CAMBRIDGE STUDIES IN
ADVANCED MATHEMATICS 16

An introduction to harmonic analysis on semisimple Lie groups

AN INTRODUCTION TO

Harmonic analysis on semisimple Lie groups

V.S. VARADARAJAN

Department of Mathematics
University of California at Los Angeles

PUBLISHED BY THE PRESS SYNDICATE OF THE UNIVERSITY OF CAMBRIDGE
The Pitt Building, Trumpington Street, Cambridge, United Kingdom

CAMBRIDGE UNIVERSITY PRESS
The Edinburgh Building, Cambridge CB2 2RU, UK www.cup.cam.ac.uk
40 West 20th Street, New York, NY 10011–4211, USA www.cup.org
10 Stamford Road, Oakleigh, Melbourne 3166, Australia
Ruiz de Alarcón 13, 28014 Madrid, Spain

First published 1989
First paperback edition (with corrections) 1999

Printed in the United Kingdom at the University Press, Cambridge

A catalogue record for this book is available from the British Library

Library of Congress Cataloguing in publication data
Varadarajan, V.S.
An introduction to harmonic analysis on semisimple Lie groups,
V.S. Varadarajan.
p. cm. – (Cambridge studies in advanced mathematics; 16)
Bibliography: p.
Includes index.
ISBN 0 521 34156 6
1. Harmonic analysis. 2. Semisimple Lie groups.
3. Representations of groups. I. Title. II. Series.
QA403. V37 1988
515′.2433–dc19 87-30911

ISBN 0 521 34156 6 hardback
ISBN 0 521 66362 8 paperback

See what a lovely shell,
Small and pure as a pearl,
Lying close to my foot,
Frail, but a work divine,
Made so fairily well
With delicate spire and whorl,
How exquisitely minute,
A miracle of design!
What is it? a learned man
Could give it a clumsy name.
Let him name it who can;
The beauty will be the same.

– Alfred, Lord Tennyson, *Maud* II, II.

Contents

Preface

A substantial part of the material covered in these notes formed the content of a course of lectures given during the Spring of 1985 in the Mathematics Institute of the University of Warwick, England. My aim was to introduce the aspiring graduate student to a beautiful and central part of mathematics, the representation theory of semisimple groups. This is of course a vast and active subject, bringing together at a fairly deep level algebra, geometry, analysis, and arithmetic. This is one of the reasons why it is difficult to get into, at least for the young student. I therefore made an attempt to keep the requirements minimal, and introduced the major themes of the subject by first working them out in the case of $SL(2, \mathbb{R})$. This approach has no claim to novelty: it has been done before, and there is the well-known book of S. Lang dealing only with $SL(2, \mathbb{R})$. I have, however, discussed a number of topics not treated by Lang, such as the Schwartz space, invariant eigendistributions, wave packets, and so on; in addition, I have included, wherever possible, indications of how these ideas may be generalized to the context of a general semisimple Lie group.

The organization of the book is not always linear because I wanted to adhere closely to the lectures and preserve their freeflowing nature. As a result, the reader will often find references to matters that are not defined or are quite advanced, especially in Chapters 1–2. This should not discourage him (or her); my advice to the reader is to ignore them and proceed ahead, and come back to the difficult points later. Chapters 4–8 are essentially linear and can be worked through by a graduate student (late first year or early second year in an American university). I have included appendices on Functional Analysis and Lie theory that offer the reader some basic definitions, explanations of some concepts, and some historical perspective.

It is a great pleasure for me to express my gratitude to the Mathematics Institute of the University of Warwick and the Science and Engineering Research Council of the United Kingdom for inviting me to Warwick, and to the staff and faculty of the Institute for their wonderful hospitality. To thank Klaus and Annelise Schmidt adequately for what they did for my wife and me is impossible; it was only through their kindness and generosity that this visit become so memorable for us. Finally I want to thank the

Cambridge University Press and their editor Dr David Tranah for being really patient while I took my time preparing the final version of this manuscript.

Pacific Palisades, October 1986 V.S. Varadarajan

1

Introduction

1.1 Aim

The aim of this book is to introduce advanced undergraduates or beginning graduate students to the subject of harmonic analysis on semisimple Lie groups. This involves doing a certain amount of representation theory for these groups, either implicitly or explicitly, because harmonic analysis is concerned mainly with expanding arbitrary functions (and generalized functions) on a group as a series or integral of functions which occur as matrix elements of irreducible representations of the group. Nevertheless this is not a book on representation theory. As long as the group is compact, the harmonic analysis point of view is not very prominent; but for noncompact groups the behaviour of the matrix elements at infinity becomes critical, and the analysis becomes decisive. Thus, although representation theory and harmonic analysis have a lot in common, the two subjects are not quite the same; and the differences will become clear to the reader when both themes have been developed to a certain extent. In this introductory chapter I shall discuss briefly a number of sources of motivation for studying representations and harmonic analysis, whose diversity and wide-ranging nature show that our subject is much more central than it seems at first sight.

1.2 Some definitions

A *representation* of a group G is a homomorphism of G into the general linear group $GL(V)$ of a complex finite-dimensional vector space V; the representation is said to be *in* (or *on*) V. If 0 and V are the only linear subspaces of V stable under the representation π (i.e., left invariant by $\pi(g)$ for all $g \in G$), then π is said to be *irreducible*. Representations π_j in V_j $(j = 1, 2)$ are *equivalent* if there is a linear isomorphism $T(V_1 \to V_2)$ such that $\pi_2(g) = T\pi_1(g)T^{-1}$ for all $g \in G$. If π_j are representations of G on V_j $(j = 1, \ldots, m)$ their direct sum $\pi = \pi_1 \oplus \cdots \oplus \pi_m$ and tensor product $\pi' = \pi_1 \otimes \cdots \otimes \pi_m$ are the representations defined respectively in $V = V_1 \oplus \cdots \oplus V_m$ and $V' = V_1 \otimes \cdots \otimes V_m$ by

$$\left.\begin{aligned} \pi(g) &= \pi_1(g) \oplus \cdots \oplus \pi_m(g) \\ \pi'(g) &= \pi_1(g) \otimes \cdots \otimes \pi_m(g) \end{aligned}\right\} \quad (g \in G)$$

The dual (or contragradient) of a representation π in V is the representation

π^* in the dual V^* of V defined by

$$(v, \pi^*(g)v^*) = (\pi(g^{-1})v, v^*) \quad (v \in V, v^* \in V^*, g \in G)$$

Actually, the category of finite-dimensional representations, with $\oplus$, $\otimes$, * defined as above, is adequate only for problems involving finite groups. For infinite groups, it is necessary to impose additional restrictions such as continuity, rationality, and so on, as well as to consider infinite-dimensional representations. We begin by looking at some of the most common examples.

Finite groups. As we remarked above, the category of representations in finite-dimensional vector spaces is the natural one to work with, but the restriction to complex vector spaces is not reasonable. Most applications, in physics and chemistry, for example, deal with complex representations; but for a general theory the underlying fields should be arbitrary [S1].

Algebraic groups. In concrete terms these are groups of matrices defined by polynomial conditions on their entries. Typical examples are the unimodular group, i.e., the group of matrices of determinant 1, the orthogonal group, the symplectic group, and so on. For such a group defined over (i.e., with entries from) an algebraically closed field k, it is natural to work only with finite-dimensional representations π which are *rational*; here rational means the entries of $\pi(g)$ relative to a basis of V are polynomial functions of the entries of g and $\det(g)^{-1}$.

Topological groups. Impulses from functional analysis and quantum physics were very much responsible for a systematic study of representations in infinite-dimensional spaces. The groups considered are topological and the vector spaces usually complete and locally convex. If G is a topological group and V is a complete locally convex space, the homomorphism π of G into the group of invertible automorphisms of V is a *representation* if the map $(g, v) \mapsto \pi(g)(v)$ of $G \times V$ into V is continuous. Important special cases are when V is a Banach space and G is locally compact. If V is a Hilbert space and each $\pi(g)$ is unitary, π is called a *unitary representation.* For a general π, irreducibility now means that 0 and V are the only *closed* linear subspaces stable under π; equivalence of π_1 and π_2 is defined as before, but with T required now to be a topological linear isomorphism; if π_1 and π_2 are unitary, and T is unitary, π_1 and π_2 are said to be *unitary equivalent.* The set of equivalence classes of irreducible unitary representations of G is written $\hat{G}$ and is called the *unitary dual* of G. One can define infinite (orthogonal) direct sums in the category of unitary representations. The definition of tensor products for infinite-dimensional representations is somewhat technically involved since tensor product is a technically complicated notion for topological vector spaces; we shall not make any serious use of this.

1.3 Classical invariant theory

The geometric invariant theory of classical geometers was one of the first examples of an important context where representation theory entered in a nontrivial way. Here $G = SL(n+1, \mathbb{C})$ and one starts with a rational representation of the algebraic group G in a complex vector space V. The action of G on V gives rise to an action on the projective space $\mathbb{P}(V)$ of V. The problem of invariant theory is that of describing the orbit space $G\backslash\mathbb{P}(V)$[MF]. This leads almost immediately to the study of the action of G on the rings of functions on $\mathbb{P}(V)$. Let R be the graded ring of polynomials on V, and $\bar{R} = R^G$ be the graded subring of G-invariant polynomials. The first step in the description of $G\backslash\mathbb{P}(V)$ is the study of the following question:

$$\text{Is } \bar{R} \text{ finitely generated?} \qquad (*)$$

Hilbert proved, at the beginning of his epoch-making work on invariant theory, that for $G = SL(n+1, \mathbb{C})$ the answer to $(*)$ is affirmative. This was eventually extended to all complex semisimple groups G by Hermann Weyl who obtained it as a consequence of his famous theorem that all rational representations of any complex semisimple group are completely reducible, i.e., direct sums of irreducible representations. Weyl's theorem is one of the deepest and most important in the finite-dimensional representation theory of semisimple groups, and we shall discuss it briefly in the next chapter.

When the questions of geometric invariant theory were examined by Mumford in the 1960s Chevalley had already developed the theory of semisimple algebraic groups over any algebraically closed field k; and Mumford's investigations led naturally to the question of finite generation of $\bar{R} = R^G$ where R is the graded ring of polynomials on a vector space V on which we have a rational representation of the algebraic semisimple group G. Unfortunately Weyl's method fails when char $(k) > 0$; representations of G are not in general completely reducible when k has positive characteristic. Nevertheless Mumford conjectured that all rational representations of the semisimple group G over an arbitrary algebraically closed field k possess the following property (M):

> if $v \neq 0$ is a vector in the space V of the given representation and v is invariant under G, there is a nonzero homogeneous polynomial f on V invariant under G such that $f(v) \neq 0$.

The property (M) is equivalent to complete irreducibility when char $(k) = 0$;

its validity for general k implies $\bar{R}$ is finitely generated. Mumford's conjecture was proved by Haboush in 1975 [Hb]. These results have been the beginning of new progress in representation theory and geometric invariant theory [MF]. For the classical theory there are of course many references; in addition to Weyl's great classic [W1] the reader may consult Schur's lectures [Sc].

1.4 Quantum mechanics and unitary representations

We now turn to a completely different source of problems in which unitary representations appear prominently. The group G is now the symmetry group of a quantum-mechanical system and one is interested in a description of the system that is covariant under G. Now, any quantum-mechanical description requires the introduction of a complex Hilbert space $\mathscr{H}$; the physical interpretation consists in identifying the ortho-complemented lattice $\mathscr{L}(\mathscr{H})$ of the closed linear subspaces of $\mathscr{H}$ with the logic of experimentally verifiable propositions of the system [V1]. The requirement of covariance means there is a homomorphism σ of G into the group of automorphisms of $\mathscr{L}(\mathscr{H})$. Now it can be proved that any automorphism σ of $\mathscr{L}(\mathscr{H})$ is induced in the obvious way by a unitary or antiunitary operator, determined uniquely up to a multiplicative constant of absolute value 1. Under mild assumptions on G and σ it can be shown that σ is induced by a *projective unitary representation* of G, i.e. a unitary representation of *an extension of G by the group T of complex numbers of absolute value* 1. This representation is obviously an important invariant of the system. For suitable G one can show that σ is induced by a unitary representation of its simply connected covering group $\tilde{G}$. This is the case when G is the group of automorphisms of Euclidean or Minkowskian affine space-time (however, this is *not* the case for the group of automorphisms of Galilean space-time).

If G is the group of automorphisms of an affine space with the structure of Minkowskian space-time, G can be written as a semidirect product $A \ltimes H$ where A is the four-dimensional group of space-time translations, $H = SL(2, \mathbb{C})$, and H acts linearly on A via the Lorentz transformations. In any description of a quantum-mechanical system consistent with special relativity there will thus appear a unitary representation of G. For instance, if the system is that of a free elementary particle, it is natural to expect this representation to be irreducible, and to expect further that it will tell us everything about this free particle. Thus the free relativistic elementary particles are in one-one correspondence with a certain subset of $\hat{G}$. Now there is a general method, due to Mackey, for determining the irreducible unitary representations of such (and even more general) locally compact semidirect products. This method, applied in the present situation, leads in a simple and natural manner to the classification of the particles in terms of their mass and spin [V1] [Ma1].

It is not always the case that the symmetry groups are locally compact. The *gauge groups* occurring in the theory of gauge fields are infinite-dimensional, and the representation theory of these and more general groups is quite active now, although not yet in any definitive state [K] [PS].

1.5 Classical Fourier analysis. Plancherel and Poisson formulae

The starting point of Fourier analysis is the idea that a more or less arbitrary function can be expanded as a 'linear combination' of the exponentials. The basic objective of the theory is to define the *Fourier transform*; the transform of a function (or a generalized function) shows how it is made up of its harmonic constituents. We shall now explain briefly the point of view of the theory of unitary representations that allows us to understand and generalize these classical themes.

Fourier series deal with functions on the torus $\mathbb{T}^n$ with coordinates $\theta = (\theta_1, \ldots, \theta_n)$. We introduce the Hilbert space $L^2(\mathbb{T}^n) = L^2(\mathbb{T}^n, d\theta)$, $d\theta = d\theta_1 \cdots d\theta_n$, $\int d\theta = 1$. For $\phi \in \mathbb{T}^n$ we define the linear operator $\lambda(\phi)$ on $L^2(\mathbb{T}^n)$ by

$$(\lambda(\phi)f)(\theta) = f(-\phi + \theta)$$

The $\lambda(\phi)$ are unitary, and it is easy to show that $\lambda(\phi \to \lambda(\phi))$ is a unitary representation of $\mathbb{T}^n$, the so-called *regular representation* of $\mathbb{T}^n$. The *irreducible* unitary representations of $\mathbb{T}^n$ are precisely all the *characters*

$$\chi_m : \theta \to \exp 2\pi i(m_1\theta_1 + \cdots + m_n\theta_n) \quad (m = (m_1, \ldots, m_n) \in \mathbb{Z}^n)$$

The functions χ_m are in $L^2(\mathbb{T}^n)$; the one-dimensional subspaces $\mathbb{C}\cdot\chi_m$ are stable under λ, and the restriction of λ to $\mathbb{C}\cdot\chi_{-m}$ is equivalent to the one-dimensional representation χ_m. The orthogonal direct sum decomposition

$$L^2(\mathbb{T}^n) = \bigoplus_m \mathbb{C}\cdot\chi_{-m}$$

shows that λ is equivalent to the infinite (orthogonal) direct sum of the χ_m ($m \in \mathbb{Z}^n$), each taken only once. We shall now see that the Fourier transform operator leads to the explicit 'diagonalization' of λ. For any $f \in L^2(\mathbb{T}^n)$ define its Fourier transform $\hat{f} = \mathscr{F} f$ by

$$\hat{f}(m) = (f, \chi_{-m}) \quad (m \in \mathbb{Z}^n)$$

where $(\cdot,\cdot)$ is the scalar product. Then $\hat{f}$ is a function on $\mathbb{Z}^n$. If we equip $\mathbb{Z}^n$ with the counting measure and introduce the Hilbert space $L^2(\mathbb{Z}^n)$, then $\hat{f} \in L^2(\mathbb{Z}^n)$ and

$$\mathscr{F} : f \mapsto \hat{f}$$

is a *unitary isomorphism* of $L^2(\mathbb{T}^n)$ with $L^2(\mathbb{Z}^n)$:

$$\|f\| = \|\hat{f}\|$$

which is the usual Parseval relation. The inverse operator $\mathscr{F}^{-1}$ is given by

$$f = \sum_m \hat{f}(m)\chi_{-m}$$

the series converging in $L^2(\mathbb{T}^n)$. If we now use $\mathscr{F}$ to carry the representation λ to a representation μ of $\mathbb{T}^n$ in $L^2(\mathbb{Z}^n)$, $\mu = \mathscr{F} \circ \lambda \circ \mathscr{F}^{-1}$, then

$$\mu(\phi)\hat{f}(m) = \chi_m(\phi)\hat{f}(m) \quad (\phi \in \mathbb{T}^n)$$

This formula shows that in the standard basis of $L^2(\mathbb{Z}^n)$ all the operators $\mu(\phi)$ are diagonal.

If f is smooth, $\hat{f}(m)$ tends to 0 very rapidly when $|m| \to \infty$; the series

$$f = \sum_m \hat{f}(m)\chi_{-m}$$

then converges very nicely: we have, for it as well as for all the series obtained by formal differentiation, uniform convergence. In particular,

$$f(0) = \sum_m \hat{f}(m) \quad (f \in C^\infty(\mathbb{T}^n)) \tag{P}$$

We shall refer to this as the *Plancherel formula*.

For $\mathbb{R}^n$ the theory is more delicate. The characters of $\mathbb{R}^n$ are the functions

$$\chi_t\colon (x_1, \ldots, x_n) \to \exp \mathrm{i}(t_1 x_1 + \cdots + t_n x_n) \quad (t \in \mathbb{R}^n)$$

The regular representation of $\mathbb{R}^n$ is defined as before; it acts on $L^2(\mathbb{R}^n)$ by

$$(\lambda(y)f)(x) = f(-y + x)$$

Proceeding as before we define the *Fourier transform* of f by

$$\hat{f}(t) = \int_{\mathbb{R}^n} f(x)\chi_t(x)\,\mathrm{d}x \quad (t \in \mathbb{R}^n) \tag{FT}$$

Then

$$(\lambda(y)f)\hat{\ }(t) = \chi_t(y)\hat{f}(t) \quad (y, t \in \mathbb{R}^n) \tag{M}$$

so that the operators $\lambda(y)$ become multiplication operators simultaneously, and thus are 'diagonalized'. However, the χ_t are of absolute value 1 and so do not lie in the Hilbert space, so that the definition of the Fourier transform in (FT) is not strictly valid for all f in $L^2(\mathbb{R}^n)$. The traditional way to overcome this difficulty is to use the definition (FT) initially for f suitably restricted, say for $f \in L^1(\mathbb{R}^n) \cap L^2(\mathbb{R}^n)$; the key step is then to prove that on this restricted domain the map $f \mapsto \hat{f}$ is essentially unitary, and then to complete its definition to all of $L^2(\mathbb{R}^n)$ by continuity, noting that $L^1(\mathbb{R}^n) \cap L^2(\mathbb{R}^n)$ is dense in $L^2(\mathbb{R}^n)$. The restricted unitarity is proved in the form

$$\int_{\mathbb{R}^n} |f(x)|^2 \, dx = (2\pi)^{-n} \int_{\mathbb{R}^n} |\hat{f}(t)|^2 \, dt \qquad (P_1)$$

It then turns out that the Fourier transform maps $L^2(\mathbb{R}^n)$ onto $L^2(\mathbb{R}^n)$ so that it is a unitary isomorphism:

$$\mathcal{F}: L^2(\mathbb{R}^n, dx) \cong L^2(\mathbb{R}^n, (2\pi)^{-n} \, dt)$$

The relations (M) are now valid rigorously and show that the representation $\mu = \mathcal{F} \circ \lambda \circ \mathcal{F}^{-1}$ acts by multiplication operators. The formula (P_1), valid for all $f \in L^2(\mathbb{R}^n)$, is the *Plancherel formula.*

The 'diagonalization' of λ effected by $\mathcal{F}$ is the classic example of a 'continuous decomposition'. Let us introduce an equivalence relation in the σ-algebra of Borel subsets of $\mathbb{R}^n$ by defining $E \sim F$ to mean $(E \backslash F) \cup (F \backslash E)$ has measure zero. If $\mathcal{B}$ is the set of equivalence classes, $\mathcal{B}$ is a σ-algebra also; but unlike the σ-algebra of Borel sets $\mathcal{B}$ has no atoms. Further, Lebesgue measure becomes a measure on $\mathcal{B}$ with the property that each nonzero element has measure strictly greater than zero. For any Borel set E let

$$S(E) = \{ f \mid f \in L^2(\mathbb{R}^n, dx)\}, \hat{f} = 0 \text{ outside } E\}$$

It is then easy to show using (M) that $S(E)$ is a closed λ-stable subspace of $L^2(\mathbb{R}^n, dx)$. Of course $S(E)$ depends only on the equivalence class of E and so we have a map $S(e \mapsto S(e))$ from the σ-algebra $\mathcal{B}$ to the orthocomplemented lattice Λ of the λ-stable closed linear subspaces of $L^2(\mathbb{R}^n, dx)$. It is not difficult to show at this stage that S is an *isomorphism*:

$$S: \mathcal{B} \cong \Lambda$$

The most elegant way to prove all of these assertions is by using the Schwartz space; this method will also bring out the duality explicitly, and will have the additional advantage of focussing on the differential aspects of the theory. The Schwartz space of $\mathbb{R}^n$ is the space $\mathcal{S}$ of all C^∞ functions f on $\mathbb{R}^n$ such that for any integers $m \geqslant 0$, $\alpha_1, \ldots, \alpha_n \geqslant 0$,

$$((\partial/\partial x_1)^{\alpha_1} \cdots (\partial/\partial x_n)^{\alpha_n} f)(x) = O((1 + x_1^2 + \cdots + x_n^2)^{-m})$$

when $x_1^2 + \cdots + x_n^2 \to \infty$. If we introduce the seminorms

$$\mu_{\alpha,m}(f) = \sup |(D^\alpha f)(x)| (1 + x_1^2 + \cdots + x_n^2)^m$$

$(D^\alpha = (\partial/\partial x_1)^{\alpha_1} \cdots (\partial/\partial x_n)^{\alpha_n})$, we can view $\mathcal{S}$ as a topological vector space also. It is easy to show that $\mathcal{S}$ is complete (so that $\mathcal{S}$ is a Fréchet space). The differential operators D^α act as continuous linear operators on $\mathcal{S}$. Also, if we write, for any smooth function g, M_g for the operator of multiplication by g, then, under suitable assumptions on g, M_g will be a continuous linear operator on $\mathscr{S}$. For instance this is true if g is a polynomial. More generally, if g is of moderate growth in the sense that for any α there is an integer $m(\alpha) \geqslant 0$ and a constant $C(\alpha) > 0$ such that

$$|(D^\alpha g)(x)| \leqslant C(\alpha)(1 + x_1^2 + \cdots + x_n^2)^{m(\alpha)}$$

for all $x \in \mathbb{R}^n$, then $M_g(f \mapsto gf)$ is a well-defined and continuous operator of $\mathscr{S}$.

The rapid decay of the elements of $\mathscr{S}$ at infinity means that $\mathscr{S} \subset L^1(\mathbb{R}^n)$ and shows at once that the Fourier transform is defined by (FT) for all $f \in \mathscr{S}$; moreover, $\hat{f}$ will be a smooth function of t, and we have the formula

$$(-\mathrm{i}\partial/\partial t_1)^{\beta_1} \cdots (-\mathrm{i}\partial/\partial t_n)^{\beta_n} \hat{f} = (M_{x_1}^{\beta_1} \cdots M_{x_n}^{\beta_n} f)^\wedge$$

(x_j is the coordinate function $(x_1, \ldots, x_n) \mapsto x_j$). Furthermore, replacing f by its derivatives and integrating by parts we find the dual formula

$$((\mathrm{i}\partial/\partial x_1)^{\alpha_1} \cdots (\mathrm{i}\partial/\partial x_n)^{\alpha_n} f)^\wedge = M_{t_1}^{\alpha_1} \cdots M_{t_n}^{\alpha_n} \hat{f}$$

The estimate

$$|\hat{g}(t)| \leqslant \|g\|_1 \quad (g \in \mathscr{S})$$

in conjunction with the above formula now shows that $\hat{f} \in \mathscr{S}$ and that

$$\mathscr{F}: f \mapsto \hat{f}$$

is a continuous linear map of $\mathscr{S}$ into $\mathscr{S}$. The basic theorem may now be formulated as the assertion that $\mathscr{F}$ is a topological linear isomorphism of $\mathscr{S}$ with itself, and that $\mathscr{F}^{-1}$ is given by the inversion formula

$$(\mathscr{F}^{-1} g)(x) = (2\pi)^{-n} \int_{\mathbb{R}^n} g(t)\chi_t(x)\,\mathrm{d}t \qquad \text{(INV)}$$

If we take $x = 0$ and $g = \hat{f}$ in (INV) we get the Plancherel formula in the form

$$f(0) = (2\pi)^{-n} \int_{\mathbb{R}^n} \hat{f}(t)\,\mathrm{d}t \quad (f \in \mathscr{S}) \qquad (\mathrm{P}_2)$$

To get the earlier version (P_1) it suffices to take $f = g * \tilde{g}$ where $\tilde{g}(x) = \overline{g(-x)}$, and $*$ is convolution, defined by

$$(h_1 * h_2)(x) = \int_{\mathbb{R}^n} h_1(y)h_2(-y + x)\,\mathrm{d}y \quad (x \in \mathbb{R}^n)$$

for $h_1, h_2 \in \mathscr{S}$; then

$$\hat{f} = |\hat{g}|^2, f(0) = \int_{\mathbb{R}^n} |g(x)|^2\,\mathrm{d}x,$$

and so (P_1) follows from (P_2).

The Plancherel formula (P_2) has an interpretation from the standpoint of Schwartz's theory of distributions [Sch] which is important for us. Let us recall that a tempered distribution on $\mathbb{R}^n$ is nothing other than a continuous linear functional on $\mathcal{S}$. If F is a measurable function such that for some constants $c > 0$, $m \geqslant 0$ we have, for all $x \in \mathbb{R}^n$,

$$|F(x)| \leqslant c(1 + x_1^2 + \cdots + x_n^2)^m$$

then

$$f \mapsto \int F(x)f(x)\,\mathrm{d}x$$

is a tempered distribution, which is usually identified with F. If δ is the Dirac measure at the origin,

$$\delta(f) = f(0) \quad (f \in \mathcal{S})$$

then (P_2) may be written as

$$\delta = (2\pi)^{-n} \int \chi_t\,\mathrm{d}t \tag{P_3}$$

interpreted as an identity of tempered distributions. It is also common to refer to (P_3) as the Plancherel formula.

The relation (P_3) is a special case of more general formulae that express how an arbitrary tempered distribution can be written as a 'linear combination' of the χ_t. Since $\mathcal{F}$ is an isomorphism of $\mathcal{S}$ with itself, it defines an isomorphism of the dual of $\mathcal{S}$ with itself. Thus, if T is a tempered distribution on $\mathbb{R}^n$, its *Fourier transform* $\hat{T}$ is the tempered distribution defined by

$$\hat{T}(f) = T(\hat{f}) \quad (f \in \mathcal{S}) \tag{FT_1}$$

The map $T \mapsto \hat{T}$ is a linear isomorphism of the space of tempered distributions with itself.

If $T = t\,\mathrm{d}x$ in the sense that $T(f) = \int tf\,\mathrm{d}x$, t being an element of $\mathcal{S}$, it is immediate that $\hat{T} = \hat{t}\,\mathrm{d}t$, so that (FT_1) is consistent with the earlier definition of Fourier transform on $\mathcal{S}$. In the general case it is usual to write (FT_1) in the symbolic form

$$T(x) = (2\pi)^{-n} \int \hat{T}(t)\chi_t(x)\,\mathrm{d}t$$

that shows how T is resolved into its harmonic constituents.

A number of classical formulae may be regarded as the computation of $\hat{T}$ for suitable T. The example that is most interesting from the point of view of arithmetic is the case when T is the counting measure on $\mathbb{Z}^n$:

$$T = \sum_{m \in \mathbb{Z}^n} \delta_m$$

Here δ_a is the Dirac measure at a, $\delta_a(f) = f(a)$. It is easy to show that T is tempered. The *Poisson (summation) formula* is the statement

$$\hat{T} = \sum_{m \in \mathbb{Z}^n} \delta_{2\pi m} \qquad (\mathrm{Po}_1)$$

which is often written as

$$\sum_{m \in \mathbb{Z}^n} f(m) = \sum_{q \in 2\pi\mathbb{Z}^n} \hat{f}(q) \quad (f \in \mathscr{S}) \qquad (\mathrm{Po}_2)$$

If we view $\mathbb{R}^n/\mathbb{Z}^n$ as the torus $\mathbb{T}^n$, then we have a unitary representation of $\mathbb{R}^n$ in $L^2(\mathbb{T}^n)$ induced by the action of $\mathbb{R}^n$ on $\mathbb{T}^n$ by translations: if γ is the natural map $\mathbb{R}^n \to \mathbb{T}^n$, the unitary representation in question, say π, is defined by

$$(\pi(y)f)(\gamma(x)) = f(\gamma(-y + x)) \quad (x, y \in \mathbb{R}^n)$$

for $f \in L^2(\mathbb{T}^n)$. The Poisson formula is entirely equivalent to the statement that

$$\pi = \bigoplus_{q \in 2\pi\mathbb{Z}^n} \chi_q$$

1.6 Fourier analysis on abelian groups

The work of Pontryagin and van Kampen in the 1920s and 1930s led to a detailed understanding of the structure of locally compact abelian groups. It then became possible to develop harmonic analysis on these groups and place the classical theory in a proper perspective. The classic work treating this is that of A. Weil [We1]. The perspective of abelian Fourier analysis would eventually prove to be the starting point of a number of related developments such as Gel'fand's theory of commutative Banach algebras, the Artin–Tate extension of Hecke's work on L-functions and their functional equations (about this we shall say more in §7), Mackey's work on unitary representations of semidirect products, the work of Weil himself [We3] that revealed the deep-lying relationship of unitary representation theory to number theory and the theory of quadratic forms, and so on.

Let G be a locally compact abelian group which we shall assume for simplicity to be second-countable. We denote by $\hat{G}$ the set of *characters* of G, i.e., the set of all continuous homomorphisms of G into T, the multiplicative group of complex numbers of absolute value 1. It follows from the spectral theorem for unitary operators that the characters of G are precisely the irreducible unitary representations of G so that $\hat{G}$ is nothing else than the unitary dual of G in this case. Under pointwise multiplication $\hat{G}$ becomes a group, and we equip it with the topology of uniform convergence on compact subsets of G. It is then a fundamental fact that $\hat{G}$ is a locally compact abelian second-countable group also. For $x \in G$, $\hat{x} \in \hat{G}$, we write

$$\langle x, \hat{x} \rangle = \hat{x}(x)$$

Then x may be viewed as a character on $\hat{G}$, namely,

$$\hat{x} \mapsto \langle x, \hat{x} \rangle$$

so that we have a natural map

$$G \to \hat{\hat{G}}$$

The Pontryagin–van Kampen duality theorem is the assertion that the above map is an *isomorphism of topological groups*; it thus allows one to identify G with $\hat{\hat{G}}$:

$$G = \hat{\hat{G}}$$

If G is compact (resp. discrete), $\hat{G}$ is discrete (resp. compact). For $G = \mathbb{Z}^n$ we have $\hat{G} = \mathbb{T}^n$ and vice versa, while $\mathbb{R}^n$ is self-dual. If H is any closed subgroup of G and its *annihilator* $\check{H}$ is the subgroup of characters $\hat{x}$ in $\hat{G}$ that are identically equal to 1 on H, we have the natural identifications

$$(G/H)^{\wedge} = \check{H}, \quad \hat{H} = \hat{G}/\check{H}$$

In particular, if H is discrete and G/H is compact, the same is true of $\check{H}$ (relative to $\hat{G}$); such a subgroup of G is often called a *lattice* in G; its annihilator is then a lattice in $\hat{G}$. For all of this, see [P].

Let $\mathrm{d}x$ be a Haar measure on G. The fundamental theorem of harmonic analysis on G is the assertion that for a suitable normalization of the Haar measure $\mathrm{d}\hat{x}$ on $\hat{G}$ the Fourier transform operator is a *unitary isomorphism* of $L^2(G, \mathrm{d}x)$ with $L^2(\hat{G}, \mathrm{d}\hat{x})$:

$$\mathscr{F}: L^2(G, \mathrm{d}x) \cong L^2(\hat{G}, \mathrm{d}\hat{x})$$

The Fourier transform is defined at first for functions which are in $L^1(G)$ rather than $L^2(G)$:

$$(\mathscr{F} f)(x) := \hat{f}(\hat{x}) = \int f(x) \langle x, \hat{x} \rangle \, \mathrm{d}x \quad (\hat{x} \in \hat{G})$$

Then $\hat{f}$ is a continuous function on $\hat{G}$, vanishes at infinity on $\hat{G}$, and

$$\sup |\hat{f}| \leqslant \|f\|_1$$

We shall say f is *admissible* if f is continuous, lies in $L^1(G)$, and $\hat{f}$ lies in $L^1(\hat{G})$, and write $A(G)$ for the space of admissible functions on G. The basic theorem then asserts that with $\mathrm{d}\hat{x}$ appropriately normalized,

$$f \mapsto \hat{f}$$

is a linear isomorphism of $A(G)$ with $A(\hat{G})$, and $\mathscr{F}^{-1}$ is given by the *inversion formula*

$$f(x) = \int \hat{f}(\hat{x}) \overline{\langle x, \hat{x} \rangle} \, \mathrm{d}\hat{x} \quad (x \in G)$$

The Haar measure $d\hat{x}$, which is uniquely determined by this requirement, is said to be *dual to* dx. It follows in the usual way from the inversion formula, replacing f by $f * \tilde{f}$, that

$$\int |f(x)|^2 \, dx = \int |\hat{f}(\hat{x})|^2 \, d\hat{x} \qquad \text{(P)}$$

for $f \in A(G) \cap L^2(G)$. As $A(G) \cap L^2(G)$ is dense in $L^2(G)$, this shows that we have a unitary isomorphism of $L^2(G)$ with $L^2(\hat{G})$ that extends Fourier transform on the space of square-integrable admissible functions. We shall continue to write $\mathscr{F}$ for this unitary extension and $\hat{f}$ for $\mathscr{F}f$, $f \in L^2(G)$. The formula (P) is the *Plancherel formula.* If G is compact and $\int dx = 1$, then $d\hat{x}$ is the counting measure on $\hat{G}$, and vice versa. Finally, dx is dual to $d\hat{x}$.

As in the classical case we can use the Fourier transform to 'diagonalize' the regular representation of G. Let λ be the regular representation,

$$(\lambda(y)f)(x) = f(-y + x) \quad (x, y \in G, f \in L^2(G))$$

If $\mu = \mathscr{F} \circ \lambda \circ \mathscr{F}^{-1}$, then

$$(\mu(y)\hat{f})(\hat{y}) = \langle y, \hat{y} \rangle \hat{f}(\hat{y}) \quad (y \in G, \hat{y} \in \hat{G}, \hat{f} \in L^2(\hat{G}))$$

showing that all the $\mu(y)$ are simultaneously multiplication operators in $L^2(\hat{G})$. We can generalize this as follows. Let $\hat{m}$ be a Borel measure on $\hat{G}$ and for $x \in G$ let $U_{\hat{m}}(x)$ be the unitary operator in $L^2(G, \hat{m})$ which is multiplication by $x(\hat{x} \mapsto \langle x, \hat{x} \rangle)$; then

$$U_{\hat{m}}: x \mapsto U_{\hat{m}}(x)$$

is a unitary representation of G in $L^2(\hat{G}, \hat{m})$. Even more generally, instead of working with a measure $\hat{m}$ and multiplication operators in $L^2(\hat{G}, \hat{m})$, we may start with a Hilbert space $\hat{H}$ and a projection-valued measure $\hat{P}$ on $\hat{G}$ whose values are projections in $\hat{H}$. Now, for any $x \in G$, if we define the operator

$$U_{\hat{P}}(x) = \int \langle x, \hat{x} \rangle \hat{P}(d\hat{x})$$

$U_{\hat{P}}$ will be a unitary representation of G in $\hat{H}$. This reduces to the representation $U_{\hat{m}}$ if we take $\hat{H} = L^2(\hat{G}, \hat{m})$ and $\hat{P}_{\hat{E}}$ to be the projection operator $f \mapsto \chi_{\hat{E}} f$ ($\chi_{\hat{E}}$ = characteristic function of $\hat{E} \subset \hat{G}$). It is a remarkable fact that every unitary representation of G is of this form and that the projection-valued measure on $\hat{G}$ is uniquely determined by the representation. We thus have a bijection between the set of equivalence classes of unitary representations of G in separable Hilbert spaces and the set of unitary equivalence classes of projection valued measures in separable Hilbert spaces defined on $\hat{G}$. If $\hat{P}$ is supported in a finite subset $\hat{F} \subset \hat{G}$, the representation $U_{\hat{P}}$ is the direct sum $\bigoplus_{\hat{x} \in \hat{F}} m(\hat{x})\hat{x}$ where $m(\hat{x})$ = dimension of the range of the projection $\hat{P}(\hat{x})$; it is thus clear that in combination with the spectral multiplicity theory of projection-valued measures the correspondence

$$\hat{P} \rightleftarrows U_{\hat{P}}$$

yields the complete theory of decomposition of unitary representations of G. The representation $U_{\hat{m}}$ may then be thought of heuristically as the 'direct integral' defined by the measure $\hat{m}$, 'each character being taken with multiplicity 1'.

The Poisson formula has a beautiful generalization to this context (see [We2], Ch. VII, §2). Let Γ be a closed subgroup of G and $\check{\Gamma}$ its annihilator in $\hat{G}$. We suppose that Γ is a lattice in G so that $\check{\Gamma}$ is a lattice in $\hat{G}$. We choose the Haar measure $\mathrm{d}x$ on G so that the induced Haar measure $\mathrm{d}\bar{x}$ on $\bar{G} = G/\Gamma$ gives volume 1 for $\bar{G}$; here, induced means

$$\int_G f(x)\,\mathrm{d}x = \int_{\bar{G}} \left(\sum_{\gamma \in \Gamma} f(x+\gamma) \right) \mathrm{d}\bar{x}$$

for all $f \in L^1(G)$. We now introduce the space $A(G, \Gamma)$ of functions admissible for (G, Γ), namely, functions f on G such that (i) f is admissible for G, i.e., $f \in A(G)$, (ii) the series

$$\sum_{\gamma \in \Gamma} f(x+\gamma), \quad \sum_{\check{\gamma} \in \check{\Gamma}} \hat{f}(\hat{x}+\check{\gamma})$$

converge uniformly when x and $\hat{x}$ vary over compact subsets of G and $\hat{G}$ respectively. It is clear that

$$f \in A(G, \Gamma) \Leftrightarrow \hat{f} \in A(\hat{G}, \check{\Gamma})$$

Moreover the measure $\mathrm{d}\hat{x}$ dual to $\mathrm{d}x$ is the one that induces the normalized Haar measure on $\hat{G}/\check{\Gamma}$, namely, the one giving volume 1 to $\hat{G}/\check{\Gamma}$. The *Poisson formula* is the relation

$$\sum_{\gamma \in \Gamma} f(\gamma) = \sum_{\check{\gamma} \in \check{\Gamma}} \hat{f}(\check{\gamma}) \qquad \text{(Po)}$$

which is valid for all $f \in A(G, \Gamma)$. In the classical applications of the Poisson formula to arithmetic, $G = \mathbb{R}^n$, $\Gamma = \mathbb{Z}^n$, and the functions f considered are of the form Pe^{-Q} where P is a polynomial and Q is a positive definite quadratic form; of course $\hat{G} = \mathbb{R}^n$ and $\check{\Gamma} = 2\pi\mathbb{Z}^n$.

1.7 Harmonic analysis and arithmetic

I would like to end this introductory chapter with a few remarks on the relationship between Fourier analysis and arithmetic.

One of the earliest nontrivial applications of Fourier analysis to number-theoretic problems is to be found in the work of Dirichlet on the distribution of primes in residue classes. Dirichlet's approach to this question involved the study of certain series of the form

$$\phi(s) = \sum_{n \geqslant 1} a_n n^{-s}$$

(nowadays known as Dirichlet series); the sequences (a_n) are not completely arbitrary; for instance they were always periodic in Dirichlet's theory. If N is an integer $\geqslant 1$ and G_N is the multiplicative group of residue classes mod N that are prime to N, Dirichlet associated to each complex-valued function f on G_N the Dirichlet series

$$\phi_f(s) = \sum_{n \geqslant 1} f(n) n^{-s}$$

Harmonic analysis on G_N (Plancherel formula) now allows one to write

$$f = \sum_{\chi \in \hat{G}_N} (f, \chi)\chi$$

so that

$$\phi_f = \sum_{\chi \in \hat{G}_N} (f, \chi)\phi_\chi \tag{$*$}$$

The functions $\phi_\chi(s)$, always written as $L(s:\chi)$, are the so-called Dirichlet L-series; as χ is *multiplicative*, we have

$$L(s:\chi) = \prod_{\substack{p \nmid N \\ p \,\text{prime}}} (1 - \chi(p)p^{-s})^{-1}$$

which is an *Euler product representation* (when $\chi = 1$ it reduces, up to a rational factor in p^{-s}, to the Euler representation

$$\zeta(s) = \prod_p (1 - p^{-s})^{-1}$$

of the Riemann zeta function). The Euler products establish a very direct link to the primes and the formula ($*$) connects the L-series to more or less arbitrary Dirichlet series with periodic coefficients.

As a second example I would like to mention the well-known application of the Poisson summation formula to the study of the Jacobi theta function. Let $f(x) = \exp(i\pi\tau x^2)$ where τ varies in the upper half-plane: $\tau \in \mathbb{C}$, $\operatorname{Im}(\tau) > 0$. It is easy to check that f lies in the Schwartz space of $\mathbb{R}$ and so one can apply the Poisson formula to it:

$$\sum_{n \in \mathbb{Z}} f(n) = \sum_{n \in \mathbb{Z}} \hat{f}(2\pi n)$$

Now $\hat{f}(t) = (-i\tau)^{-1/2} e^{-it^2/4\pi\tau}$ where the square root is the branch (as a function of τ) that is 1 when $-i\tau = 1$. So, if

$$\theta(\tau) = \sum_{n \in \mathbb{Z}} e^{i\pi\tau n^2}$$

we have

$$\theta(\tau) = (-i\tau)^{-1/2}\theta(-1/\tau)$$

This relation, together with the obvious periodicity $\theta(\tau+2)=\theta(\tau)$, expresses the fact that θ is a *modular form*. On the other hand, we can use the theory of Fourier transforms on the multiplicative group of positive reals (known classically as the *Mellin transforms*) to introduce the Mellin transform

$$M\theta(s)=\int_0^\infty (\theta(iv)-1)v^s\frac{dv}{v}$$

and find

$$M\theta(s)=2\Gamma(s)\pi^{-s}\zeta(s)\quad (\mathrm{Re}\,(s)>\tfrac{1}{2})$$

so that the fact that θ is a modular form is essentially equivalent to the analytic continuation and functional equation of the zeta function.

This is only a special (but typical) case of the relationship between modular forms and Dirichlet series in which the Poisson formula and Mellin transform play critical roles. These connections were explored by Hecke in the 1930s and the 1940s. Hecke's ideas were generalized in a far-reaching manner in the 1950s by Artin and Tate, and by Selberg, Weil and Langlands in the 1960s. Their work revealed the remarkable fact that the arithmetic theory of Galois extensions of number fields and modular forms were intimately related to fundamental questions of harmonic analysis and representation theory of certain locally compact groups and their homogeneous spaces, namely, the groups $G(R)$ where G is the group $GL(n)$ or more generally an algebraic group defined over a number field k, and R is a locally compact ring containing the number field. The examples characteristic of arithmetic are: $R=\mathbb{R}, \mathbb{C}$, or the p-adic fields $\mathbb{Q}_p$ and their finite extensions, as well as $\mathbb{A}_k$, the ring of adéles attached to k. The subgroup $G(k)$ of $G(\mathbb{A}_k)$ is *discrete*, and it is the spectral decomposition of the representation of $G(\mathbb{A}_k)$ in $L^2(G(\mathbb{A}_k)/G(k))$ that has critical importance for the central questions of modern arithmetic. The foundations for doing harmonic analysis in these contexts were laid by Gel'fand, Selberg, Langlands, and above all, Harish-Chandra. The far-reaching and profound nature of Harish-Chandra's work has made the representation-theoretic approach one of the most powerful in number theory.

It may not be out of place to go into a little more detail into this. Let $k=\mathbb{Q}$ so that the adéle ring $\mathbb{A}_k=\mathbb{A}$ is the so-called *restricted direct product* $\mathbb{R}\times\prod'_p\mathbb{Q}_p$, the prime referring to the fact that only those sequences (x_p) are admitted for which x_p is in the ring $\mathbb{Z}_p$ of p-adic integers for all but finitely many primes p. If $G=GL(n)$, the group $G(\mathbb{A})$ is then the restricted direct product $GL(n,\mathbb{R})\times\prod'_p GL(n,\mathbb{Q}_p)$, with the prime corresponding to the fact that only those sequences (g_p) are admitted for which $g_p\in GL(n,\mathbb{Z}_p)$, a compact open subgroup of $GL(n,\mathbb{Q}_p)$ (even maximal) for all but finitely many p. The irreducible unitary representations π of $GL(n,\mathbb{A})$ are then of the form $\pi_\infty \otimes \bigotimes'_p \pi_p$ where π_p (resp. π_∞) is an irreducible unitary representation of $GL(n, \mathbb{Q}_p)$ (resp. $GL(n, \mathbb{R})$), and $\bigotimes'_p$ means that for all but finitely many p π_p has a unit vector v_p in its space left fixed by $GL(n, \mathbb{Z}_p)$ (π_p is then called *spherical*), and the Hilbert space of $\bigotimes'_p$ is the completion of the linear span of tensors $\bigotimes'_p u_p$ where $u_p = v_p$ for all but finitely many p. One can associate to π an Euler product

$$L(s:\pi) = \phi_\infty(s) \prod_p \phi_p(s), \quad \phi_p(s) = P_p(p^{-s})^{-1}$$

where $P_p(T)$ is a polynomial $1 + c_{p,1}T + \cdots + c_{p,r}T^r$ (r independent of p). Actually only the ϕ_p for those p for which π_p is spherical are defined in a relatively direct fashion, but we shall finesse this point. However, when π 'occurs' in the spectral decomposition of $L^2(GL(n, \mathbb{A})/GL(n, \mathbb{Q}))$, $L(s:\pi)$ converges in a half-plane, extends meromorphically to the whole s-plane, and satisfies a functional equation relating $L(s:\pi)$ to $L(1 - s:\check{\pi})$, $\check{\pi}$ being the contragradient of π; moreover, these properties are true essentially only for such π. The central arithmetic significance of these Dirichlet series $L(s:\pi)$ is the following circumstance (checked in many cases and believed to be true in general): among these are to be found all the Dirichlet series that one can build from arithmetic data coming from the Galois groups of the extensions of $\mathbb{Q}$. For $n = 1$, the π are the Grössencharacters attached to $\mathbb{Q}$, and the $L(s:\pi)$ are the corresponding *Hecke L-series*; for $n = 2$ some of the π may be identified with modular forms vanishing at the cusps and the $L(s:\pi)$ are the Dirichlet series associated to them by Hecke.

The study of problems of harmonic analysis and representation theory whose origin is in these ideas is a very exciting programme (the 'Langlands philosophy') and is currently attracting a tremendous amount of interest. No matter where this programme will eventually go, there is no doubt that a deeper understanding of arithmetic is not possible without the use of Fourier analysis in its most sophisticated form, on semisimple groups and their homogeneous spaces.

Notes and comments

The expositions of Mackey [Ma1] [Ma2] are excellent for getting an overview of the subject. For group theory and quantum mechanics the great classical source is Weyl [W3]. For the arithmetic connections Serre's book [S2] is a good starting point. For a nice account of Dirichlet's work see Davenport [Da]. The 'Langlands philosophy' originated in two famous lectures of Langlands [L1] [L2]. There are many sources for learning about modular forms; for a recent one see Terras [Te]. The classic sources for getting a view of arithmetic from the perspective of harmonic analysis are Tate's famous thesis [Ta] and Weil's book [We2].

Problems

1. Let k be a field of prime characteristic $p > 1$; $M_n =$ the algebra of $n \times n$ matrices with entries from k; $G = GL(n, k)$; G acts on M_n by $(g, m) \mapsto gmg^{-1}$. If M_0 is the subspace of scalar matrices and M_1 the subspace of matrices of trace 0, prove the following:

(a) if n is prime to p, M_1 is the unique G-stable complementary subspace to M_0;

(b) if p divides n, $M_0 \subset M_1$ and M_0 has no G-stable complementary subspace.

2. If $k = \mathbb{C}$ prove that complete reducibility of representations is a consequence of Mumford's conjecture.

3. The following steps lead to an 'algebraic' proof of the Plancherel formula for $\mathbb{R}^n$; $\mathscr{S} = \mathscr{S}(\mathbb{R}^n)$.

(a) Let $f \in \mathscr{S}$ and $f(0) = 0$. Prove that there are $g_1, \ldots, g_n \in \mathscr{S}$ such that $f = x_1 g_1 + \cdots + x_n g_n$.

(b) Let $L(f) = \int \hat{f}\, dx_1 \cdots dx_n$. Deduce from (a) that if $f(0) = 0$ then $L(f) = 0$. Hence conclude that $L(f) = cf(0)$ for some $c \in \mathbb{C}$.

(c) Prove that $c = (2\pi)^n$ by computing that for $f(x) = (2\pi)^{-n/2} e^{-(x^2 + \cdots + x^2)/2}$, $\hat{f} = (2\pi)^{n/2} f$.

4. Let $L_i (i = 1, 2)$ be two irreducible unitary representations of a group G. If T is a closed densely defined operator such that $L_1(g) T L_2(g)^{-1} = T$ for all $g \in G$, prove that $T = cU$ where U is unitary and c is a constant. (Hint: consider $T^\dagger T$).

2

Compact groups: the work of Weyl

2.1 Characters. Orthogonality relations

The theory of compact groups is dominated by ideas and results of Hermann Weyl. Weyl's work had a profound impact on the entire course of development of harmonic analysis, and it is important to have a good understanding of his ideas. It is, however, not my intention to prove everything.

Let G be a compact group which for simplicity will be assumed to satisfy the second axiom of countability. It has a unique positive measure, written dx, invariant under both left and right translations, normalized by the condition

$$\int_G dx = 1$$

We shall refer to it as *the* Haar measure on G (see Appendix 2). Averaging with respect to it is a powerful technique in harmonic analysis on G.

Any finite-dimensional representation of G is equivalent to a unitary representation. This is a consequence of group averaging. In fact, if π is a (continuous) representation of G in V and $(\cdot|\cdot)_0$ is any positive-definite scalar product for V, we define a new scalar product $(\cdot|\cdot)$ by

$$(u|v) = \int_G (\pi(g)u|\pi(g)v)_0 \, dg \quad (u, v \in V)$$

It is then easy to prove that $(\cdot|\cdot)$ is also a positive-definite scalar product for V; the invariance of dg implies that $(\cdot|\cdot)$ is invariant under π so that π is unitary with respect to $(\cdot|\cdot)$.

It is thus enough to consider only unitary representations, at least when working with finite-dimensional representations. The theory may then be developed along the same lines as for finite groups. The central notion is that of the character of a finite dimensional unitary representation; if π is the representation, its *character* θ_π is the function on G defined by

$$\theta_\pi(x) = \mathrm{tr}\,(\pi(x)) \quad (x \in G)$$

The function θ_π has the following properties:

(i) θ_π is an invariant function, i.e., $\theta_\pi(yxy^{-1}) = \theta_\pi(x)$ for all $x, y \in G$; in classical language, θ_π is a *class function*;

(ii) $\theta_\pi(1) = \dim(\pi) := d(\pi)$;
(iii) $\theta_\pi = \theta_{\pi'} \Leftrightarrow \pi \cong \pi'$; if ω is the equivalence class of π we often write θ_ω for θ_π;
(iv) θ_π is additive, i.e., $\theta_{\pi_1 \oplus \cdots \oplus \pi_m} = \theta_{\pi_1} + \cdots + \theta_{\pi_m}$;
(v) $\theta_{\pi \otimes \pi'} = \theta_\pi \theta_{\pi'}$;
(vi) if $\omega, \omega' \in \hat{G}$,† and $(\cdot|\cdot)$ is the scalar product in $L^2(G)$, $(\theta_\omega|\theta_{\omega'}) = \delta_{\omega\omega'}$ (Kronecker delta);
(vii) if $\omega = \bigoplus_{1 \leqslant i \leqslant r} m_i \omega_i$ where the m_i are integers $\geqslant 1$ and the ω_i are irreducible classes, i.e., $\omega_i \in \hat{G}$, then $(\theta_\omega|\theta_\omega) = m_1^2 + \cdots + m_r^2$; in particular, ω irreducible $\Leftrightarrow (\theta_\omega|\theta_\omega) = 1$.

The relations (vi) are *the orthogonality relations for the characters.* They show that the irreducible characters form an orthonormal family in the closed subspace $L^2(G)^{\text{inv}}$ of $L^2(G)$ consisting of functions in $L^2(G)$ invariant under all inner automorphisms.

In addition to the character we can associate with a finite dimensional unitary representation π its *matrix elements*, namely, the functions

$$x \to (\pi(x)u|v) \quad (u, v \in V)$$

The orthogonality relations (vi) are consequences of orthogonality relations between matrix elements which are formulated as follows. Let π (resp. π') be irreducible unitary; then

$$\int_G (\pi(x)u|v)\overline{(\pi'(x)u'|v')}\,dx = \begin{cases} 0 & \text{if } [\pi] \neq [\pi'] \\ d(\pi)^{-1}(u|v)\overline{(u'|v')} & \text{if } \pi = \pi' \end{cases}$$

where $[\pi]$ (resp. $[\pi']$) is the class of π (resp. π'), u, v (resp. u', v') are vectors in the space of π (resp. π'). In particular, if $A(\omega)$ is the linear span of the matrix elements of π ($[\pi] = \omega, d(\pi) = d(\omega)$), and $(e_i)_{1 \leqslant i \leqslant d(\omega)}$ is an orthonormal basis in the space of π, then the functions

$$f_{ij:\omega}: x \to d(\omega)^{1/2}(\pi(x)e_j|e_i)(1 \leqslant i, j \leqslant d(\omega))$$

form an orthonormal basis for $A(\omega)$; of course,

$$A(\omega) \perp A(\omega') \quad (\omega \neq \omega')$$

2.2 The Peter–Weyl theorem

All these remarks hang in the air unless one can produce finite-dimensional unitary representations. For finite groups this is done by decomposing the regular representation. Note that if π is a finite-dimensional unitary representation in V and $U \subset V$ is an invariant subspace for π, $U^\perp$ is also π-invariant and $V = U \oplus U^\perp$. This procedure may be continued till we obtain

†See §1.2 for definition of $\hat{G}$

a decomposition of π as a direct sum of irreducible unitary representations. In other words, π is *completely reducible.* For finite G, the regular representation is finite-dimensional; but it is not obvious that $L^2(G)$ has finite-dimensional invariant subspaces when G is infinite but compact.

Let us consider the right regular representation ρ (the left and right regular representations are equivalent):

$$(\rho(x)f)(y) = f(yx) \quad (x, y \in G)$$

The key idea is to look for self-adjoint operators that commute with ρ; their spectral subspaces will then be ρ-stable. If the self-adjoint operator is compact,[†] it will have *finite-dimensional eigenspaces.* Now, integral operators defined by kernels k are compact (even Hilbert–Schmidt); the condition for self-adjointness is $k(y,x) = \overline{k(x,y)}$, and the condition for commuting with ρ is $k(xg, yg) = k(x,y)$, for all $g, x, y \in G$. The last condition is equivalent to saying that $k(x,y)$ depends only on xy^{-1}. So if a is a continuous function such that $a(x) = \overline{a(x^{-1})}$ $(x \in G)$ the kernel $k_a(x,y) = a(xy^{-1})$ defines an integral operator K_a which is self-adjoint, of Hilbert–Schmidt class, and commutes with ρ. It is now easy to show that ρ splits as an orthogonal direct sum of finite-dimensional subrepresentations if and only if the linear span of the finite-dimensional ρ-invariant subspaces of $L^2(G)$ is dense in $L^2(G)$. Suppose $f \in L^2(G)$ is orthogonal to all finite-dimensional ρ-invariant subspaces of $L^2(G)$. Then f is orthogonal to all the eigenspaces of K_a corresponding to its nonzero eigenvalues and hence $K_a f = 0$, i.e.,

$$\int_G a(xy^{-1}) f(y)\,dy = \int_G a(z) f(z^{-1}x)\,dz = 0$$

for almost all x in G. In other words, with $*$ denoting convolution as usual,

$$a * f = 0$$

Select a sequence $(a_n)_{n \geqslant 1}$ of functions such that

(i) a_n is real, continuous, and $\geqslant 0$,
(ii) $\int a_n\,dx = 1$,
(iii) $a_n(x) = a_n(x^{-1})$,
(iv) $\operatorname{supp}(a_n) \searrow (1)$ in the sense that if U is any neighbourhood of 1 the support of a_n is contained in U for all sufficiently large n.

This is easy to do. We leave it to the reader to show that for $g \in L^2(G)$

$$a_n * g \to g \quad \text{in } L^2(G), n \to \infty$$

[†]See Appendix 1

So $f = \lim (a_n * f) = \lim 0 = 0$. What we have given is the classical argument of Peter–Weyl.

This is the fundamental result of the general theory. It leads to the *completeness theorem* of Peter–Weyl.

Theorem 1. *The irreducible unitary representations of G are all finite-dimensional and they separate the points of G. The irreducible characters form an orthonormal basis of $L^2(G)^{\mathrm{inv}}$, and $L^2(G)$ is the orthogonal direct sum of matrix elements*:

$$L^2(G)^{\mathrm{inv}} = \bigoplus_{\omega \in \hat{G}} \mathbb{C}\theta_\omega, \quad L^2(G) = \bigoplus_{\omega \in \hat{G}} A(\omega)$$

where the sums are orthogonal.

Let $f \in C(G)$, the space of continuous functions on G. For any $\omega \in \hat{G}$ let $d(\omega)^{1/2}(\pi(x)e_j|e_i) = u_{ij;\omega}(x)$ where π is in the class ω and $(e_i)_{1 \leqslant i \leqslant d(\omega)}$ is an orthonormal basis in the space of π. Then

$$(u_{ij;\omega})_{1 \leqslant ij \leqslant d(\omega)}, \quad \omega \in \hat{G}$$

is an orthonormal basis for $L^2(G)$ and hence

$$\| f \|^2 = \sum_{\omega \in \hat{G}} \sum_{1 \leqslant ij \leqslant \omega} |(f | u_{ij;\omega})|^2$$

But if we write

$$\pi(g) = \int_G g(x)\pi(x)\,\mathrm{d}x \quad (g \in C(G))$$

then

$$\sum_{ij} |(f | u_{ij;\omega})|^2 = d(\omega) \sum_{ij} |(\pi(\bar{f})e_j | e_i)|^2 = d(\omega)\,\mathrm{tr}\,(\pi(\bar{f})^\dagger \pi(\bar{f}))$$

where † denotes adjoint. This leads to

$$\| f \|^2 = \sum_{\omega \in \hat{G}} d(\omega)\theta_\omega(\tilde{f} * f) \quad (\tilde{f}(x) = \overline{f(x^{-1})}) \tag{P}$$

which may be viewed as the Plancherel formula. As in the case of Fourier analysis on $\mathbb{R}^n$ we can rewrite it on observing that $\| f \|^2 = (\tilde{f} * f)(1)$. Let $L(G)$ be the linear span of all the $\tilde{f} * f$, $f \in C(G)$. Then

$$h(1) = \sum_{\omega \in \hat{G}} d(\omega)\theta_\omega(h) \quad (h \in L(G))$$

Let

$$A(\omega) = \bigoplus_{1 \leqslant i \leqslant d(\omega)} A_i(\omega), \quad A_i(\omega) = \bigoplus_{1 \leqslant j \leqslant d(\omega)} \mathbb{C} f_{ij,\omega}$$

Then it is easy to show that the $A_i(\omega)$ are stable under the right regular representation ρ and the restriction of ρ to $A_i(\omega)$ is a representation of the class ω. Thus the restriction of ρ to $A(\omega)$ is equivalent to the multiple $d(\omega)\,\rho$, and

$$\rho \cong \sum_{\omega \in \hat{G}} d(\omega)\omega$$

This is the representation-theoretic version of the Plancherel formula. Thus, for a compact group, the decomposition of the regular representation yields all the irreducible representations of the group.

We have formulated the completeness theorem only in $L^2(G)$. We actually have the following.

Theorem 2. *The algebraic sum*

$$A = \sum_{\omega \in \hat{G}} A(\omega)$$

is an algebra stable under complex conjugation, and is dense in $C(G)$ in the topology of uniform convergence.

Proof. If $\pi_j \in \omega_j \in \hat{G}$ $(j = 1, 2)$, the matrix elements of $\pi_1 \otimes \pi_2$ are in $A(\omega_1) \cdot A(\omega_2)$; since $\pi_1 \otimes \pi_2$ is completely reducible it follows that $A(\omega_1) \cdot A(\omega_2) \subset A$. Similarly if $\bar{\omega}$ is the class contragradient to ω, $A(\bar{\omega}) = \overline{A(\omega)}$. So A is an algebra and $A = \bar{A}$. Now use the Stone–Weierstrass theorem.

It follows from the completeness theorem that given a fundamental sequence of neighbourhoods of 1 in G, say (U_n), we can find finite-dimensional unitary representations π_n of G such that $\ker(\pi_n) \subset U_n$, ker denoting the kernel. Now we can view π_n as a continuous homomorphism

$$\pi_n: G \to L_n, \quad \pi_n(G) = L_n,$$

where L_n is a closed subgroup of the unitary group $U(d_n)$, $d_n = \dim(\pi_n)$, hence a *Lie group*. Hence we can exhibit G as *a limit of Lie groups*:

$$G = \varprojlim L_n \quad L_n \subset U(d_n)$$

This was the starting point of von Neumann's classical solution of Hilbert's fifth problem (Is a locally Euclidean group a Lie group?). Note however that the L_n need not be connected; in fact they may all be finite, which must be the case if G is totally disconnected ([MZ], §2.20).

It is usual to say that a topological group H has *no small closed (normal) subgroups* if there is a neighbourhood U of 1 in H such that (1) is the only closed (normal) subgroup of H that is entirely contained in U. If G is a compact group and we assume that G has no small closed normal subgroups, then G is a closed subgroup of some $U(n)$. Now one can show that a Lie group has no small subgroups so that a compact Lie group can always be regarded as a closed subgroup of some $U(n)$. In other words, *every compact Lie group has a faithful (finite-dimensional) representation.*

If $f \in C(G)^{\text{inv}}$, we may view the series

$$\sum_{\omega \in \hat{G}} (f, \theta_\omega)\theta_\omega$$

as the (invariant) *Fourier series of f*. The reader should note that Theorem 2 does not assert that this converges to f in the topology of uniform convergence. If G is a compact *Lie group* and $C^\infty(G)$ is the space of infinitely differentiable functions, and if we regard $C^\infty(G)$ as a Fréchet space in which convergence is equivalent to uniform convergence of all derivatives, then we can prove that

$$f = \sum_{\omega \in \hat{G}} (f, \theta_\omega)\theta_\omega$$

the series converging in $C^\infty(G)$. The proof of this requires much preparation and will not be given here ([V3], II, §7.6). We may even replace $C^\infty(G)$ by $C^r(G)$, $r \gg 1$. For $G = U(n)$, see ex. 5.

2.3 Weyl's character formula for $U(n)$

It is natural to ask whether one can supplement the general theory with the explicit description of $\hat{G}$ when G is one of the commonly encountered compact groups, in particular when $G = U(n)$, the group of $n \times n$ unitary matrices. We shall now derive Hermann Weyl's famous formula for the irreducible characters of $U(n)$. Actually Weyl's arguments carry over to the case of any compact connected Lie group; but the general case needs considerable background in the structure theory of semisimple groups and is beyond this initial introduction (see §5). The case of $U(n)$ is nevertheless typical and we shall therefore look at it in detail.

So let $G = U(n)$ and let $D \subset G$ be the subgroup of diagonal matrices

$$D = \{\text{diag}(t_1, \ldots, t_n) \,|\, |t_j| = 1, 1 \leqslant j \leqslant n\}$$

We denote by W the group of permutations $\{1, 2, \ldots, n\}$. Then W acts naturally on D as well as on the dual group $\hat{D}$, D being viewed as a compact abelian group; $D \cong \mathbb{T}^n$, $\hat{D} \cong \mathbb{Z}^n$. If γ is a conjugacy class in G, it follows from the diagonalizability of unitary matrices that γ meets D and that $\gamma \cap D$ is a single W-orbit, namely the W-orbit of diagonal matrices whose diagonal entries are the eigenvalues $\lambda_1, \ldots, \lambda_n$ (in some order) of the matrices in γ. So if we write, for any class function f on G, f_D for its restriction to D, then f_D is W-invariant and the correspondence $f \rightleftarrows f_D$ is one-one. If now π is a finite dimensional unitary representation of G, the restriction of π to D is a direct sum of characters of D; so we have

$$(\theta_\pi)_D = \sum_j m_j \chi_j$$

where the sum is finite, the χ_j are in D, and m_j are integers $\geqslant 0$, i.e., $(\theta_\pi)_D$ is a finite Fourier series on D with nonnegative integer coefficients; it is moreover *symmetric*, i.e., W-invariant. Weyl's theorem gives a formula for $(\theta_\pi)_D$ when π is irreducible.

The idea behind Weyl's method is to start with the orthogonality relations

$$\int_G \theta_\omega \theta_{\omega'} \, dx = \delta_{\omega\omega'} \quad (\omega, \omega' \in \hat{G})$$

Although the integration is over G, the integrand is a class function and so the integration reduces to integrating over D. The key to Weyl's calculation is thus the result that exhibits the integral of any function (not necessarily a class function) as a repeated integral – an integral over a typical conjugacy class, and then an integral over D whose elements may be viewed as parametrizing the conjugacy classes. This calculation, given in the lemma below and its corollary, assumes some familiarity with differential methods.

Note first that we have an action of G on itself by inner automorphisms, and that the conjugacy classes are the corresponding orbits. However, there are some 'exceptional orbits' around which this partition of G into orbits exhibits some unpleasant features. A good analogy is the situation where the rotation group acts on $\mathbb{R}^3$; all the *nonzero* orbits are of dimension 2 but the origin is a zero-dimensional orbit all by itself. This sudden drop in dimension at 0 corresponds to the analytical fact that the spherical coordinate system works only on $\mathbb{R}^3\backslash(0)$. In the present situation, for any point $x \in G$, one should view the points of D in the conjugacy class of x as representing the *angular* coordinates of x, while the points of D where the conjugacy class of x meets it should be thought of as representing the *radial* coordinates of x. From this point of view, Weyl's integral formula appears as the analogue, for $G = U(n)$, of the classical formula which gives the transformation from Cartesian to polar or spherical coordinates. The formula works only on the part of G which remains after all the orbits of less than the maximum dimension have been removed from G. To find out the condition $x \in G$ so that the class of x has maximum possible dimension, observe that this class may be identified with the coset space G/G_x where G_x is the centralizer of x in G i.e., the subgroup of G of elements that commute with x; for, the map $y \mapsto yxy^{-1}$ maps G onto the class of x, and $y_1 x y_1^{-1} = y_2 x y_2^{-1} \Leftrightarrow y_1 \in y_2 G_x$. If $\lambda_1, \ldots, \lambda_r$ are the distinct eigenvalues of x with respective multiplicities $m_1, \ldots, m_r$, then the matrices in G_x are precisely those of the block diagonal form

$$\begin{pmatrix} * & & & 0 \\ & * & & \\ & & \ddots & \\ 0 & & & * \end{pmatrix}$$

where the blocks are of sizes $m_1, \ldots, m_r$ respectively and correspond to the eigenspaces of x associated to $\lambda_1, \ldots, \lambda_r$. So

$$\dim(G_x) = m_1^2 + \cdots + m_r^2$$

Since $m_1 + \cdots + m_r = n$, we have

$$m_1^2 + \cdots + m_r^2 \geqslant n, = n \Leftrightarrow r = n \quad \text{and} \quad m_1 = \cdots = m_n = 1$$

Since $\dim(G/G_x)$ is a maximum if and only if $\dim(G_x)$ is a minimum, we get

$$\dim(G/G_x)\,\text{maximum} \Leftrightarrow x \text{ has } n \text{ distinct eigenvalues}$$

The condition that x has n distinct eigenvalues is the same as the condition that the characteristic polynomial of x has distinct roots, and so, as the coefficients of the characteristic polynomials of x are polynomials in the entries of x, there is a polynomial function δ of the entries of x such that $\delta(x) \neq 0$ is equivalent to x having n distinct eigenvalues. It is also quite simple to exhibit such a δ explicitly. Indeed, if y is a variable $n \times n$ complex matrix with eigenvalues $\lambda_1, \ldots, \lambda_n$,

$$\begin{aligned} \det(T \cdot 1 - y) &= T^n - c_1(y)T^{n-1} + c_2(y)T^{n-2} - \cdots \pm c_n(y) \\ &= \prod_1^n (T - \lambda_j) \end{aligned}$$

where T is an indeterminate. Now

$$\prod_{i \neq j} (\lambda_i - \lambda_j)$$

is invariant under the group of permutations of the variables $\lambda_1, \ldots, \lambda_n$ and so, by Newton's theorem, can be written as a polynomial $P(\sigma_1, \ldots, \sigma_n)$ where σ_j are the elementary symmetric functions. If

$$\delta(y) = P(c_1(y), \ldots, c_n(y))$$

it is immediate that δ, called the *discriminant* of G, is a polynomial in the coefficients of y and

$$\delta(y) \neq 0 \Leftrightarrow y \text{ has } n \text{ distinct eigenvalues}$$

We shall call $x \in G$ *regular* if x has n distinct eigenvalues, and write G' for the set of regular elements of G. We see at once from the preceding remarks that G' is a dense open subset of G. Let

$$D' = G' \cap D$$

It is usual to call $G\backslash G'$ the *singular set*. It is an analytic subvariety of G since it is the set of zeros of the analytic function δ. We now observe that G is an analytic manifold, and that in any system of local coordinates the Haar measure is absolutely continuous with respect to Lebesgue measure (see Appendix 3, no. 4). It follows from this that any analytic subvariety of G that is not equal to G has Haar measure 0. In particular, $G\backslash G'$ has measure 0.

Lemma 3. *For $x \in G$, $t \in D$, xtx^{-1} depends only on the coset xD. If $\psi(xD, t) = xtx^{-1}$,*

$$\psi: G/D \times D' \to G'$$

is an analytic map whose differential $\mathrm{d}\psi$ is bijective everywhere. It is actually a covering map each of whose fibers has exactly $n!$ elements. Moreover, if

$$\Delta(t) = \prod_{1 \leqslant i < j \leqslant n} (t_i - t_j) \quad (t = \mathrm{diag}(t_1, \ldots, t_n))$$

and $\mathrm{d}x, \mathrm{d}t, \mathrm{d}\bar{x}$ are the exterior differential forms that respectively define the normalized invariant measures on $G, D, \bar{G} = G/D$, then

$$\psi^*(\mathrm{d}x) = \Delta\bar{\Delta}\,\mathrm{d}\bar{x}\,\mathrm{d}t$$

Proof. We will have $\psi^*(\mathrm{d}x) = \omega(\cdot, \cdot)\,\mathrm{d}\bar{x}\,\mathrm{d}t$ where ω is an analytic function on $\bar{G} \times D'$. Now, writing $\bar{x}$ for xD, we have $\psi(g[x], t) = g\psi(\bar{x}, t)g^{-1}$ for all $g \in G$, where we have written $g[\bar{x}]$ for gxD; as the form $\mathrm{d}x$ is invariant under inner automorphisms, the function $\omega(x, t)$ does not change when $\bar{x}$ is changed to $g[\bar{x}]$, and so depends only on t. We can thus determine it by a differential calculation at $(\bar{1}, t)$. To compute $\mathrm{d}\psi_{(\bar{1},t)}$ it is convenient to identify† the tangent spaces to G at 1 and D at t respectively with the Lie algebra $\mathfrak{g}$ of skew Hermitian matrices and the Lie algebra $\mathfrak{d}$ of diagonal matrices. Then $\mathfrak{g} = \mathfrak{d} \oplus \mathfrak{q}$ where $\mathfrak{q}$ is the subspace of $\mathfrak{g}$ of matrices with zeros in the diagonal; we identify $\mathfrak{q}$ with the tangent space to $\bar{G}$ at $\bar{1}$. Now for $s \in \mathbb{R}, Z \in \mathfrak{q}$,

$$\psi(\exp sZ \cdot D, t) = \exp sZt \cdot \exp(-sZ) = t \exp(s(t^{-1}Zt)) \cdot \exp(-sZ)$$

so that, taking derivatives at $s = 0$ we obtain

$$\mathrm{d}\psi_{(\bar{1},t)}(Z, 0) = t^{-1}Zt - Z$$

Hence, for $Z \in \mathfrak{q}$, $H \in \mathfrak{d}$, noting that $\psi(\bar{1}, t) = t$ and

$$\mathrm{d}\psi_{(\bar{1},t)}(Z, H) = \mathrm{d}\psi_{(\bar{1},t)}(Z, 0) + \mathrm{d}\psi_{(\bar{1},t)}(0, H)$$

we get

$$\mathrm{d}\psi_{(\bar{1},t)}(Z, H) = (t^{-1}Zt - Z) + H$$

†See Appendix 3 for the Lie theoretic concepts used in this discussion

For any $u \in D$ let us write $A(u)$ for the linear map $Z \mapsto u^{-1}Zu - Z$ of $\mathfrak{q}$ onto itself. It then follows from the above formula for $d\psi_{(\bar{1},t)}$ and the definition of ω that

$$\omega(t) = \det A(t)$$

To calculate $\det A(t)$, we go over to the complexification $\mathfrak{q}_{\mathbb{C}}$ of $\mathfrak{q}$ and view $A(t)$ as the linear map $X \mapsto t^{-1}Xt - X$ of the space of $n \times n$ complex matrices with zero diagonal entries. If E_{rs} $(r \neq s)$ is the matrix with 1 as its (rs)th entry and 0 for its other entries, the E_{rs} form a basis for $\mathfrak{q}_{\mathbb{C}}$ and

$$A(t)\cdot E_{rs} = \left(\frac{t_s}{t_r} - 1\right)\cdot E_{rs}$$

Hence

$$\det A(t) = \prod_{r \neq s}\left(\frac{t_r}{t_s} - 1\right) = \prod_{r<s}\left(\frac{t_r}{t_s} - 1\right)\left(\frac{\bar{t}_r}{\bar{t}_s} - 1\right)$$

since $|t_r| = 1$ for all r. Hence

$$\omega = \Delta\bar{\Delta}$$

For $t \in D'$, $\omega(t) \neq 0$ so that $d\psi_{(\bar{1},t)}$ is bijective. Also, as the t_i are distinct, the $n!$ permutations of $(t_1, \ldots, t_n)$ are distinct so that each point of G' has $n!$ conjugates in D', proving that the fibers of ψ all have cardinality $n!$. To conclude that ψ is a covering map it is enough, using a simple topological argument, to show that ψ is *proper*, i.e., for any compact set $E \subset G'$, $\psi^{-1}(E)$ is compact in $\bar{G} \times D'$. This is an easy consequence of the compactness of G.

We can now prove the integral formula that follows from this lemma. If f is any Borel function on G we write

$$g_f(t) = \int_G f(xtx^{-1})\,dx \quad (t \in D)$$

provided the integral exists.

Proposition 4. *A Borel function f on G lies in $L^1(G)$ if and only if g_f lies in $L^1(D, \Delta\bar{\Delta}\,dt)$; in this case*

$$\int_G f\,dx = \frac{1}{n!}\int_D g_f \Delta\bar{\Delta}\,dt$$

In particular, an invariant function f on G lies in $L^1(G)$ if and only if its restriction f_D to D lies in $L^1(D, \Delta\bar{\Delta}\,dt)$, and then

$$\int_G f\,dx = \frac{1}{n!}\int_D f_D \Delta\bar{\Delta}\,dt$$

Proof. It follows from the lemma that, if f is a continuous function on G'

with compact support,

$$\begin{aligned}\int_G f\,\mathrm{d}x &= \frac{1}{n!}\int_{\bar{G}\times D'} f(\psi(\bar{x},t))\Delta\bar{\Delta}\,\mathrm{d}\bar{x}\,\mathrm{d}t \\ &= \frac{1}{n!}\int_D \Delta\bar{\Delta}\,\mathrm{d}t \int_{\bar{G}} f(xtx^{-1})\,\mathrm{d}x \\ &= \frac{1}{n!}\int_D g_f\Delta\bar{\Delta}\,\mathrm{d}t\end{aligned}$$

From this, standard measure theoretic arguments lead to the first assertion. The second is an immediate consequence of the first because, for an invariant f, $g_f = f_D$.

We can now give a proof of Weyl's formula for the irreducible characters of $U(n)$ ([W1] Theorem 7.5A).

***Theorem 5** **(Weyl's character formula).** The irreducible characters of $U(n)$ are in natural one-one correspondence with n-tuples of integers $(m_1,\ldots,m_n)$ with $m_1 > m_2 > \cdots > m_n$. The character $\Theta_{m_1,\ldots,m_n}$ is given on D by*

$$(\Theta_{m_1,\ldots,m_n})_D = \left(\sum_{w\in W} \varepsilon(w)\chi^w_{m_1,\ldots,m_n}\right)\Big/\Delta$$

Here $\chi_{m_1,\ldots,m_n}$ is the character $t = \mathrm{diag}(t_1,\ldots,t_n) \mapsto t_1^{m_1}\cdots t_n^{m_n}$ of D, w refers to the action of the permutation w on D, and $\varepsilon(w)$ is the signature of the permutation w.

It must be noted that, for any character Θ, Θ_D is a finite Fourier series, so that the right side of the above formula should be a finite Fourier series. Actually it is not difficult to give a direct argument showing that the numerator on the right side is divisible by Δ in the ring of finite Fourier series with integer coefficients.

Proof. If π is an irreducible unitary representation let us define $(\Theta_\pi)_D\cdot\Delta = \Phi_\pi$. Then the Φ_π have the following properties:

(i) Φ_π is a finite Fourier series with integer coefficients,
(ii) Φ_π is skew symmetric, i.e.,

$$\Phi_\pi^w = \varepsilon(w)\Phi_\pi \quad (w\in W)$$

(iii) $\displaystyle \frac{1}{n!}\int_D \Phi_\pi\bar{\Phi}_{\pi'}\,\mathrm{d}t = \delta_{[\pi][\pi']}$

Indeed, since

$$\Delta = \prod_{i<j}(t_i - t_j)$$

is a finite Fourier series with integer coefficients which is skew-symmetric while Θ_π is a symmetric finite Fourier series with integer coefficients, (i) and (ii) are clear; (iii) is just the orthogonality relation. We shall now determine all functions satisfying (i) and (ii). For any $\chi \in \hat{D}$ the alternating sum

$$\tilde{\chi} := \sum_{w \in W} \varepsilon(w) \chi^w$$

satisfies (i) and (ii) above; if χ_1 and χ_2 are in different W-orbits, $(\tilde{\chi}_1 | \tilde{\chi}_2) = 0$ while, for $\chi_2 = \chi_1^{w_0}$, we have $\tilde{\chi}_2 = \varepsilon(w_0)\tilde{\chi}_1$. If $\chi = \chi_{m_1,\ldots,m_n}$ and $m_i = m_j$, then, for the element w in W that interchanges i and j and leaves everything else fixed, we have $\tilde{\chi} = \tilde{\chi}^w = \varepsilon(w)\tilde{\chi} = -\tilde{\chi}$, so that $\tilde{\chi} = 0$. On the other hand, if $\chi = \chi_{m_1,\ldots,m_n}$ where m_i are distinct, the χ^w are all distinct and we have

$$\tilde{\chi} = \pm \tilde{\chi}_{m'_1,\ldots,m'_n} (m'_1 > \cdots > m'_n), \quad \| \chi \|^2 = n!$$

where $(m'_1, \ldots, m'_n)$ is the rearrangement of $(m_1, \ldots, m_n)$ in decreasing order. So any Φ satisfying (i) and (ii) has the form

$$\Phi = \sum_{\substack{m_1 > \cdots > m_n \\ m_i \in \mathbb{Z}}} c(m_1, \ldots, m_n) \tilde{\chi}_{m_1,\ldots,m_n}$$

where the sum if finite, the coefficients $c(m_1, \ldots, m_n)$ are integers, and

$$\| \Phi \|^2 = \sum | c(m_1, \ldots, m_n) |^2$$

Since $\| \Phi_\pi \|^2 = 1$, it follows that in the above expression for $\Phi = \Phi_\pi$ only one $(m_1, \ldots, m_n)$ will occur, and moreover $c(m_1, \ldots, m_n) = \pm 1$. Hence there are integers $m_i(\pi) = m_i$, $m_1 > \cdots > m_n$ such that

$$\Phi_\pi = \varepsilon(\pi) \tilde{\chi}_{m_1,\ldots,m_n} \quad \varepsilon(\pi) = \pm 1 \qquad (*)$$

To complete the proof of the theorem we must show that all possible $(m_1, \ldots, m_n)$ (with $m_1 > \cdots > m_n$) occur in $(*)$ as π varies, and that $\varepsilon(\pi) = 1$ always. If $(q_1, \ldots, q_n)$ does not occur, we consider the function $\Theta_D = \tilde{\chi}_{q_1,\ldots,q_n}/\Delta$ on D'. It is invariant under W and hence, by the lemma above, it is the restriction to D' of an invariant function on G', say Θ. Then the proposition above shows that $\Theta \in L^2(G)^{\text{inv}}$ and

$$(\Theta, \Theta_\pi)_G = (n!)^{-1} (\tilde{\chi}_{q_1,\ldots,q_n}, \tilde{\chi}_{m_1,\ldots,m_n}) = 0$$

Since π is arbitrary, this contradicts the completeness of the Θ_π in $L^2(G)^{\text{inv}}$. Finally, to prove $\varepsilon(\pi) = 1$, let us order $\mathbb{Z}^n$ *lexicographically* and write

$$(\Theta_\pi)_D = c\chi_{r_1,\ldots,r_n} + \sum_{(s_1,\ldots,s_n) \prec (r_1,\ldots,r_n)} d(s_1, \ldots, s_n) \chi_{s_1,\ldots,s_n}$$

where $c, d(s_1, \ldots, s_n)$ are integers $\geqslant 0$ and denote the multiplicites of $\chi_{r_1,\ldots,r_n}$ and $\chi_{s_1,\ldots,s_n}$ in the restriction of π to D. We now observe that

$$\Delta = \tilde{\chi}_{n-1,n-2,\ldots,0} = \chi_{n-1,n-2,\ldots,0} + \sum_{(s_1,\ldots,s_n) \prec (n-1,\ldots,0)} d'(s_1, \ldots, s_n) \chi_{s_1,\ldots,s_n}$$

where the $d'(s_1, \ldots, s_n)$ are integers, and hence

$$\Phi_\pi = c\chi_{r_1+n-1, r_2+n-2, \ldots, r_n} + \sum_{(s_1,\ldots,s_n) \prec (r_1+n-1,\ldots,r_n)} e(s_1, \ldots, s_n)\chi_{s_1,\ldots,s_n}$$

with $e(s_1, \ldots, s_n) \in \mathbb{Z}$. On the other hand, we conclude from the formula $\Phi_\pi = \varepsilon(\pi)\tilde{\chi}_{m_1,\ldots,m_n}$ that $(m_1, \ldots, m_n)$ is the *highest* element of $\mathbb{Z}^n$ for which the corresponding character occurs in Φ_π, and that the coefficient of $\chi_{m_1,\ldots,m_n}$ is $\varepsilon(\pi)$. Hence

$$c = \varepsilon(\pi), \quad r_1 + n - 1 = m_1, \quad r_2 + n - 2 = m_2, \ldots, r_n = m_n$$

In particular, $\varepsilon(\pi) > 0$ so that $\varepsilon(\pi) = 1$, completing the proof. Note that $r_1 \geqslant r_2 \cdots \geqslant r_n$.

For any irreducible π, the characters of D which occur in the restriction of π to D are called the *weights* of π. We now transfer the lexicographic order from $\mathbb{Z}^n$ to $\hat{D}$ and conclude that there is always a unique highest weight and that this necessarily has the form $\chi_{r_1,\ldots,r_n}$ with $r_1 \geqslant \cdots \geqslant r_n$. Moreover, it occurs with multiplicity 1 and the irreducible character is just $\Theta_{r_1+n-1, r_2+n-2, \ldots, r_n}$. Hence we have obtained the following.

Theorem 6 (Cartan–Weyl description via highest weights). *If $r_1, \ldots r_n$ are integers $r_1 \geqslant \cdots \geqslant r_n$ there is (up to equivalence) a unique irreducible representation of $U(n)$ with highest weight $\chi_{r_1,\ldots,r_n}$ and this weight has multiplicity 1 in the representation. Its character is $\Theta_{m_1,\ldots,m_n}$ where $m_1 = r_1 + n - 1, m_2 = r_2 + n - 2, \ldots, m_n = r_n$, and all irreducible representations are thus obtained.*

Unlike the character formula this is a description of the *irreducible representations themselves*; it plays a fundamental role in all questions which involve the representations themselves as well as in constructions which explicitly realize the representations.

The one-dimensional representation det $(x \mapsto \det(x))$ corresponds to the highest weight $(1, \ldots, 1)$ and hence changing π to $\pi \otimes (\det)^r$ corresponds to the changes

$$(r_1, \ldots, r_n) \to (r_1 + r, \ldots, r_n + r), \quad (m_1, \ldots, m_n) \to (m_1 + r, \ldots, m_n + r)$$

in the two-parameter systems we have introduced. It follows easily from this that the above theorems for $U(n)$ remain valid for $SU(n)$ with the sole modification that the highest weights are of the form

$$\chi_{r_1,\ldots,r_n} \quad (r_1 \geqslant r_2 \geqslant \cdots \geqslant r_n = 0)$$

and the characters are of the form

$$\Theta_{m_1,\ldots,m_n} \quad (m_1 > m_2 > \cdots > m_n = 0)$$

The simplest case is when $G = SU(2)$. D is then the group

$$D = \{u_\theta \mid \theta \in \mathbb{R}\}, \quad u_\theta = \begin{pmatrix} e^{i\theta} & 0 \\ 0 & e^{-i\theta} \end{pmatrix}$$

The irreducible characters are the functions Θ_m $(m = 1, 2, \ldots)$ where

$$\Theta_m(u_\theta) = \frac{e^{im\theta} - e^{-im\theta}}{e^{i\theta} - e^{-i\theta}} = e^{i(m-1)\theta} + e^{i(m-3)\theta} + \cdots + e^{-i(m-1)\theta}$$

Of course

$$\Delta(u_\theta) = e^{i\theta} - e^{-i\theta}$$

and W consists of the identity and the reflection $u_\theta \to u_{-\theta}$. The highest weight is $\chi_{m-1}(u_\theta \to e^{i(m-1)\theta})$ and the dimension is m, as can be seen from the expansion of Θ_m, or else from the relation

$$\Theta_m(1) = \dim(\pi_m)$$

However, as both the numerator and denominator in the formula for Θ_m vanish at $u_\theta = 1$ we have to apply the differential operator $-\mathrm{i}\,d/d\theta$ to get the value at $u_\theta = 1$ (L'Hospital's rule). The same applies in the general case for $G = U(n)$ or $SU(n)$. Let

$$\omega_{m_1,\ldots,m_n} \quad (m_1 > \cdots > m_n)$$

be the irreducible class with character $\Theta_{m_1,\ldots,m_n}$. We introduce the differential operator

$$\partial(\textstyle\prod) = \prod_{r<s}\left(t_r\frac{\partial}{\partial t_r} - t_s\frac{\partial}{\partial t_s}\right) = \prod_{r<s}\left(\frac{1}{i}\frac{\partial}{\partial\theta_r} - \frac{1}{i}\frac{\partial}{\partial\theta_s}\right) \quad (t_r = e^{i\theta_r})$$

on D. Exactly like Δ, $\partial(\prod)$ is skew-symmetric with respect to W. We now apply $\partial(\prod)$ to both sides of the relation

$$\sum_w \varepsilon(w)\chi^w_{m_1,\ldots,m_n} = \Delta(\Theta_{m_1,\ldots,m_n})_D$$

and evaluate at the identity element of D. Since

$$\begin{aligned}\partial(\textstyle\prod)(\varepsilon(w)\chi^w_{m_1,\ldots,m_n}) &= \varepsilon(w)\varepsilon(w^{-1})(\partial(\textstyle\prod)\chi_{m_1,\ldots,m_n})^w \\ &= \prod_{r<s}(m_r - m_s)\cdot\chi^w_{m_1,\ldots,m_n}\end{aligned}$$

we have

$$\partial(\textstyle\prod)\tilde{\chi}_{m_1,\ldots,m_n} = \prod_{1\leqslant r<s\leqslant n}(m_r - m_s)\cdot\chi^{\#}_{m_1,\ldots,m_n}$$

where

$$\chi^{\#}_{m_1,\ldots,m_n} = \sum_w \chi^w_{m_1,\ldots,m_n}$$

In particular,

$$(\partial(\textstyle\prod)\tilde{\chi}_{m_1,\ldots,m_n})(1) = n!\prod_{1\leqslant r<s\leqslant n}(m_r - m_s)$$

The computation of the right side is an immediate consequence of the following.

Lemma 7. *For any C^∞ function u on D,*

$$\partial(\prod)(\Delta u) = \left(\prod_{1 \leqslant r < s \leqslant n} (s-r)\right)\chi^{\#}_{n-1,n-2,\ldots,1,0} \cdot u$$

modulo the ideal in $C^\infty(D)$ generated by the $(t_r - t_s)$ $(r \neq s)$. In particular,

$$(\partial(\prod))(\Delta u)(1) = n! \prod_{1 \leqslant r < s \leqslant n} (s-r) \cdot u(1)$$

Proof of lemma. It is obvious that $\partial(\prod)(\Delta u) = (\partial(\prod)\Delta)u + \sum_\alpha v_\alpha$ where each v_α is of the form $(\partial(\prod_\alpha)\Delta)u_\alpha$, with $u_\alpha \in C^\infty(D)$ and $\partial(\prod_\alpha)$ is a product $\prod_\alpha(t_r(\partial/\partial t_r) - t_s(\partial/\partial t_s))$, the suffix α indicating that the product excludes a *nonempty* set α of pairs (r, s) with $r < s$. A downward induction on $|\alpha|$ will show that each $\partial(\prod_\alpha)\Delta$ is in the ideal generated by the $(t_r - t_s)(r \neq s)$, and in fact in the ideal generated by the m-fold products of the $(t_r - t_s)$, where $m = |\alpha|$. To complete the proof of the lemma it remains to show that

$$\partial(\prod) = \prod_{1 \leqslant r < s \leqslant n} (s-r) \cdot \chi^{\#}_{n-1,n-2,\ldots,1,0}$$

This is immediate from the previous calculation if we remember that

$$\Delta = \tilde{\chi}_{n-1,n-2,\ldots,1,0}$$

Applying this lemma to $u = (\Theta_{m_1,\ldots,m_n})_D$ we get *Weyl's dimension formula*:

$$d(\omega_{m_1,\ldots,m_n}) = \frac{\prod_{1 \leqslant r < s \leqslant n} (m_r - m_s)}{\prod_{1 \leqslant r < s \leqslant n} (s-r)}$$

2.4 Plancherel formula and orbital integrals for $U(n)$

Weyl's method of integrating over the conjugacy classes is a very powerful one. It leads to the explicit formula for the irreducible characters, and, as a close examination of the proof shows, it links harmonic analysis on $U(n)$ to that on D. A further exploration of this connection shows that one can get the Plancherel formula for $U(n)$ from the Plancherel formula for D. Like the character and dimension formulae, this method works in general, and when properly generalized is the cornerstone of Fourier analysis on all semi-simple Lie groups.

For a continuous function f on $G = U(n)$ we define its *orbital integral* as the function on D given by

$$F_f(t) = (-1)^{n(n-1)/2}(t_1 \cdots t_n)^{-(n-1)}\Delta(t) \int_G f(xtx^{-1})\,dx \quad (t \in D)$$

If we ignore the factors in front, this is essentially the average M_f of f over the conjugacy class of t which is viewed as the orbit of t for the action of G on itself by inner automorphisms:

$$M_f(t) = \int_G f(xtx^{-1})\,dx \quad (t \in D)$$

Since the map $(x, t) \mapsto xtx^{-1}$ from $G \times D$ to G is C^∞ and the domain of integration is a compact manifold, it follows that M_f is continuous and, more generally, is of class C^r if f is of the class C^r $(0 \leqslant r \leqslant \infty)$. Thus the same is true for F_f, and hence

$$f \mapsto F_f$$

is a linear map from $C^\infty(G)$ to $C^\infty(D)$. We shall now formulate its key properties in the following theorem.

Theorem 8. (i) *F_f is skew-symmetric, i.e.,*

$$F_f(t^w) = \varepsilon(w) F_f(t) \quad (t \in D, w \in W)$$

(ii) *The irreducible characters are the Fourier transforms of the orbital integrals, i.e., for integers $m_1 > \cdots > m_n$, $f \in C^\infty(G)$*

$$\Theta_{m_1,\ldots,m_n}(f) := \int_G \Theta_{m_1,\ldots,m_n} f\,dx = \hat{F}_f(m_1, \ldots, m_n)$$

where $\hat{F}_f$ is the Fourier transform of F_f on D.

(iii) (*Limit formula*): *The delta function at the identity element can be calculated by applying the differential operator $\partial(\prod)$ to the orbital integrals and calculating the limit as $t \to 1$:*

$$(\partial(\textstyle\prod) F_f)(1) = (-1)^{n(n-1)/2} \cdot n! \prod_{r<s} (s-r) f(1) \quad (f \in C^\infty(G))$$

Proof. (i) For any $w \in W$ there is a permutation matrix $\tilde{w}$ in $U(n)$ which has the same action as w on D; if w is the permutation $1 \to 1', \ldots, n \to n'$, then $\tilde{w}$ is the linear transformation $(x_i) \to (x_{i'})$. Then

$$\begin{aligned} M_f(t^w) &= \int_G f(x\tilde{w}t\tilde{w}^{-1}x^{-1})\,dx \\ &= \int_G f(xtx^{-1})\,dx \end{aligned}$$

since $dx = d(x\tilde{w})$. As $(t_1, \ldots, t_n)^{-(n-1)}$ is symmetric and Δ is skew-symmetric, we have (i).

(ii) This is an immediate consequence of the integration formula. We have

$$\Theta_{m_1,\ldots,m_n}(f) = \frac{1}{n!}\int_D \bar{\Delta}(t)\Delta(t)\Theta_{m_1,\ldots,m_n}(t)\left(\int_G f(xtx^{-1})\,dx\right)$$

$$= \frac{1}{n!}\int_D F_f(t)\left(\sum_w \varepsilon(w)\chi^w_{m_1,\ldots,m_n}(t)\right)dt$$

Here we have used the character formula and the easily verified identity

$$\bar{\Delta}(t) = (-1)^{n(n-1)/2}(t_1\cdots t_n)^{-(n-1)}\Delta(t)$$

So

$$\Theta_{m_1,\ldots,m_n}(f) = \frac{1}{n!}\sum_w \varepsilon(w)\int_D F_f\chi^w_{m_1,\ldots,m_n}\,dt$$

$$= \frac{1}{n!}\sum_w \varepsilon(w)\int_D F_f^{w^{-1}}\chi_{m_1,\ldots,m_n}\,dt$$

$$= \int_D F_f\chi_{m_1,\ldots,m_n}\,dt$$

in view of (i). This proves (ii).

(iii) We have, writing $u(t) = (-1)^{n(n-1)/2}(t_1\cdots t_n)^{-(n-1)}M_f(t)$,

$$(\partial(\textstyle\prod)F_f)(1) = (\partial(\textstyle\prod)(\Delta u))(1)$$

$$= n!\prod_{r<s}(s-r)(-1)^{n(n-1)/2}M_f(1)$$

$$= n!\prod_{r<s}(s-r)(-1)^{n(n-1)/2}f(1)$$

To obtain the Plancherel formula for G we use the Plancherel formula on D for the function F_f. If $u\in C^\infty(D)$ we have

$$\left(\frac{1}{i}\frac{\partial}{\partial\theta_r}u\right)^\wedge(m_1,\ldots,m_n) = -m_r\hat{u}(m_1,\ldots,m_n)$$

so that

$$(\partial(\textstyle\prod)u)^\wedge(m_1,\ldots,m_n) = (-1)^{n(n-1)/2}\prod_{r<s}(m_r-m_s)\hat{u}(m_1,\ldots,m_n)$$

Hence, calculating $(\partial(\prod)F_f)$ (1) by the Plancherel formula on D, we get

$$(-1)^{n(n-1)/2}n!\prod_{r<s}(s-r)f(1)$$

$$= (\partial(\textstyle\prod)F_f)(1)$$

$$= \sum_{m_1,\ldots,m_n}(\partial(\textstyle\prod)F_f)^\wedge(m_1,\ldots,m_n)$$

$$= (-1)^{n(n-1)/2}\sum_{m_1,\ldots,m_n}\left(\prod_{r<s}(m_r-m_s)\right)\hat{F}_f(m_1,\ldots,m_n)$$

We now observe that in this sum only the terms with *distinct* $(m_1,\ldots,m_n)$

contribute because $\prod_{r<s}(m_r - m_s)$ vanishes whenever there are coincidences. Moreover, since F_f is skew-symmetric, $\hat{F}_f$ is also skew-symmetric so that $\prod_{r<s}(m_r - m_s)$. $\hat{F}_f(m_1, \ldots, m_n)$ is symmetric. Hence,

$$n! \prod_{r<s} (s-r)\cdot f(1) = n! \sum_{m_1 > m_2 > \cdots > m_n} \prod_{r<s} (m_r - m_s)\hat{F}_f(m_1, \ldots, m_n)$$

This gives the Plancherel formula

$$f(1) = \sum_{m_1 > \cdots > m_n} d(\omega_{m_1,\ldots,m_n}) \quad \Theta_{m_1,\ldots,m_n}(f)$$

in view of the dimension formula and the calculation (ii) of the theorem above.

2.5 Weyl's theory

In a series of path-breaking memoirs [W2] Hermann Weyl developed the representation theory of *all compact semisimple Lie groups.* It is not an exaggeration to say that his work changed the entire landscape of the subject. Among other things he obtained the formulae for the character and dimension of the irreducible representations of which the results for $SU(n)$ described in the preceding section are special cases. Furthermore he used the character formula to obtain the characterization of the irreducible representations as highest-weight modules, thereby unifying the global and infinitesimal points of view. We shall now sketch, in this and in the next section, an outline of Weyl's ideas. For a full treatment the reader should consult [V2]; it is really a part of the theory of semisimple Lie algebras.

Weyl's method is based on the structure theory of compact connected Lie groups. If G is such a Lie group, and we write C for the connected component of the centre of G, the group G/C is semisimple, and if we denote by (G, G) the commutator subgroup of G, namely the smallest closed subgroup of G containing all the commutators $xyx^{-1}y^{-1}$ $(x, y \in G)$, the map $G \to G/C$ restricts on (G, G) to an isogeny, i.e., to a homomorphism which is surjective and has finite kernel. In other words, (G, G) is semisimple and $C \times (G, G) \to G$ is also an isogeny. In this way the structure and representation theory of G reduces to that of a semisimple group. For $G = U(n)$, $(G, G) = SU(n)$ and $C = \{t\cdot 1 \,|\, |t| = 1\}$.

The fundamental result on the structure of G is that G has maximal tori, that all such are mutually conjugate in G, and that, if T is one such, every element of G is conjugate to some element of T. Let us fix a maximal torus T. If $N_G(T)$ is the normalizer of T in G, then it can be shown that T is its own centralizer in G, and

$$W = N_G(T)/T$$

is finite; this finite group is known as the *Weyl group* of (G, T). The action by inner automorphisms of $N_G(T)$ on T is trivial when the acting element is in T and so gives rise to an action of W on T. Every conjugacy class of G meets T and the intersection is always a single orbit of W. If $\hat{T}$ is the group dual to T, W acts naturally on $\hat{T}$.

For any $x \in G$ we write $\mathrm{Ad}(x)$ for the action of x on $\mathfrak{g}$, the Lie algebra of G, so that $\mathrm{Ad}\,(x \mapsto \mathrm{Ad}\,(x))$ is the adjoint representation of G. The centralizer of x in $\mathfrak{g}$, say $\mathfrak{g}_x$, is by definition the subalgebra of $\mathfrak{g}$ of elements fixed by $\mathrm{Ad}\,(x)$. If $\dim(T) = l$, one can show that $\dim(\mathfrak{g}_x) \geqslant l$ for all x and that there is equality when x lies in a dense open subset of G. The integer l is known as the rank of G and the elements x for which $\dim(\mathfrak{g}_x) = l$ are said to be *regular*: G' is the set of regular elements, and we write T' for $T \cap G'$.

For describing Weyl's formula in this general context we now need the root-space decomposition of $\mathfrak{g}$. Let $\mathfrak{t} = \mathrm{Lie}(T)$ be the Lie algebra of T, viewed as usual as a subalgebra of $\mathfrak{g}$. Then $\mathfrak{t}$ is abelian and the compactness of T shows that the linear transformations

$$\mathrm{ad}\,H = (\mathrm{d}/\mathrm{d}t)_{t=0}(\mathrm{Ad}\,(\exp tH)) \quad (H \in \mathfrak{t})$$

are completely reducible. So if $\mathfrak{g}_{\mathbb{C}}$ is the complexification of $\mathfrak{g}$, and $\mathfrak{t}_{\mathbb{C}}$ that of $\mathfrak{t}$, we can write $\mathfrak{g}_{\mathbb{C}}$ as the direct sum of subspaces on each of which $\mathrm{ad}\,(\mathfrak{t}_{\mathbb{C}})$ acts through a linear function on $\mathfrak{t}_{\mathbb{C}}$. These linear functions are the *roots* of $(\mathfrak{g}, \mathfrak{t})$ and we write R for the (finite) set of all nonzero roots. For $\alpha \in R$ let

$$\mathfrak{g}_\alpha = \{X \in \mathfrak{g}_{\mathbb{C}} \mid [H, X] = \alpha(H)X\}$$

Then the $\mathfrak{g}_\alpha$ are the *root spaces, they are all of dimension 1*, and the decomposition we have been speaking about is

$$\mathfrak{g}_{\mathbb{C}} = \mathfrak{t}_{\mathbb{C}} \oplus \bigoplus_{\alpha \in R} \mathfrak{g}_\alpha$$

Each root space is stable under $\mathrm{Ad}\,(T)$ and $\mathrm{Ad}\,(T)$ acts on it through an element of $\hat{T}$; this element, written ξ_α, is called the *global root*, and we have

$$\mathrm{Ad}\,(t)\cdot X = \xi_\alpha(t)\cdot X \quad (t \in T, X \in \mathfrak{g}_\alpha)$$

Of course,

$$\xi_\alpha(\exp H) = \mathrm{e}^{\alpha(H)} \quad (H \in \mathfrak{t})$$

In particular, it follows from this formula (and the compactness of T) that each root is pure-imaginary-valued on $\mathfrak{t}$, showing that going over to the complexification is essential. We also note the following commutation rules which are very useful and quite easy to prove:

$$[\mathfrak{g}_\alpha, \mathfrak{g}_\beta] \begin{cases} = 0 & (\alpha + \beta \neq 0, \alpha + \beta \notin R) \\ \in \mathfrak{g}_{\alpha+\beta} & (\alpha + \beta \in R) \\ \in \mathfrak{t}_{\mathbb{C}} & (\alpha + \beta = 0) \end{cases}$$

If we remove from (it) the finite number of hyperplanes which are the zeros of the roots, we get a dense open set whose connected components are called *chambers* in it. Each chamber is a convex cone, and the subset of R of roots which are positive on a given chamber is called the *positive system of roots corresponding to that chamber*. Now the action of the group W on T induces an action on (it) and it can be shown that W acts *simply transitively* on the set of chambers, i.e., if C_1, C_2 are two chambers, there is exactly one element $w \in W$ such that $C_2 = wC_1$; in particular, there are exactly $|W|$ chambers. Let us write W again for the image of W in GL(it) via this action. Then W has the following beautiful geometric description in terms of the root system R; there is a natural structure of a Euclidean space for (it) such that if s_α is the reflection in the hyperplane of zeros of α W is the group generated by the s_α, $\alpha \in R$. The Euclidean structure actually exists on $\mathfrak{g}$ itself; if $\mathfrak{g}$ is semisimple, it is the structure associated with the invariant quadratic form

$$(X, X) = \mathrm{tr}((\mathrm{ad}\, X)^2) \quad (X \in \mathfrak{g})$$

the *Cartan–Killing* of $\mathfrak{g}$, and its restriction to $\mathfrak{t}$ is

$$(H, H) = \sum_{\alpha \in R} \alpha(H)^2 \quad (H \in \mathfrak{t})$$

Since each $\alpha \in R$ takes only pure imaginary values on $\mathfrak{t}$ and iR spans $\mathfrak{t}^*$, the positive definiteness of $(\cdot,\cdot)$ on (it) is clear. If $\mathfrak{g}$ is not semisimple, a simple modification is needed in this definition.

If $(\text{it})^+$ is any chamber and R^+ is the corresponding positive system of roots, the *simple roots of* R are those roots in R whose zeros form a part of the boundary of $(\text{it})^+$; if $S \subset R^+$ is the set of simple roots of R, then S has exactly l' elements where $l' = l - \dim(\text{centre}(\mathfrak{g}))$, these are linearly independent, and every element of R is a (unique) linear combination of elements of S with coefficients which are all integers $\geqslant 0$. Finally, $R = R^+ \coprod (-R^+)$. If $G = U(n)$, the rank is n, and the root spaces are the one-dimensional spaces $\mathbb{C} \cdot E_{rs}$ $(r \neq s)$ where E_{rs} are the matrix units, E_{rs} being the matrix with $\delta_{ir}\delta_{js}$ as its ijth entry. The roots are the linear functions

$$\alpha_{rs}: \mathrm{diag}(a_1, \ldots, a_n) \mapsto a_r - a_s$$

the global roots are the characters

$$\xi_{\alpha_{rs}}: \mathrm{diag}(t_1, \ldots, t_n) \mapsto t_r / t_s$$

and

$$R^+ = \{\alpha_{rs} | r < s\}$$

is a positive system; $l' = n - 1$, and

$$\alpha_{12}, \alpha_{23}, \ldots, \alpha_{n-1,n}$$

is the set of simple roots in R^+.

We are now in a position to introduce the alternating sums entering Weyl's formula. For any character χ of T let

$$\tilde{\chi} := \sum_{w} \varepsilon(w)\chi^{w} \quad (\varepsilon(w) = \det(w))$$

where $\det(w)$ is the determinant of the linear transformation of $\mathfrak{t}$ induced by w. Then $\tilde{\chi} = 0$ unless χ is such that all the transforms χ^{w} $(w \in W)$ are distinct; such χ are called *regular*, and, for regular χ, $\tilde{\chi}^{w} = \varepsilon(w)\tilde{\chi}$. To understand the meaning of regularity for a character χ observe that the surjective homomorphism $H \mapsto \exp H$ of $\mathfrak{t}$ onto T allows us to imbed T as a lattice in the real vector space $i\mathfrak{t}^*$ by identifying any character χ with the linear function $\lambda(\chi)$ such that

$$\chi(\exp H) = e^{\lambda(\chi)(H)} \quad (H \in \mathfrak{t})$$

So χ is regular if and only if all $w \cdot \lambda(\chi)$ $(w \in W)$ are distinct. But the interpretation of W (acting on $\mathfrak{t}$) as the group generated by the reflections in the root hyperplanes now allows us to prove that

$$\chi \text{ regular} \Leftrightarrow \alpha(H_{\lambda(\chi)}) \neq 0 \quad \text{for all } \alpha \in R,$$

where $H_{\lambda(\chi)} \in i\mathfrak{t}$ is defined by

$$(H_{\lambda(\chi)}, H) = \lambda(\chi)(H) \quad (H \in \mathfrak{t})$$

If R^+ is a positive system of roots, we say that χ is positive *with respect to* R^+ if and only if

$$H_{\lambda(\chi)} \in (i\mathfrak{t})^+$$

i.e., if and only if

$$\alpha(H_{\lambda(\chi)}) > 0 \quad \text{for all } \alpha \in R^+$$

Clearly positive χ are regular, and every regular χ can be moved to a positive one by applying a uniquely determined element of W. For $G = U(n)$ and for the positive system $R^+ = \{\alpha_{rs} | r < s\}$, the character $\chi_{m_1,\ldots,m_n}$ is positive if and only if $m_1 > \cdots > m_n$.

The function Δ, which is a special alternating sum, is more subtle to define. The correct way to do it is to begin by introducing the function

$$D(t) = \det((\mathrm{Ad}(t^{-1}) - 1)_{\mathfrak{g}/\mathfrak{t}}) \quad (t \in T)$$

where the suffix refers to the linear map induced on $\mathfrak{g}/\mathfrak{t}$. Clearly

$$D(t) = \prod_{\alpha \in R} (\xi_{\alpha}(t) - 1) \quad (t \in T)$$

and D is W-invariant; what we want to do is to define Δ as a finite Fourier series so that it is skew-symmetric and $\Delta\bar{\Delta} = D$. It turns out that this imposes a restriction on the fundamental group of G but that (fortunately) it can always be ensured by passing to a finite covering of G. For instance, when $G = SU(2)$ and $G_1 = PSU(2) = SU(2)/(\pm 1)$, $\Delta(u_\theta) = e^{i\theta} - e^{-i\theta}$ cannot be defined on the image of the diagonal group in G_1. To see this even more clearly, let us go over to $\mathfrak{t}$ via the exponential map $\mathfrak{t} \to T$ so that

$$D(\exp H) = \prod_{\alpha \in R} (e^{\alpha(H)} - 1) \quad (H \in \mathfrak{t})$$

If we now define

$$\Delta_{\mathfrak{t}}^+(H) = \prod_{\alpha \in R^+} (e^{\alpha(H)/2} - e^{-\alpha(H)/2}) \quad (H \in \mathfrak{t})$$

where R^+ is some positive system, we can easily show that

$$\Delta_{\mathfrak{t}}^{+w} = \varepsilon(w)\Delta_{\mathfrak{t}}^+, \quad D \circ \exp = \Delta_{\mathfrak{t}}{}^+ \bar{\Delta}_{\mathfrak{t}}{}^+$$

The 'square roots' of the global roots ξ_α that enter the definition of Δ^+ point to the difficulty of making Δ^+ go over to T. Actually it is enough to require that just a single appropriately chosen 'square root' be defined on T; for we can rewrite $\Delta_{\mathfrak{t}}^+$ as

$$\Delta_{\mathfrak{t}}^+ = e^{-\rho} \prod_{\alpha \in R^+} (e^\alpha - 1)$$

where

$$\rho = \frac{1}{2} \sum_{\alpha \in R^+} \alpha$$

is the famous half-sum of positive roots introduced by Weyl. We shall say that G is *acceptable* (Harish-Chandra's terminology) if there is a character ξ_ρ of T such that

$$\xi_\rho(\exp H) = e^{\rho(H)} \quad (H \in \mathfrak{t})$$

Since

$$w^\rho = \rho - \sum_{\substack{\beta \in R^+ \\ w^{-1}\beta \in -R^+}} \beta,$$

it is clear that all the characters $\xi_{w\rho}$ are also well defined, showing that acceptability is independent of the choice of R^+. For an acceptable G, we can define Δ^+ corresponding to R^+ by

$$\Delta^+ = \xi_\rho^{-1} \prod_{\alpha \in R^+} (\xi_\alpha - 1)$$

so that

$$(\Delta^+)^w = (-1)^{l(w)} \Delta^+$$

where

$$l(w) = \text{number of roots } \alpha \in R^+ \text{ for which } w^{-1}\alpha \in -R^+$$

and

$$D = \Delta^+ \overline{\Delta^+}$$

It can be shown that

$$\varepsilon(w) = (-1)^{l(w)}$$

so that Δ^+ is skew-symmetric. The key point in all this discussion is Weyl's discovery that a suitable covering group of G is always acceptable (and compact). Weyl's fundamental theorem is now the following ([V2], Theorem 4.14.3).

Theorem 9 (Weyl's character formula). *Let G be an acceptable compact connected Lie group and let R^+ be a positive system. Then the irreducible characters of G are in one-one correspondence with the positive characters of T; if $\chi \in \hat{T}$ is a positive character, the corresponding irreducible character Θ_χ of G is given by*

$$(\Theta_\chi)_T = \left(\sum_w \varepsilon(w) \chi^w \right) \Big/ \Delta^+$$

where

$$\Delta^+ = \xi_\rho^{-1} \prod_{\alpha \in R^+} (\xi_\alpha - 1)$$

Moreover, the character ξ_ρ is positive, Θ_{ξ_ρ} is the trivial character, and

$$\Delta^+ = \sum_w \varepsilon(w) \xi_\rho^w$$

To formulate the result giving Weyl's dimension formula we need one more notation. For any $\mu \in \mathfrak{it}^*$ let $H_\mu \in \mathfrak{it}$ be defined by

$$(H_\mu, H) = \mu(H) \quad (H \in \mathfrak{t})$$

(cf. the definition earlier of $H_{\lambda(\chi)}$). In particular, the H_α are all well-defined elements of it ($\alpha \in R$), and $\alpha(H_\alpha) > 0$. We now have the following.

Corollary 10 (Weyl's dimension formula). *Let χ be a positive character of T, $\lambda = \lambda(\chi)$, and let ω_λ be the irreducible class with character Θ_χ. If $d(\omega_\lambda)$ is the dimension of ω_λ, then*

$$d(\omega_\lambda) = \frac{\prod_{\alpha \in R^+} \lambda(H_\alpha)}{\prod_{\alpha \in R^+} \alpha(H_\alpha)}$$

In general a compact connected Lie group will have an infinite fundamental group; for instance, the fundamental group of $U(n)$ is $\mathbb{Z}$. But $SU(n)$ is simply connected and one can show that for the group $SO(n)$ the fundamental group is $\mathbb{Z}_2 = (\pm 1)$. It was a major discovery of Weyl that for a general compact connected semisimple Lie group its fundamental group is finite, so that its universal covering group is still compact. Weyl also proved that if G is compact and simply connected, and $T \subset G$ is a maximal torus, the character group $\hat{T}$, which we have identified naturally with a lattice in it^*, consists precisely of all those $\lambda \in it^*$ such that

$$\frac{2\lambda(H_\alpha)}{\alpha(H_\alpha)}$$

is an *integer* for every root α. Since one can easily check that ρ satisfies this condition, it follows that G is acceptable. Thus we get the following ([V2], Theorems 4.11.6 and 4.14.4).

Theorem 11 (Weyl). *The universal covering group of a compact connected semisimple Lie group is compact and acceptable.*

2.6 The infinitesimal approach

The fact that we are working with *Lie groups* allows us to approach their representation theory via the theory of representations of Lie algebras. For semisimple Lie groups this method was pioneered by Elie Cartan. It led him to the characterization of the finite-dimensional irreducible representations as highest-weight modules. Subsequently, when Harish-Chandra began his epoch-making work on infinite-dimensional representations of real semisimple Lie groups, it was to the infinitesimal method that he turned first for proving the fundamental results concerning the representations.

The idea behind the method is extremely simple. For every Lie group G we have its Lie algebra $\mathrm{Lie}(G) = \mathfrak{g}$, identified with the Lie algebra of all left-invariant vector fields on G. The assignment

$$G \to \mathrm{Lie}(G)$$

is functorial; if

$$\pi : G \to H$$

is a morphism of Lie groups, it gives rise to a morphism of Lie algebras

$$d\pi : \mathrm{Lie}(G) \to \mathrm{Lie}(H)$$

in a natural manner. If $X \in \mathrm{Lie}(G)$ and X_1 is the tangent vector at the identity element of G defined by the vector field X, then $d\pi$ is determined uniquely by the relation

$$(d\pi(X))_1 = (d\pi)_1(X_1)$$

The morphism $d\pi$ determines π on the connected component of the identity by the formula

$$\pi(\exp X) = \exp(d\pi(X)) \quad (X \in \mathfrak{g})$$

In general not every morphism from $\mathrm{Lie}(G)$ to $\mathrm{Lie}(H)$ is of the form $d\pi$ for some morphism π from G to H; this is the case however if G is connected and simply connected. The application to finite-dimensional representation theory arises now from the remark that an analytic representation of G in V is the same as a morphism $G \to GL(V)$; here, analytic refers to the property that the matrix elements are analytic functions on G. If $\mathfrak{gl}(V)$ is the Lie algebra of endomorphisms of V, we thus arrive at the result that for a connected Lie group G the assignment

$$\pi \to d\pi$$

is injective from the set of analytic representations of G in V into the set of Lie algebra representations of $\mathfrak{g}$ in V, and is even bijective when G is simply connected. To get $d\pi$ from π one uses the formula

$$d\pi(X) = (\mathrm{d}/\mathrm{d}t)_{t=0}\, \pi(\exp tX) \quad (X \in \mathfrak{g})$$

and to go from $d\pi$ to π one uses the formula

$$\pi(\exp X) = \exp(d\pi(X)) \quad (X \in \mathfrak{g})$$

It is clear from these formulae that π is irreducible if and only if $d\pi$ is irreducible, and in fact π and $d\pi$ have the same set of invariant subspaces; moreover

$$\mathrm{Hom}_G(\pi_1, \pi_2) = \mathrm{Hom}_{\mathfrak{g}}(d\pi_1, d\pi_2)$$

If our Lie groups are *complex*, analyticity is a substantial restriction on the representation; analytic representations are often called *holomorphic*. For instance, for $G = GL(n, \mathbb{C})$,

$$g \to g^{\mathrm{conj}} \quad (\mathrm{conj} = \text{complex conjugation})$$

is a representation which is *not* holomorphic. For *real* Lie groups there is no restriction. In fact let G be a connected real Lie group and π a representation of G in a finite-dimensional complex vector space V. If $GL(V)_{\mathbb{R}}$ (resp. $\mathfrak{gl}(V)_{\mathbb{R}}$) denotes $GL(V)$ regarded as a real Lie group (resp. $\mathfrak{gl}(V)$ regarded as a real Lie algebra), π can be proved to be a morphism from G to $GL(V)_{\mathbb{R}}$, so that $d\pi$ is a representation of $\mathfrak{g}$ in V. If $\mathfrak{g}_{\mathbb{C}}$ is the complexification of $\mathfrak{g}$, $d\pi$ extends to a representation, also denoted by $d\pi$, of $g_{\mathbb{C}}$ in V. Thus the assignment

$$\pi \mapsto d\pi$$

is injective from the set of representations of G in V to the set of representations of $\mathfrak{g}_{\mathbb{C}}$ in V, and bijective if G is in addition simply connected. If G is not simply connected and $\tilde{G}$ is the universal covering group of G with covering morphism $\gamma(\tilde{G} \to G)$, the map $\pi \mapsto d\pi$ has as its range the set of those representations of $\mathfrak{g}_{\mathbb{C}}$ in V such that the corresponding representation of $\tilde{G}$ is trivial on $\ker(\gamma)$; here $\mathrm{Lie}(\tilde{G}) = \mathrm{Lie}(G)$ and $d\gamma = $ identity.

The determination of the representations of the Lie algebra $\mathfrak{g}_{\mathbb{C}}$ is a *linear* problem and can be attacked by the methods of linear algebra. If $\{X_1,\ldots,X_n\}$ is a basis of $\mathfrak{g}_{\mathbb{C}}$ with commutation rules

$$[X_i,X_j]=\sum_k c_{ijk}X_k \quad (c_{ijk}\in\mathbb{C})$$

then giving a representation of $\mathfrak{g}_{\mathbb{C}}$ in V means the specifying of endomorphisms $L_1,\ldots,L_n$ of V such that

$$[L_i,L_j]:=L_iL_j-L_jL_i=\sum_k c_{ijk}L_k$$

which is in general a very tractable question. Let us illustrate this method for the group $G=SU(2)$. The Lie algebra $\mathfrak{g}$ is the Lie algebra of 2×2 skew-Hermitian matrices of trace 0 and so $\mathfrak{g}_{\mathbb{C}}$ may be identified with the Lie algebra of all 2×2 complex matrices of trace 0; indeed, if A is any $n\times n$ complex matrix of trace 0, we have the decomposition

$$A=(A-A^{\dagger})/2+(A+A^{\dagger})/2\mathrm{i}$$

so $A=S_1+\mathrm{i}S_2$ where $S_1=(A-A^{\dagger})/2$ and $S_2=(A+A^{\dagger})/2\mathrm{i}$ are skew-Hermitian. For $\mathfrak{g}_{\mathbb{C}}$ we take the so-called *standard basis*

$$H=\begin{pmatrix}1 & 0\\ 0 & -1\end{pmatrix},\quad X=\begin{pmatrix}0 & 1\\ 0 & 0\end{pmatrix},\quad Y=\begin{pmatrix}0 & 0\\ 1 & 0\end{pmatrix}$$

with the commutation rules

$$[H,X]=2X,\quad [H,Y]=-2Y,\quad [X,Y]=H$$

Let now V be a finite-dimensional complex vector space and τ a representation of $\mathfrak{g}_{\mathbb{C}}$ in V. Assume τ to be irreducible and let λ be an eigenvalue of $\tau(H)$ with the property that $\lambda+2$ is not an eigenvalue. To understand the significance of this choice let us notice the following lemma.

Lemma 12. *If $v\in V$ is an eigenvector of $\tau(H)$ for the eigenvalue μ, then $\tau(X)v$ (resp. $\tau(Y)v$) is an eigenvector for the eigenvalue $\mu+2$ (resp. $\mu-2$).*

Proof. This follows immediately from the relations

$$\tau(H)\tau(X)v=\tau(X)\tau(H)v+2\tau(X)v$$

$$\tau(H)\tau(Y)v=\tau(Y)\tau(H)v-2\tau(Y)v$$

So if we write S for the set of eigenvalues of $\tau(H)$ and V_μ $(\mu\in S)$ for the eigenspaces, then

$$V'=\bigoplus_{\mu\in S}V_\mu$$

is stable under $\tau(H)$, $\tau(X)$, $\tau(Y)$, hence stable under τ. The irreducibility of τ now shows that $V'=V$, i.e.,

$$V=\bigoplus_{\mu\in S}V_\mu$$

In other words, $\tau(H)$ is diagonalizable. If we choose λ as above, and $v \neq 0$ is an element of V_λ, we must have

$$\tau(H)v = \lambda v, \quad \tau(X)v = 0$$

We now consider the sequence of vectors

$$v_\lambda, v_{\lambda-2}, \ldots, v_{\lambda-2q}$$

where

$$v_\lambda = v, \quad v_{\lambda-2r} = \tau(Y)^r v$$

Since $v_{\lambda-2r} \in V_{\lambda-2r}$ we know that this sequence eventually becomes zero and so we can define q by the requirement that

$$v_{\lambda-2q} \neq 0, \quad \tau(Y)v_{\lambda-2q} = 0$$

Since the eigenvalues $\lambda, \lambda-2, \ldots, \lambda-2q$ are distinct, the vectors $v_\lambda, v_{\lambda-2}, \ldots, v_{\lambda-2q}$ are linearly independent. The vectors of V_μ are said to have *weight* μ, so that for our sequence $(v_{\lambda-2r})$ the subscripts are just the weights. The key to the entire calculation is now the following lemma.

Lemma 13 *If σ is any representation of $\mathfrak{g}_{\mathbb{C}}$, then for any integer $m \geqslant 0$*

$$\sigma(X)\sigma(Y)^{m+1} = \sigma(Y)^{m+1}\sigma(X) + (m+1)\sigma(Y)^m(\sigma(H) - m\cdot 1)$$

Proof. For $m = 0$ this is just the relation

$$\sigma(X)\sigma(Y) = \sigma(Y)\sigma(X) + \sigma(H)$$

which follows from the commutation rule $[X, Y] = H$. For $m \geqslant 1$ we use induction on m and the relation $\sigma(H)\sigma(Y) = \sigma(Y)(\sigma(H) - 2\cdot 1)$ to get

$$\begin{aligned}\sigma(X)\sigma(Y)^{m+1} &= \sigma(X)\sigma(Y)^m\sigma(Y)\\ &= \{\sigma(Y)^m\sigma(X) + m\sigma(Y)^{m-1}(\sigma(H) - (m-1)\cdot 1)\}\sigma(Y)\\ &= \sigma(Y)^m(\sigma(Y)\sigma(X) + \sigma(H)) + m\sigma(Y)^{m-1}\sigma(Y)(\sigma(H) - (m+1)\cdot 1)\\ &= \sigma(Y)^{m+1}\sigma(X) + (m+1)\sigma(Y)^m(\sigma(H) - m\cdot 1).\end{aligned}$$

Resuming our discussion and taking $\sigma = \tau$ in the lemma, we get, on applying both sides of the identity to the vector v,

$$\tau(X)v_{\lambda-2(m+1)} = (\lambda - m)(m+1)v_{\lambda-2m}$$

This formula shows that the linear span of $(v_{\lambda-2m})_{0\leqslant m\leqslant q}$ is stable under τ, so that, once again, by irreducibility,

$$V = \bigoplus_{0\leqslant m\leqslant q} \mathbb{C}v_{\lambda-2m}$$

At the same time, taking $m = q$, we see that

$$0 = (\lambda - q)(q+1)v_{\lambda-2q}$$

which shows that $\lambda = q$. In other words, *λ itself is an integer $\geqslant 0$ and the spectrum of $\tau(H)$ is*

$$S_\lambda = \{\lambda, \lambda - 2, \ldots, -\lambda\} \quad (\lambda \text{ integer} \geqslant 0)$$

These calculations lead to the following result.

Theorem 14. *The irreducible representations of $\mathfrak{g}_{\mathbb{C}} = \mathfrak{sl}(2, \mathbb{C})$ are parametrized (up to equivalence) by integers $\lambda \geqslant 0$. For any integer $\lambda \geqslant 0$ the corresponding irreducible representation τ_λ has dimension $\lambda + 1$; and there is a basis $(v_{\lambda - 2m})_{0 \leqslant m \leqslant \lambda}$ such that $\tau_\lambda(H)$, $\tau_\lambda(Y)$ and $\tau_\lambda(X)$ are given by the following formulae:*

$$\tau_\lambda(H) v_{\lambda - 2m} = (\lambda - 2m) v_{\lambda - 2m}$$
$$\tau_\lambda(Y) v_{\lambda - 2m} = v_{\lambda - 2(m+1)}, \quad \tau_\lambda(Y) v_{-\lambda} = 0$$
$$\tau_\lambda(X) v_{\lambda - 2(m+1)} = (\lambda - m)(m+1) v_{\lambda - 2m}, \quad \tau_\lambda(X) v_\lambda = 0$$

We can represent the vectors v schematically by

$$\begin{array}{ccc} & \bullet\lambda & \\ & \bullet\lambda_{-2} & Y \\ \uparrow & & \downarrow \\ X & & \\ & \bullet -\lambda & \end{array}$$

with the weights alongside. Then λ is the 'highest' weight, $-\lambda$ is the 'lowest' weight; the action of X (or rather $\tau(X)$) moves the vectors upwards, one step at a time, while the action of Y moves them downwards, one step at a time. The vector v_λ (resp. $v_{-\lambda}$) is characterized up to a scalar multiple by the equation

$$\tau(X) v_\lambda = 0 \quad (\text{resp. } \tau(Y) v_{-\lambda} = 0)$$

The corresponding group representation can be realized in the space of polynomials of two variables. Since $SU(2)$ is simply connected, representations of $\mathfrak{sl}(2, \mathbb{C})$ are the same as representations of $SU(2)$ and so are completely reducible. The reader should try to work out an algebraic proof of this result.

With the help of the structure theory of semisimple Lie algebras one can carry over the arguments given above to obtain a view of all the representations of a compact Lie group from the infinitesimal standpoint. Let us go back to the discussion of section 5 and the notations used there. Let R be the set of roots of $(\mathfrak{g}, \mathfrak{t})$, $\mathfrak{g}$ being the Lie algebra of the compact connected Lie group G. We select a chamber $(i\mathfrak{t})^+$ and denote by R^+ the corresponding set of positive roots. Let π be an irreducible representation of $\mathfrak{g}_{\mathbb{C}}$ in a finite-dimensional vector space V. For any $\mu \in \mathfrak{t}_{\mathbb{C}}^*$ put

$$V_\mu = \{v \in V \mid \pi(H)v = \mu(H)v \quad \text{for all } H \in \mathfrak{t}_{\mathbb{C}}\}$$

μ is called a *weight* of π if $V_\mu \neq 0$; $W(\pi)$ is the (finite) set of weights of π. The commutation rules

$$[H, X_\alpha] = \alpha(H)X_\alpha \quad (H \in \mathfrak{t}_{\mathbb{C}}, \alpha \in R,\ X_\alpha \in \mathfrak{g}_\alpha)$$

now imply that

$$\pi(X_\alpha)V_\mu \subset V_{\mu+\alpha}$$

Here, as π is irreducible,

$$V = \bigoplus_{\mu \in W(\pi)} V_\mu$$

(if π arises from a representation of G, which we also denote by π, the restriction of π to T is a direct sum of characters of T and so V is the direct sum of weight spaces. Actually V is always the direct sum of the weight spaces but this needs a little more effort to prove). We call a weight μ *maximal* if $\mu + \alpha$ is not a weight for any $\alpha \in R^+$. Since any π has only finitely many weights, there are always maximal weights. If $\lambda \in W(\pi)$ is maximal and $0 \neq v_\lambda \in V_\lambda$, then

$$\pi(X_\alpha)v_\lambda = 0 \quad (\alpha \in R^+)$$

It can then be shown that the linear span of all vectors of the form

$$\pi(X_{-\alpha_1})\pi(X_{-\alpha_2}) \cdots \pi(X_{-\alpha_m})(v_\lambda)$$

where $m \geqslant 0$, $\alpha_1, \alpha_2, \ldots, \alpha_m \in R^+$ (if $m = 0$, this vector is v_λ by convention) is stable under π and hence, as π is irreducible, must coincide with V; at the same time this shows that any weight of π must have the form

$$\lambda - \sum_{\alpha \in R^+} m(\alpha)\alpha \quad (m(\alpha) \text{ integers} \geqslant 0)$$

In other words, if we introduce a partial ordering on $\mathfrak{t}_{\mathbb{C}}^*$, $\leqslant$, by defining

$$\mu_1 \leqslant \mu_2 \Leftrightarrow \mu_2 - \mu_1 = \sum_{\alpha \in R^+} m(\alpha)\alpha \quad (m(\alpha) \geqslant 0)$$

then λ is *the highest weight of* π.

It is obvious that the highest weight of an irreducible representation π is uniquely determined by π. It is less obvious, although true, that conversely the highest weight determines the representation. So the problem of explicit determination of the irreducible representations reduces to the problem of determining which linear functions on $\mathfrak{t}_{\mathbb{C}}$ can be highest weights (of irreducible representations of $\mathfrak{g}_{\mathbb{C}}$). It turns out that if $\mathfrak{g}_{\mathbb{C}}$ (or G) is *semisimple* a linear function λ on $\mathfrak{t}_{\mathbb{C}}$ is a highest weight if and only if it satisfies the following remarkable *integrality condition*:

$$\frac{2\lambda(H_\alpha)}{\alpha(H_\alpha)}\in\mathbb{Z}^+ \quad \text{for all } \alpha\in R^+$$

The reader will recognize this as a generalization of the integrality condition encountered in the representation theory of $\mathfrak{sl}(2,\mathbb{C})$. If $\tilde{G}$ is the universal covering group of G, the representations of $\tilde{G}$ thus have their highest weights in a *lattice* – namely the lattice of all $\lambda\in\mathfrak{t}_{\mathbb{C}}^*$ with $2\lambda(H_\alpha)/\alpha(H_\alpha)\in\mathbb{Z}$ for all $\alpha\in R$.

It is now quite easy to describe how the representations of G are singled out; they are the ones *whose highest weights define characters of T*. If G is simply connected this is automatically satisfied (for λ satisfying the integrality condition); and the character of the irreducible representation with highest weight λ is then $\Theta_{\lambda+\rho}$, with $\rho=(\sum_{\alpha\in R^+}\alpha)/2$; this can be deduced from Weyl's character formula exactly as in the case of the unitary group.

This last result makes it clear why the irreducible representation is determined by the knowledge of its highest weight. This result, as well as the characterization of the highest weights, emerges very naturally in the treatment of Weyl, but becomes quite difficult to establish in the purely algebraic development; in fact Cartan had to resort to tedious calculations and the classification of simple Lie algebras to prove them. It was only in the early 1950s, when Chevalley and Harish-Chandra made a deep study of the representations of the universal enveloping algebra of a semisimple Lie algebra, that a purely algebraic and entirely conceptual approach to these results was found ([V2] Ch. 4, §§6–7).

For groups $SU(n)$ where everything is quite explicit, it is not difficult to prove some of these statements. The roots are the linear forms

$$\alpha_{rs}\colon \operatorname{diag}(a_1,\ldots,a_n)\mapsto a_r-a_s \quad (r\neq s)$$

and R^+ consists of the α_{rs} with $r<s$. The Cartan–Killing form is given on $\mathfrak{t}_{\mathbb{C}}$ by

$$(\operatorname{diag}(a_1,\ldots,a_n),\ \operatorname{diag}(b_1,\ldots,b_n))=2n\sum_j a_jb_j$$

so that

$$H_{\alpha_{rs}}=(2n)^{-1}(\ldots,\overset{\overset{r}{\downarrow}}{1},\ldots,\overset{\overset{s}{\downarrow}}{-1},\ldots) \quad (r<s)$$

(zero for all other entries). If

$$H_{rs}=(\ldots,1,\ldots,-1,\ldots)$$

the elements

$$H_{rs}, X_{rs} = E_{rs}, \quad Y_{rs} = E_{sr}$$

form a standard basis of a subalgebra isomorphic to $\mathfrak{sl}(2, \mathbb{C})$, and the representation theory of $\mathfrak{sl}(2, \mathbb{C})$ allows us to conclude that, if

$$\mu: \operatorname{diag}(a_1, \ldots, a_n) \mapsto \mu_1 a_1 + \cdots + \mu_n a_n$$

is a weight, we must have the integrality condition

$$\mu_r - \mu_s \in \mathbb{Z}$$

and that, if $\lambda \in W(\pi)$ is maximal,

$$\lambda_r - \lambda_s \quad \text{is an integer } \geqslant 0 \quad \text{for } r < s$$

Let $0 \neq v_\lambda \in V_\lambda$, λ maximal and let U be the linear span of v_λ and the vectors

$$u = \pi(Y_{r_1 s_1})\pi(Y_{r_2 s_2}) \cdots \pi(Y_{r_m s_m})v_\lambda \quad (r_j < s_j, m \geqslant 1) \tag{*}$$

U is stable under $\pi(\mathfrak{t})$ and $\pi(Y_{rs})$ $(r < s)$. The commutation rules

$$[X_{rs}, X_{st}] = X_{rt} \quad (r < s < t)$$

show that the elements

$$X_{12}, X_{23}, \ldots, X_{n-1,n}$$

generate the Lie algebra $\sum_{r<s} \mathbb{C} X_{rs}$ and so, to prove that U is stable under $\pi(\mathfrak{g})$ it suffices to prove stability under all $\pi(X_{r,r+1})$. We can do this by an induction on the integer m occurring in (*). For $m = 0$ the vector u is v_λ by convention, and $\pi(X_{rs})u = 0$ by maximality of λ. Let then $m \geqslant 1$. We need the following lemma.

Lemma 15. *If $r_1 < s_1$, the commutator $[X_{r,r+1}, Y_{r_1 s_1}]$ is $H_{r,r+1}$ when $r_1 = r$, $s_1 = r+1$, $Y_{r_1 r}$ when $r_1 < r < r+1 = s_1$, $-Y_{r+1,s_1}$ when $r_1 = r < r+1 < s_1$, 0 in all other cases.*

Proof. The easiest way to see this is to remark that the commutator lies in $[\mathfrak{g}_\alpha, \mathfrak{g}_\beta]$ where $\alpha = \alpha_{r,r+1}$, $\beta = \alpha_{s_1 r_1}$, so that we may assume that $\alpha + \beta$ is a root. Since $\alpha + \beta$ is the linear form $\operatorname{diag}(a_1, \ldots, a_n) \mapsto a_r - a_{r+1} + a_{s_1} - a_{r_1}$, it is a root only if either $s_1 = r+1$, $r \neq r_1$ so that $r_1 < r < r+1 = s_1$, or $r = r_1$, $r+1 \neq s_1$, so that $r_1 = r < r+1 < s_1$, coinciding with $\alpha_{r r_1}$ $(r_1 < r)$ or $\alpha_{s_1, r+1}$ $(r+1 < s_1)$, so that the commutator is $Y_{r_1 r}$ or $-Y_{r+1,s_1}$.

Continuing the discussion, we have

$$\begin{aligned} \pi(X_{r,r+1})u &= \{\pi(Y_{r_1 s_1})\pi(X_{r,r+1}) + \pi([X_{r,r+1}, Y_{r_1 s_1}])\}\pi(Y_{r_2 s_2}) \cdots \pi(Y_{r_m s_m})v_\lambda \\ &= u_1 + u_2 \quad \text{(say)} \end{aligned}$$

But

$$u_1 = \pi(Y_{r_1s_1})\pi(X_{r,r+1})\pi(Y_{r_2s_2})\cdots\pi(Y_{r_ms_m})v_\lambda$$
$$\in\pi(Y_{r_1s_1})U \quad \text{(induction)}$$
$$\subset U$$

and, by the lemma

$$u_2\in(\pi(\mathfrak{t}_{\mathbb{C}}) + \sum_{p<q} \mathbb{C}\pi(Y_{pq}))\pi(Y_{r_2s_2})\cdots\pi(Y_{r_ms_m})v_\lambda$$
$$\subset(\pi(\mathfrak{t}_{\mathbb{C}}) + \sum_{p<q} \mathbb{C}\pi(Y_{pq}))(U)$$
$$\subset U$$

At this stage we know that maximal weights are highest weights. The theory can be completed now without difficulty. The argument is the same in the general case also; instead of $X_{r,r+1}$ one uses X_α where $\alpha\in R^+$ is simple; the key lemma is then the assertion that, for any $\beta\in R^+$ for which $\alpha-\beta$ is a root, $\alpha-\beta\in -R^+$.

2.7 Compact groups and complex groups. Unitarian trick

Our development has led us naturally to the close relationship that exists between compact groups and complex groups. We begin by discussing this in the special case when $G = SU(n)$ and $G_{\mathbb{C}} = SL(n, \mathbb{C})$.

Let V be a vector space and π a *holomorphic* representation of $G_{\mathbb{C}}$ in V The restriction of π to G is then a representation of G in V in the usual sense; we write π^0 for this. We thus have an assignment

$$\pi\mapsto\pi^0$$

from the category of holomorphic representations of $G_{\mathbb{C}}$ to the category of *continuous representations of G. It is a functor and it is a remarkable fact that this functor establishes an isomorphism of categories.* In more detail we have

Theorem 16. *Every continuous representation G is of the form π^0, where π is a holomorphic representation of $G_{\mathbb{C}}$ and is uniquely determined by π^0. π and π^0 have the same collection of invariant subspaces; if π_j $(j = 1,2)$ is a holomorphic representation of $G_{\mathbb{C}}$ in V_j and $L(V_1 \to V_2)$ is a linear map, then L intertwines π_1 and π_2 if and only if it intertwines π_1^0 and π_2^0. Finally all holomorphic representations of $G_{\mathbb{C}}$ are rational and they are all completely reducible.*

The key fact in the proof is the following lemma which is a special case of the celebrated *unitarian trick* of Hermann Weyl ([V2], Lemma 4.11.13).

Lemma 17. *If F is a holomorphic function defined on a connected open subset N of $G_{\mathbb{C}}$ containing G, and $F = 0$ on G, then $F = 0$.*

Proof. It is a question of proving that all derivatives of F vanish on G (or even at a single point of G). The complex tangent space to $G_{\mathbb{C}}$ at a point $u \in G$ may be identified with $\mathfrak{g}_{\mathbb{C}} =$ the space of complex $n \times n$ matrices of trace 0, while the real tangent space to G at u may be identified with $\mathfrak{g} =$ the space of $n \times n$ skew-Hermitian matrices of trace 0; for $X \in \mathfrak{g}_{\mathbb{C}}$, the map $s \mapsto u \exp sX$ $(s \in \mathbb{C})$ is a (complex) curve in $G_{\mathbb{C}}$ through u which has X as the tangent vector at u, and, when $X \in \mathfrak{g}$ and s is restricted to $\mathbb{R}$, the same map defines a curve in G through u with X as its tangent vector at u. We now observe that $\mathfrak{g}_{\mathbb{C}}$ is the complex *linear span* of $\mathfrak{g}$. As a consequence,

$$\begin{aligned} F = 0 \text{ on } G &\Rightarrow \partial(X)F = 0 \text{ on } G \quad \text{for all } X \in \mathfrak{g} \\ &\Rightarrow \partial(X)F = 0 \text{ on } G \quad \text{for all } X \in \mathfrak{g}_{\mathbb{C}} \end{aligned}$$

Thus, for any $X \in \mathfrak{g}_{\mathbb{C}}$, $\partial(X)F$ satisfies the same assumptions as F. It follows that $\partial(X_1)\partial(X_2)\cdots\partial(X_r)F = 0$ on G for all $r \geqslant 1$, $X_1, \ldots, X_r \in \mathfrak{g}_{\mathbb{C}}$.

Proof of theorem. If π_1 and π_2 are holomorphic representations of $G_{\mathbb{C}}$ and $L(V_1 \to V_2)$ a linear map, the condition $L\pi_1(g) = \pi_2(g)L$ for all $g \in G_{\mathbb{C}}$ is a set of polynomial identities, and the lemma above shows that they are valid as soon as they are satisfied for $g \in G$. The same argument shows that π and π^0 leave the same subspaces invariant. It therefore only remains to show that every continuous representation of G is of the form π^0 for a *rational* representation π; the uniqueness of π will follow from the above lemma. Let σ be the representation $u \mapsto u$ $(u \in G_{\mathbb{C}})$ and $\check{\sigma}$ the representation $u \mapsto (u^{-1})^t$ contragradient to σ. Obviously σ and $\check{\sigma}$ are rational and $(\check{\sigma})^0 = (\sigma^0)^{\text{conj}}$. Since σ^0 is already faithful, the algebra generated by the matrix elements of σ^0 and $(\sigma^0)^{\text{conj}}$ is closed under complex conjugation and separates the points of G, hence dense in $C(G)$. This implies (cf. ex.2) that *every* irreducible representation of G occurs as a direct summand in the tensor products $\tau_1 \otimes \tau_2 \otimes \cdots \otimes \tau_N$ where each τ_j is either σ^0 or $(\sigma^0)^{\text{conj}}$. Now $\tau_j = \pi_j^0$ where π_j is either σ or $\check{\sigma}$, and any subspace in the space of $\tau_1 \otimes \cdots \otimes \tau_N$, stable under G, is necessarily stable under $G_{\mathbb{C}}$, by our earlier remarks. Thus any irreducible representation of G is of the form π^0 where π is a rational representation of $G_{\mathbb{C}}$ and occurs as a direct summand of some $\pi_1 \otimes \cdots \otimes \pi_N$. The rest of the theorem follows easily.

These arguments can be carried over to the general context. The fundamental ideas remain the same but the technical details are somewhat less trivial. I shall limit myself to just a brief description of the outline of the proofs ([V2], Theorem 4.11.14). One starts with a compact simply connected semisimple Lie group G with the Lie algebra $\mathfrak{g}$, and introduces the complex simply connected Lie group $G_{\mathbb{C}}$ corresponding to $\mathfrak{g}_{\mathbb{C}}$; then $\mathfrak{g} \subsetneq \mathfrak{g}_{\mathbb{C}}$ defines a morphism $G \to G_{\mathbb{C}}$ of *real* Lie groups. It is not obvious that this latter morphism is injective, but this can be proved to be the case, so that we may view G as naturally imbedded in $G_{\mathbb{C}}$, $G \subset G_{\mathbb{C}}$. The unitarian trick for the pair $(G, G_{\mathbb{C}})$ can now be formulated and proved exactly as we did for the pair $(SU(n), SL(n, \mathbb{C}))$. On the other hand, suppose we start with a complex semisimple Lie algebra $\mathfrak{l}$; then one can construct a *real form* $\mathfrak{g}$ of $\mathfrak{l}$ with the property that the Cartan–Killing form of $\mathfrak{g}$ is real and negative-definite on $\mathfrak{g}$. Then the adjoint group of $\mathfrak{g}$ is compact and so its universal covering group is compact, showing that any complex simply connected semisimple group occurs as the second member of a pair $(G, G_{\mathbb{C}})$ considered above. Finally G can be characterized as a maximal compact subgroup of G and shown to be unique up to conjugation by an inner automorphism of $G_{\mathbb{C}}$.

This discussion has been entirely transcendental and it is not at all clear how one can view $G_{\mathbb{C}}$ as an *algebraic group*; it is obviously necessary to do this before one can speak of *rational* representations. One of the most direct ways of doing this is to introduce the algebra A generated by the matrix elements of the irreducible holomorphic representations of $G_{\mathbb{C}}$; A is nothing more than the *linear span* of the matrix elements of all the (not just the irreducible) holomorphic representations of $G_{\mathbb{C}}$. We now use the fact that G has a faithful representation to conclude that A is *finitely generated*. A is thus an *affine algebra* in the language of algebraic geometry. Moreover, A has a canonical *involution* which we write as the map $a \mapsto a^*$; it is an automorphism of period 2 of the complex algebra A, and is completely determined by the requirement

$$a^*|_G = (a|_G)^{\mathrm{conj}}$$

Since A is an affine algebra we can consider its *spectrum* Spec(A), namely, the set of homomorphisms of A into $\mathbb{C}$, and view it as an algebraic variety. The central result is now the following.

Theorem 18 (Tannaka–Chevalley). *We have*

$$\mathrm{Spec}(A) = G_{\mathbb{C}}$$

and, if for any $\xi \in \mathrm{Spec}(A)$ *we denote by* ξ^* *the element* $a \mapsto \xi(a^*)$ *of* Spec(A),

$$G = \{\xi \in \mathrm{Spec}(A) \mid \xi = \xi^*\}$$

In particular, the natural structure of an algebraic variety on Spec(A) *converts* $G_{\mathbb{C}}$ *into an affine algebraic group with algebraic group structure compatible with the holomorphic structure, and with the property that the holomorphic representations are precisely the rational ones.*

For a detailed treatment that leads to this theorem see [C] (cf. also [Y]).

To conclude this discussion let me show how the complete reducibility of the rational representations of $SL(n, \mathbb{C})$ leads to the fundamental theorem of the theory of invariants asserting that the algebra of invariants is finitely generated. Let π be a rational representation of $SL(n, \mathbb{C})$ in a vector space V; let $P = P(V)$ be the algebra of polynomials on V and $I = P^G$ the subalgebra of P of elements invariant under $SL(n, \mathbb{C})$ acting through π. More generally let us consider the action of $SL(n, \mathbb{C})$ on P, and, for any stable subspace L of P let us consider the subspaces L^ω 'transforming like ω', ω being any element of the set $\mathscr{E}$ of equivalent classes of irreducible rational representations of $SL(n, \mathbb{C})$; in more detail, L^ω is the linear span of subspaces $F \subset L$ with the property that F is stable under $SL(n, \mathbb{C})$ and the representation on F is irreducible and belongs to the class ω. If P_d $(d \geqslant 0)$ are the homogeneous components of P, the action of $SL(n, \mathbb{C})$ leaves each P_d invariant and the representation of $SL(n, \mathbb{C})$ on P_d is also rational. So, by the complete reducibility of the rational representations, $P_d = \bigoplus_\omega P_d^\omega$, so that, summing over d, we have

$$P = \bigoplus_{\omega} P^\omega, \quad P^\omega = \bigoplus_{d \geqslant 0} P_d^\omega$$

The P_d^ω are all finite-dimensional, but the P^ω are not so in general. If ω_0 is the class of the trivial representation, $P^{\omega_0} = I$. We now have the following.

Lemma 19. *The P^ω are stable under multiplication by elements of I, i.e., the P^ω are I-modules. If E is the projection operator $P \to I$ modulo $\sum_{\omega \neq \omega_0} P^\omega$ then E has the following properties:*

(i) *E maps P into I and is the identity on I,*
(ii) *if $f \in P_d, Ef \in I \cap P_d$,*
(iii) *E commutes with I: $E(gf) = gEf$ if $g \in I$, $f \in P$,*
(iv) *E commutes with $SL(n, \mathbb{C})$: $E(f^x) = Ef (x \in SL(n, \mathbb{C}))$.*

Proof. For the first statement we must show that if $F \subset P$ is irreducibly invariant and defines a representation of the class ω, and $g \in I$, gF has the same property as F. If (f_j) is a basis for F and $f_j^x = \sum_i a_{ij}(x) f_i$ $(x \in SL(n, \mathbb{C}))$ we have $(gf_j)^x = \sum_i a_{ij}(x)(gf_i)$, and our assertion is obvious. The properties of E are now immediate.

Once we have this lemma, the classical argument of Hilbert goes through without change. Let me recount it here for the sake of completeness. Let

I_+ be the set of all $f \in I$ with $f(0)=0$ and let $P_+ = PI_+$. Since P_+ is an ideal in P we can find an ideal basis for it; a little thought will show that we can arrange matters so that the elements $i_1, \ldots, i_m$ of the ideal basis are homogeneous elements of I_+. Then the claim is that I is *generated as an algebra over* $\mathbb{C}$ by $i_1, \ldots, i_m$. This is proved by showing that $I \cap P_d := I_d \subset \mathbb{C}[i_1, \ldots, i_m]$ for all $d \geqslant 0$, $\mathbb{C}[i_1, \ldots, i_m]$ being the algebra over $\mathbb{C}$ generated by $i_1, \ldots, i_m$. Write $d_j = \deg(i_j)$; $d_j \geqslant 1$ obviously. This last statement is proved by induction on d. For $d=0$ it is obvious and so let $d \geqslant 1$ and let us suppose that $I_e \subset \mathbb{C}[i_1, \ldots, i_m]$ for $e<d$. Let $f \in I_d$ and write $f = \sum_j g_j i_j$, $g_j \in P$; we may clearly assume that the g_j are homogeneous, that $\deg(g_j) = d - d_j$, and that the sum is only over those j for which $d_j \leqslant d$. We now apply the operator E to f and get

$$Ef = f = \sum_j (Eg_j) i_j$$

But $Eg_j \in I_{d-d_j}$ and $d - d_j < d$, so that $Eg_j \in \mathbb{C}[i_1, \ldots, i_m]$ by the induction hypothesis. Hence $f \in \mathbb{C}[i_1, \ldots, i_m]$.

Remark. Since E is the projection on I it can be constructed much more directly by the formula

$$Ef = \int_{SU(n)} f^u \, du$$

The properties (i)–(iv) are immediate (for (iv) we use the unitarian trick). However, the P^ω constitute an important aspect of the situation and so it was not unreasonable to use them to define E. For the actual argument of Hilbert, properties (i)–(iv) are enough; they do not characterize the E we have used. In fact Hilbert worked with a different operator (Cayley's 'Ω-process'). It is clear that the construction of E, and hence the finite generation theorem also, goes over to the case of rational representations of complex semisimple groups [MF].

Notes and comments

For a thorough study of compact groups and their representations Zhelebenko's book [Z] is a good reference. For the semisimple theory from both the global and infinitesimal points of view the reader might consult my book [V2], for the infinitesimal theory, the book of Serre [S3].

Problems

1. Fill in the details in the proof of Theorem 1.
2. Let A be as in Theorem 2 and $B \subset A$ a subalgebra containing 1 and closed under complex conjugation. Prove that, if B is dense in A, then

$B = A$ (Hint: use the orthogonality relations). Deduce that, if L is a faithful irreducible unitary representation of G, then all the irreducible representations of G occur as constituents in some $L \otimes \cdots \otimes L \otimes \check{L} \otimes \cdots \otimes \check{L}$ ($\check{L}$= contragradient of L).

3. Prove that any compact Lie group G has a faithful unitary representation (Hint: use the fact that G has no small subgroups).

4. Let $G = SU(2)$ and let L_m be the irreducible representation of dimension $m+1$. Use the character formula to verify the *Clebsch–Gordon formula*:

$$L_{m_1} \otimes L_{m_2} = L_{|m_1 - m_2|} \oplus L_{|m_1 - m_1|+2} \oplus \cdots \oplus L_{m_1 + m_2}$$

Similarly, for $G = SU(3)$, let L_{m_1, m_2} be the irreducible representation with character $\Theta_{m_1+2, m_2+1, 0}$ ($m_1 \geqslant m_2$); prove that

$$L_{m,0} \otimes L_{0,m} = L_{0,0} \oplus L_{1,1} \oplus \cdots \oplus L_{m,m}$$

5. Let $G = U(n)$, $\mathfrak{g} = \mathfrak{u}(n)$, and for $Z, Z' \in \mathfrak{g}$ let $(Z, Z') = -\operatorname{tr}(ZZ')$. For $Z \in \mathfrak{g}$, $\partial(Z)$ denotes the corresponding left-invariant vector field viewed as differential operator so that $(\partial(Z)f)(x) = ((\mathrm{d}/\mathrm{d}t)f(x \exp tZ))_{t=0}$, $f \in C^\infty(G)$.

(a) Prove that $(\cdot,\cdot)$ is positive-definite and G-invariant

(b) Let E_{rs} be the matrix units and let $H_r = \mathrm{i}E_{rr}$, $Z_{rs} = E_{rs} - E_{sr}$, $Z'_{rs} = \mathrm{i}(E_{rs} + E_{sr})$ $(r < s)$. Verify that $(H_r, 2^{-1/2}Z_{rs}, 2^{-1/2}Z'_{rs})$ form an orthonormal basis of $\mathfrak{g}$. Deduce that

$$\square = \sum_r \partial(H_r)^2 + \frac{1}{2}\left(\sum_{r<s} (\partial(Z_{rs})^2 + \sum_{r<s} \partial(Z'_{rs})^2) \right)$$

is a differential operator invariant under both left and right translations.

(c) Rewriting $\square$ in the form

$$\square = \sum_r \partial(H_r)^2 + \sum_{r<s} (\partial(E_{rs})\partial(E_{sr}) + \partial(E_{sr})\partial(E_{rs}))$$

prove that, if f is a matrix element of the representation of G of highest weight $k_1, k_2, \ldots, k_n$ $(k_1 \geqslant \cdots \geqslant k_n)$, then

$$\square f = \left(\sum_r k_r^2 + \sum_{r<s} (k_r - k_s) \right) f$$

(Hint: compute when f is highest weight vector.)

(d) Use (c) to prove that if $g \in C^\infty(G)$, its Fourier series converges to g in the topology of $C^\infty(G)$ (Hint: if $\mu(k_1, \ldots, k_n)$ is the eigenvalue computed in (c), observe that $f = \mu^{-r}\square^r f$ and that

$$\sum (\mu(k_1, \ldots, k_n) + 1)^{-r} \text{ is convergent if } r \gg 1).$$

6. Let $G = U(n)$ and f a C^∞ function on D which is skew-symmetric. Show that we can write $f = \Delta g$ where $g \in C^\infty(D)$ and symmetric.

3

Unitary representations of locally compact groups

3.1 Brief history

The representation theory of groups which are either compact or abelian was essentially complete by the 1930s. Nevertheless, except for isolated studies by physicists such as Dirac and Wigner, there was no serious attempt to develop the theory of unitary representations of groups which were neither compact nor abelian. This is not surprising; for the theory of unitary representations of general locally compact groups leads very quickly to questions involving algebras of operators in infinite-dimensional Hilbert spaces which are very nonclassical. However, the situation changed dramatically in the 1940s; for by then von Neumann had completed his epoch-making work (with Murray) on operator algebras and it became clear that there existed a reasonable framework for constructing a theory of infinite-dimensional unitary representations. It was in 1943 that Gel'fand and Raikov showed that every locally compact group possesses sufficiently many irreducible unitary representations to separate its points.

3.2 Haar measures on groups and homogeneous spaces

I shall begin by recalling some general facts concerning locally compact groups and their homogeneous spaces, and measure theory on them (see [V1], Ch. V); the groups will always be assumed to be second-countable, and we shall only deal with representations in a separable Hilbert space; the latter will not be a restriction because irreducible unitary representations of second-countable groups necessarily act in separable Hilbert spaces. A locally compact second-countable group G always has a left-invariant Haar measure dx; dx is unique up to a multiplicative constant >0 and $L^2(G{:}dx)$ is separable. In general, dx will not be invariant under right translations; if it is so, G will be called *unimodular*. In this case dx will be invariant under all inner automorphisms $x \mapsto axa^{-1}$ and the inversion $x \mapsto x^{-1}$. The classical Lie groups such as $GL(n, \mathbb{C})$, $SL(n, \mathbb{C})$, $SO(n, \mathbb{C})$, $Sp(n, \mathbb{C})$ and their real forms are unimodular. If G is a Lie group with Lie algebra $\mathfrak{g}$, G is unimodular if and only if, for each $X \in \mathfrak{g}$, the endomorphism $\operatorname{ad} X (Y \mapsto [X, Y])$ of $\mathfrak{g}$ has trace 0; this is the case if $\mathfrak{g} = [\mathfrak{g}, \mathfrak{g}]$, or, at the other extreme, if $\mathfrak{g}$ is nilpotent. Compact and abelian groups are unimodular. The triangular group B consisting of $n \times n$ matrices of the form

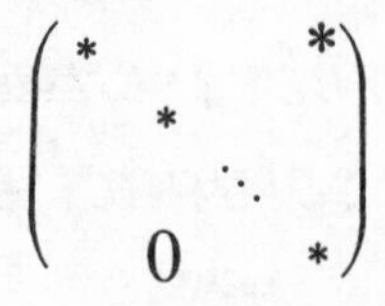

is not unimodular. If K is a nonarchimedean local field, the groups such as $GL(n, K)$ or $SL(n, K)$ are locally compact, second-countable, and unimodular, but are not Lie groups; as we remarked in the introduction, they play an important role in modern number theory. We shall always work with unimodular G although on occasion we may consider subgroups that are not.

If H is a closed subgroup of a unimodular G, the homogeneous space G/H is locally compact Hausdorff and second-countable, and G acts on it as a group of homeomorphisms, the action being denoted by $g, x \mapsto g[x]$ $(g \in G, x \in G/H)$. Up to a multiplicative constant >0 there will be at most one σ-finite measure invariant under this action, and there will be one if and only if H is also unimodular. The invariant measures are then Borel measures on G/H; if $\mathrm{d}x$ and $\mathrm{d}\xi$ are left Haar measures on G and H respectively, there is a unique invariant measure $\mathrm{d}\bar{x}$ on G/H such that, for all $\alpha \in C_c(G)$ (= the space of continuous functions on G with compact supports),

$$\int_G \alpha \, \mathrm{d}x = \int_{G/H} \mathrm{d}\bar{x} \left(\int_H \alpha(x\xi) \, \mathrm{d}\xi \right)$$

This relationship between $\mathrm{d}x$, $\mathrm{d}\xi$ and $\mathrm{d}\bar{x}$ is usually abbreviated as $\mathrm{d}x = \mathrm{d}\bar{x}\,\mathrm{d}\xi$; the uniqueness comes from the fact that the map that takes α to the function $\bar{x} \mapsto \int_H \alpha(x\xi)\,\mathrm{d}\xi$ is *surjective* from $C_c(G)$ to $C_c(G/H)$. Notice that if H is compact or discrete G/H always has an invariant measure. Even when G/H does not have an invariant σ-finite measure, it always has quasi-invariant measures, even finite ones. Here, μ on G/H is said to be *quasi-invariant* if the collection of μ-null sets is stable under the action of G, i.e., if the transforms of μ by elements of G are mutually absolutely continuous. One can show that up to mutual absolute continuity there is a unique quasi-invariant measure. If μ is one such, then for any $g \in G$ we have a strictly positive Borel function $\rho(g:\cdot)$ on G/H such that, for all $\alpha \in C_c(G/H)$, $g \in G$,

$$\int_G \alpha(g[x])\rho(g:x)^{-1} \, \mathrm{d}\mu(x) = \int_G \alpha(x) \, \mathrm{d}\mu(x)$$

Obviously $\rho(1:x)=1$ almost everywhere and it is easy to check that

$$\rho(g_1g_2:x)=\rho(g_1:g_2[x])\,\rho(g_2:x)$$

for each $g_1, g_2 \in G$ for almost all x. Note that, for any $g \in G$, $\rho(g:\cdot)$ is only determined up to a null set; it can be shown nevertheless that there is a Borel function ρ' on $G\times(G/H)$ such that, for almost all $g\in G$, $\rho'(g:x)=\rho(g:x)$ for almost all $x\in G/H$.

The classical theory of representations of compact and abelian groups focuses attention on two very general and related questions. The first is to determine, for a given G, the unitary dual $\hat{G}$ in as explicit a manner as possible. The second is to decompose the regular representation of G into its irreducible components. More generally, if H is a closed subgroup such that G/H has an invariant measure, the operators of left translation by elements of G are unitary in $L^2(G/H)$, giving rise to a unitary representation of G, often called the *quasiregular representation* of G/H; and one may ask for its decomposition into irreducible constituents. The case of abelian G already shows that the decompositions of the regular or the quasiregular representation need not be discrete, and that one should expect their decomposition to be intimately related to, perhaps even effected by, a Fourier transform theory for G.

3.3 Induced representations

The quasiregular representations are special cases of *induced representations*. The concept of induced representations goes back to Frobenius who developed their theory when the groups involved are finite. For compact groups they were studied by Weil. The definition and systematic treatment of induced representations for general locally compact groups are due to Mackey [Ma3]. For Lie groups, where the theory of distributions becomes available, it is possible to study the theory of induced representations at a much more profound level; this was done by Bruhat [Br]. I would like to introduce them in our general context (see [V1], Ch. VI).

Let $H\subset G$ be a closed subgroup, and let σ be a unitary representation of H in a (separable) Hilbert space K, whose norm and scalar product are denoted by $|\cdot|$ and $(\cdot|\cdot)$ respectively. We select a quasi-invariant measure μ on G/H and consider the Hilbert space $\mathscr{H}$ of Borel functions (or rather μ-equivalence classes of Borel functions)

$$f: G\to K$$

such that

(i) $f(x\xi)=\sigma(\xi)^{-1}f(x) \quad (x\in G,\ \xi\in H)$,
(ii) $\|f\|^2 := \int_{G/H}|f(x)|^2\,\mathrm{d}\mu(x)<\infty$.

In (ii) we should note that, in view of (i), the function $gH \mapsto |f(g)|^2$ is well defined on G/H and it is this function that is required to be μ-summable. The condition (i) does not change when f is replaced by a left translate. So, if μ is invariant, this Hilbert space is stable under left translations by elements of G and we obtain a unitary representation of G. In the general case when μ is assumed to be only quasi-invariant, we define the action of $g \in G$ on $\mathscr{H}$ by

$$(\pi(g)f)(g') = \rho(g:g^{-1}[g'])^{1/2} f(g^{-1}g') \quad (g' \in G)$$

where $\bar{g}' = g'H$ and $\rho(g:\cdot)$ is the Radon–Nikodym derivative of μ with respect to its transform by g^{-1}. The adjustment by the Radon–Nikodym derivatives is exactly what is needed to make $\pi(g)$ unitary and π a representation; here and elsewhere, when dealing with separable groups and Hilbert spaces, it is often convenient to use the principle: a homomorphism τ into the unitary group is a representation if the function $(\tau(\cdot)u|v)$ is a measurable function of G for any to vectors u, v in the space of τ.

The representation π defined above is called the representation of G *induced by* σ and is denoted by

$$\mathrm{Ind}_H^G \sigma$$

Strictly speaking, we should write it as π_μ, acting on the Hilbert space $\mathscr{H}_\mu$; however, if ν is another quasi-invariant measure on G/H, the map

$$f \mapsto \left(\frac{d\mu}{d\nu}\right)^{1/2} f$$

is a unitary isomorphism of $\mathscr{H}_\mu$ with $\mathscr{H}_\nu$ taking π_μ to π_ν, so that the equivalence class of π_μ depends only on that of σ. If μ is invariant and σ the trivial representation, we obtain the quasiregular representation on $L^2(G/H)$, which is the regular representation when $H = (1)$.

There is an alternative description of induced representations that uses the notion of *cocycles*. Given a unitary representation σ of H in K, let $U(K)$ be the group of unitary operators on K; then a σ-*cocycle* is a map

$$\gamma: G \times G/H \to U(K)$$

such that

(i) $\gamma(g_1 g_2 : x) = \gamma(g_1 : g_2[x])\gamma(g_2 : x)$ for all $g_1, g_2 \in G$, $x \in G/H$,
(ii) $\gamma(\xi : \bar{1}) = \sigma(\xi)$ for $\xi \in H, \bar{1}$ being the coset H,
(iii) γ is Borel, i.e., for all $u, v \in K$, the map $g, x \mapsto (\gamma(g:x)u|v)$ is Borel.

We now select a quasi-invariant measure μ as before and a σ-cocycle γ, and define the operators $\pi(g)$ $(g \in G)$ on $L^2(G/H:K:\mu)$ by

$$(\pi(g)f)(x) = \rho(g:g^{-1}[x])^{1/2}\gamma(g:g^{-1}[x])f(g^{-1}[x]) \quad (x \in G/H)$$

The cocycle identities and the identities satisfied by ρ ensure that π is a representation, and one can show that it is equivalent to the induced representation. To see this, let us first define a *Borel section* for G/H to be a Borel map

$$s: x \mapsto s(x) \quad (x \in G/H)$$

of G/H into G such that

(i) $s(1) = 1$,
(ii) $s(x)H = x \quad (x \in G/H)$.

It is known ([V1], Theorem 5.11) that Borel sections always exist. If s is a Borel section, then, for $g \in G$, $x \in G/H$,

$$\xi(g:x) = s(g[x])^{-1} g s(x)$$

fixes 1 and hence lies in H; thus, if we set

$$\gamma_s(g:x) = \sigma(\xi(g:x))$$

it is almost immediate that γ_s is a σ-cocycle. If γ is any σ-cocycle,

$$\begin{aligned} \gamma(g:x) &= \gamma(g:s(x)[\bar{1}]) \\ &= \gamma(gs(x):\bar{1})\gamma(s(x):\bar{1})^{-1} \\ &= \gamma(s(g[x])\xi(g:x):\bar{1})\gamma(s(x):\bar{1})^{-1} \\ &= \gamma(s(g[x]:\bar{1})\gamma_s(g:x)\gamma(s(x):\bar{1})^{-1} \end{aligned}$$

So if we put

$$a(x) = \gamma(s(x):\bar{1})$$

then

$$\gamma(g:x) = a(g[x])\gamma_s(g:x)a(x)^{-1}$$

The map

$$f \mapsto af$$

is then a unitary operator in $L^2(G/H:K:\mu)$ which takes the representation defined by γ to the one defined by γ_s. If we now take $\gamma = \gamma_s$, the map

$$f \mapsto Uf, Uf(g) = \sigma(\xi(g:\bar{1}))f(g[\bar{1}]) \quad (g \in G)$$

is unitary and takes the representation defined by γ_s to the induced representation $\mathrm{Ind}_H^G \sigma$ defined earlier on $\mathscr{H}$.

Let us go back to $\mathscr{H}$ and introduce the projection-valued measure P on G/H where, for any Borel set $E \subset G/H$, P_E is the operator of multiplication by the characteristic function of $E \cdot H$; $E \cdot H$ is of course a Borel set, being the pre-image of E under the map $G \to G/H$. Writing π^σ to emphasize that π depends on σ we then have

$$\pi^\sigma(g)P_E\pi^\sigma(g)^{-1} = P_{g[E]}$$

More generally, if τ is a unitary representation of G in some Hilbert space and Q is a projection valued measure based on G/H in the same Hilbert space such that

$$\tau(g)Q_E\tau(g)^{-1} = Q_{g[E]}$$

for all $g \in G$ and (Borel) $E \subset G/H$, (τ, Q) is called a *system of imprimitivity* for $(G, G/H)$; heuristically, one may think of it as a representation of $(G, G/H)$. Thus (π^σ, P) is a system of imprimitivity. If G and hence G/H are finite, they were considered first by Frobenius; in this case, the restriction of τ to H leaves the range of the projection $Q_{(\bar{1})}$ invariant; and, if σ is the representation of H defined on this range (τ, Q) is isomorphic to (π^σ, P). For general G it was Mackey who first considered systems of imprimitivity and proved the *imprimitivity theorem* (see [V1], Ch. VI, §61): *any system of imprimitivity for $(G, G/H)$ is isomorphic to one of the form (π^σ, P); the assignment*

$$\sigma \mapsto (\pi^\sigma, P)$$

commutes with taking direct sums, and σ is irreducible if and only if (π^σ, P) is so; and it induces a bijection between the set of equivalence classes of (separable) unitary representations of H and (separable) systems of imprimitivity for $(G, G/H)$.

If we fix a Borel section s and a quasi-invariant measure μ and define the induced representation on $L^2(G/H:K:\mu)$ using the cocycle γ_s, one can show that the assignment

$$\sigma \mapsto (\pi^\sigma, P)$$

is functorial; and for any two unitary representations σ_1, σ_2 we have a natural bijection from $\mathrm{Hom}_H(\sigma_1, \sigma_2)$ to $\mathrm{Hom}_G((\pi^{\sigma_1}, P), (\pi^{\sigma_2}, P))$. Thus the technique of induced representations gives a very general and essentially functorial way of going from representations of H to representations of G.

The key step in the proof of the imprimitivity theorem is the argument that associates to a system (τ, Q) a unitary representation σ of H so that $(\tau, Q) \cong (\pi^\sigma, P)$. In fact, the *transitivity* of the action of G on G/H allows us to conclude that Q is unitarily equivalent to $P = P^{K,\mu}$ for some K where P_E is the projection $f \mapsto \chi_E f$ in $L^2(G/H:K:\mu)$, χ_E being the characteristic function of $E \subset G/H$. We may thus assume that $Q = P$. If τ^0 is the representation given by

$$(\tau^0(g)f)(x) = \rho(g:g^{-1}[x])^{1/2} f(g^{-1}[x])$$

then (τ^0, P) and (τ, P) are both systems of imprimitivity for $(G, G/H)$; and if $t(g) = (\tau^0(g))^{-1}\tau(g)$, $t(g)$ commutes with all P_E and so can be written as

$$(t(g)f)(x) = T(g:x)f(x)$$

where $T(g:\cdot)$ is a Borel map of G/H into $U(K)$. Moreover

$$t(g_1g_2) = (\tau^0(g_2))^{-1}t(g_1)(\tau^0(g_2))\cdot t(g_2)$$

which implies that

$$T(g_1g_2:x) = T(g_1:g_2[x])T(g_2:x)$$

for almost all x, for each $g_1, g_2 \in G$. The map T may not be a cocycle, but this is only a technical difficulty (although nontrivial), and one can construct a unitary representation σ in K and a σ-cocycle γ such that for almost all $g \in G$ $T(g:x) = \gamma(g:x)$ for almost all x. It is then clear that $\tau = \pi^\sigma$.

The equations defining a system of imprimitivity make sense in much wider contexts than those of coset spaces. For instance if X is a locally compact Hausdorff second-countable space and G acts on X as a group of homeomorphisms, the action $g, x \mapsto g[x]$ being continuous, one can define a system of imprimitivity for (G, X) in a Hilbert space $\mathscr{H}$ as a pair (π, P) where π is a unitary representation of G in $\mathscr{H}$, P, a projection valued measure based on X in $\mathscr{H}$, and

$$\pi(g)P_E\pi(g)^{-1} = P_{g[E]}$$

for all $g \in G$, Borel sets $E \subset X$. A very general assumption is that P is *ergodic* for the action of G; this means that, if $E \subset X$ is a G-invariant Borel set, then either $P_E = 0$ or $P_{X \setminus E} = 0$. We then say that (π, P) is an *ergodic system of imprimitivity*. The reader who is familiar with the theory of measure-preserving transformations will recognize the similarity of this definition to that of a measure being ergodic for a group of transformations. Indeed, if μ is an ergodic G-invariant measure on X, $\mathscr{H} = L^2(X:\mu)$, and P_E is multiplication by the characteristic function of E, then for the natural representation π given by $(\pi(g)f)(x) = f(g^{-1}[x])$ $(f \in \mathscr{H}, x \in X, g \in G)$, (π, P) is an ergodic system of imprimitivity. For an ergodic system it is no longer possible to determine in a simple manner all the cocycles, so that no classification theorem for ergodic systems of imprimitivity can be proved.

Each G-orbit in X is a Borel set so that we have a partition of X into G-invariant Borel sets. The characteristic feature that distinguishes the ergodic context from the nonergodic one is that the quotient space of X relative to this partition, which can be constructed in the category of Borel spaces, is a very bad one. The classical example, going back to the famous construction by Lebesgue of nonmeasurable sets, is that of $\mathbb{Z}$ acting on the circle T by $(n, t) \mapsto e^{2i\pi n\alpha}t$ where $\alpha \in \mathbb{R}$ is irrational. Let us say that the action of G on X is *regular* if the σ-algebra of G-invariant Borel subsets of X is generated by a countable collection of sets. The Lebesgue example is an irregular action; in counterpoint to this is the fact that if there is a

Borel section for the action, i.e., a Borel set that meets each G-orbit exactly once, then the action is regular. For a regular action it can be shown that, if (π, P) is an ergodic system of imprimitivity, there is an *orbit* X_0 such that $P_{X \setminus X_0} = 0$. The restriction of (π, P) to X_0 may then be viewed, after choosing a point $x_0 \in X_0$ and identifying X_0 with G/H, where H is the stabilizer of x_0 in G (i.e., the subgroup of elements g in G such that $g[x_0] = x_0$), as a system of imprimitivity for $(G, G/H)$.

If G is a connected Lie group, one often encounters induced representations of G in a *differential* setting. Let V be a C^∞ complex vector bundle on G/H and let us suppose that there is an action of G on V satisfying the following conditions:

(i) the map $G \times V \to V$ is C^∞ and compatible with the action $G \times G/H \to G/H$;
(ii) each element of G acts as a vector-bundle automorphism of V;
(iii) there is a positive-definite scalar product $(\cdot|\cdot)_{\bar{1}}$ on the fibre of V at $\bar{1}$ that is invariant under H;
(iv) there is a quasi-invariant measure μ on G/H and a C^∞ map $\rho(G \times G/H \to$ positive reals) such that $\int_{G/H} \alpha(g[x])\rho(g{:}x)^{-1}\,\mathrm{d}\mu(x) = \int_{G/H} \alpha(x)\,\mathrm{d}\mu(x)$ for all $\alpha \in C_c^\infty(G/H)$.

The condition (iii) implies that there is a family of scalar products $(\cdot|\cdot)_x$ on the fibre at x, depending smoothly on x, such that

$$(g[u]|g[v])_{g[x]} = (u|v)_x$$

The group G then operates linearly on the space of sections of V. Let $\mathscr{H} = \mathscr{H}_{V,\mu}$ be the Hilbert space of all sections f such that

$$\|f\|^2 \int_{G/H} (f(x)|f(x))_x\,\mathrm{d}\mu(x) < \infty$$

We then obtain a representation of G in $\mathscr{H}$ by setting

$$(\pi(g)f)(x) = \rho(g{:}g^{-1}[x])^{1/2} f(g^{-1}[x])$$

In particular, if μ is invariant, π is the natural representation of G on the space of square-integrable sections of V. It can be shown that $\pi \cong \mathrm{Ind}_H^G\,\sigma$ where σ is the finite-dimensional unitary representation of H on the fibre of V at $\bar{1}$, obtained by restricting the action of G to H (see [V1], pp. 233–5).

The concept of induced representation is a natural way to pass from unitary representations of H to unitary representations of G. It is in a certain sense dual to the restriction functor Res_H^G that associates with any unitary representation of G its restriction to H. When G is compact the relationship between these two processes leads to the *Frobenius reciprocity theorem: if σ (resp. τ) is any irreducible unitary representation of H (resp. G),* *the number of times τ occurs (as a direct summand) in* $\mathrm{Ind}_H^G \sigma$ *is exactly equal to the number of times σ* occurs in $\mathrm{Res}_H^G \tau$.

The technique of induced representations is a very versatile one. We shall see later that quite often induced representations are irreducible, so that one has a very explicit method of constructing irreducible unitary representations. Indeed, for certain semidirect products and nilpotent Lie groups, *all* irreducible unitary representations are induced from suitable subgroups. At the other extreme, for semisimple groups, the inducing process does not in general lead to the construction of all the irreducible unitary representations; nevertheless, even here, once certain basic (discrete series) representations are constructed, the others, at least those needed for the Plancherel formula, can be obtained by inducing from the so-called parabolic subgroups. Finally, for (suitable) subgroups H and unitary representations σ of H, the decomposition of Ind_H^G into irreducible constituents is a very central problem.

3.4 Semidirect products

([Ma1], [Ma3]; [V1], Ch. VI, §8). One of the most elegant illustrations of these ideas is Mackey's theory of unitary representations of certain semidirect products. Let A be a locally compact abelian group and L a locally compact group acting on A as a group of automorphisms. Both L and A are second-countable and we write

$$(h, a) \mapsto h[a] \quad (h \in L,\ a \in A)$$

for the action of L on A. The semidirect product $A \rtimes L$ is then the locally compact group whose underlying topological space is $A \times L$ with the group operation defined by

$$(a, h)\cdot(a', h') = (ah[a'], hh')$$

Let G be the semidirect product. We identify A with $A \times (1)$ and L with $(1) \times L$ so that A and L are closed subgroups of G, with A normal and $G = AL$, $L \cap A = (1)$. L acts naturally on $\hat{A}$ and we write $h, \hat{a} \mapsto h[\hat{a}]$ for this action.

If π is a unitary representation of G and $\pi^A = \pi|_A$, $\pi^L = \pi|_L$, π^A (resp. π^L) is a unitary representation of A (resp. L) and

$$\pi^A(hah^{-1}) = \pi^L(h)\pi^A(a)\pi^L(h)^{-1} \quad (a \in A,\ h \in L)$$

Conversely, given unitary representations π^1 (resp. π^2) of A (resp. L), the condition $\pi^1(hah^{-1}) = \pi^2(h)\pi^1(a)\pi^2(h)^{-1}$ $(a \in A,\ h \in L)$ is sufficient to ensure that $\pi^1 = \pi|_A$, $\pi^2 = \pi|_L$ for a unitary representation π of G. Now, the unitary representations π^1 of A are in natural one-one correspondence with the projection valued measures P^1 on $\hat{A}$ by Fourier transform theory on A:

$$\pi^1(a) = \int_{\hat{A}} \langle a, \hat{a} \rangle \, \mathrm{d}P^1(\hat{a}) \quad (a \in A)$$

The relation between π^1 and π^2 translates to the condition

$$\pi^2(h) P^1_{\hat{E}} \pi^2(h)^{-1} = P^1_{h[\hat{E}]} \quad (h \in L, \ \hat{E} \subset \hat{A})$$

In other words, unitary representations of G are in natural one-one correspondence with systems of imprimitivity (π^2, P^1) for $(L, \hat{A})$. Although the action of L on A need not be transitive, it may happen that there is an *orbit* $X_0 \subset \hat{A}$ such that $P^1_{\hat{A} \backslash X_0} = 0$. Then (π^2, P^1) may be viewed as an irreducible system of imprimitivity for $(L, L/L_0)$ when L_0 is the subgroup of L which fixes an arbitrary but fixed element $\hat{x}_0 \in X_0$; it is thus isomorphic to the system induced by an irreducible unitary representation of L_0. The corresponding representation of G is quite easy to describe directly. Let $H(\hat{x}_0) = AL_0$, and let $\sigma_{\hat{x}_0}(ah) = \langle a, \hat{x}_0 \rangle \sigma(h)$ $(a \in A, \ h \in L_0)$. Then $H(\hat{x}_0)$ is a closed subgroup of G, $\sigma_{\hat{x}_0}$ is an irreducible unitary representation of $H(\hat{x}_0)$, and our representation is just $\mathrm{Ind}^G_{H(\hat{x}_0)} \sigma_{\hat{x}_0}$. Changing $\hat{x}_0$ to a different point in $X_0 = L[\hat{x}_0]$ does not change the equivalence class of this representation. If the semidirect product is not regular, or more precisely in the presence of genuinely ergodic phenomena, there will in general be many more irreducible unitary representations of G, corresponding to ergodic measures and cocycles associated with them. For the *regular* semidirect products the equivalence classes of the irreducible unitary representations of G obtained above exhaust G, so that we can say that the irreducible unitary representations of G are classified up to equivalence by pairs $(\hat{x}_0, \sigma)$ where $\hat{x}_0$ varies in a subset of $\hat{A}$ meeting each L-orbit in $\hat{A}$ exactly once, and for each $\hat{x}_0$ σ varies over a complete set of mutually inequivalent irreducible unitary representations of the stabilizer of $\hat{x}_0$ in L. If A is Minkowski space-time and L the homogeneous Lorentz group (or rather its universal covering group), we have a regular semidirect product, and the above considerations lead to the classification of the free relativistic particles in terms of their mass and spin [V1] [Ma1].

3.5 Factor representations

In spite of its initial promise and excitement, the project of generalizing classical harmonic analysis and representation theory to the setting of *arbitrary* locally compact second-countable groups gradually turned out to be a hopeless one. The work of von Neumann had already given indications of this. Let me now try to explain some of the reasons for this.

Let G be a second-countable locally compact group and π a unitary representation of G in a Hilbert space $\mathscr{H}$. One can associate to π two algebras $R(\pi)$ and $R'(\pi)$: $R'(\pi)$ is the algebra of all bounded operators of $\mathscr{H}$ commuting with π, and $R(\pi)$ is the algebra of all bounded operators of $\mathscr{H}$ commuting with $R'(\pi)$. Both are *von Neumann algebras*, and $R(\pi)$ is the closure, in the weak operator topology, of the linear combinations of all the $\pi(g)$, $g \in G$. The significance of $R'(\pi)$ arises from the fact that for a projection to belong to $R'(\pi)$ it is necessary and sufficient that its range define an invariant subspace of π. If $\mathscr{H}$ is finite-dimensional, one can show that π is a *multiple* of an irreducible representation if and only if $R'(\pi) \cap R(\pi)$ consists only of multiples of the identity, i.e., the centre of $R'(\pi)$ (or $R(\pi)$) consists of scalars. The representation π is then called a *factor representation* because we can factorize $\mathscr{H}$ and π as $\mathscr{H} = \mathscr{H}_0 \otimes \mathscr{H}'_0$, $\pi = \pi_0 \otimes 1$, where π_0 is an irreducible representation of G acting on $\mathscr{H}_0$. In the general case it is still natural to call π a *factor representation* if $R(\pi) \cap R'(\pi) = \mathbb{C} \cdot 1$; the algebras $R(\pi)$ and $R'(\pi)$ are also called *factors*. *If* we can write $\mathscr{H} = \mathscr{H}_0 \otimes \mathscr{H}'_0$, $\pi = \pi_0 \otimes 1$ where π_0 is an irreducible representation of G in $\mathscr{H}_0$, π is a factor representation; in this case $R'(\pi) \cong$ the algebra of bounded operators on $\mathscr{H}'_0$, and π is said to be of *type 1*. To von Neumann we owe the fundamental discovery that there exist factor representations which are *not* of type 1. He proved for instance that, if G is a countable discrete group all of whose conjugacy classes other than (1) contain infinitely many elements (for example the free groups with 2 generators or the group of permutations of the integers which move only finitely many of them), the regular representation of G is a factor representation which is *not* of type 1. If we compare this with the case of a finite group whose classes of irreducible representations correspond bijectively with the set of conjugacy classes, and all of which occur as direct summands of its regular representation, we can see how remote from classical intuition von Neumann's examples are. A systematic development of von Neumann's ideas led eventually (in the 1950s) to a deep understanding of the decomposition theory of unitary representations and to results which implied more or less that a reasonable generalization of classical Fourier analysis and representation theory could be expected only for the so-called *type I groups*; i.e., *groups all of whose factor representations are of type I*.

This is the historical background of our subject. I must emphasize that it is very sketchy and not always in strict chronological order. Nevertheless its thrust is very clear; *classical Fourier analysis is a very uncertain guide in predicting how noncommutative harmonic analysis should be done*. It is thus not surprising that already in the middle and late 1940s some people were turning their attention to some of the simplest noncompact non-abelian groups like $SL(2, \mathbb{R})$ and $SL(2, \mathbb{C})$, to get a better insight into the new phenomena which arise. The work of Gel'fand–Naimark, Bargmann and Harish-Chandra dates to this period and marks, in my opinion, the real beginning of modern noncommutative Fourier analysis and representation theory.

4

Parabolic induction, principal series representations, and their characters

4.1 Early work

The work of Bargmann, Gel'fand–Naimark, and Harish-Chandra in the 1940s and early 1950s brought out several features of the representation theory of semisimple Lie groups which were completely nonclassical. The most striking of these were the following.

(1) There are irreducible unitary representations that 'do not occur' in the regular representation.
(2) The regular representation may have both continuous and discrete parts in its decomposition.
(3) Even at the level of infinite-dimensional representations there is an intimate connection between the representations of the Lie algebra and the corresponding Lie group. This allows one to introduce the notion of 'arbitrary' (not just unitary) irreducible representations and bring in a completely algebraic point of view of the representation theory.
(4) One can associate to any irreducible representation a *character* which is a class function on the group, which determines the representation up to equivalence, which is given by explicit formulae analogous to those of Weyl in the compact case, and in terms of which an explicit Plancherel formula can be proved.
(5) The matrix elements of the irreducible representations are eigenfunctions for the (bi-)invariant differential operators of the group, and the spectral theorem for these operators may be interpreted as another version of the Plancherel formula.

Gel'fand and Naimark treated the *complex classical groups*, and their investigations were collected together in their great monograph [GN]. By the vastness of its reach and the classic beauty of its ideas, this book exerted a profound influence for a very long time and still continues to be one of the great sources of motivation and inspiration of the subject. The fundamental method of constructing irreducible unitary representations, namely inducing from parabolic subgroups (*parabolic induction*), as well as the introduction of characters and the discovery of the deep link between the Plancherel formula and the limit formula for orbital integrals, goes back to this work.

Bargmann [Ba] confined himself to $SL(2, \mathbb{R})$; nevertheless, his discovery of the discrete series for $SL(2, \mathbb{R})$ and the orthogonality relations satisfied by its matrix elements, his formulation of the Plancherel formula in terms of the spectral theory of the Casimir operator, and the determination of the Plancherel measure in terms of the leading terms of the asymptotics of the matrix elements of the continuous spectrum, were absolutely fundamental. Moreover his construction of the discrete series in terms of Hilbert spaces of holomorphic functions and forms on the unit disc underscored the deep relationship between discrete series and holomorphic structures on homogeneous spaces of the group.

Harish-Chandra [H1] [H2] [H3] worked from the very beginning with *general* semisimple Lie groups and began with a penetrating study of the representations of the group from the Lie algebraic point of view. His most profound insight was however his rather early realization [H4] that the method of *parabolic induction will not work for real groups unless one has already constructed the discrete series.* From this perspective it was inevitable that the construction of the discrete series was the central focus of his efforts. His eventual explicit determination of the discrete series [H5] [H6] and the Plancherel formula [H7] must be regarded as one of the most profound achievements of the mathematics of our time. Furthermore, his ideas and methods reached far beyond the Plancherel formula for real semisimple groups, and have played a fundamental role in the (still incomplete) extension of harmonic analysis to semisimple groups over local and global fields. One can say without undue exaggeration that the history of harmonic analysis in the years 1950–80 is essentially the development of the Harish-Chandra programme, carried out almost single-handedly by him.

4.2 Parabolic subgroups and principal series for complex groups

The simplest context for parabolic induction, which was also historically the first to be considered, is induction from *Borel subgroups* of complex semisimple Lie groups. If G is a connected complex algebraic matrix group, its *parabolic subgroups* are by definition the algebraic subgroups P for which G/P is a *projective variety.* The *minimal* ones among them are the *Borel subgroups*; these are all mutually conjugate in G. For $G = SL(n, \mathbb{C})$, the parabolic subgroups are, up to conjugacy, the *block-upper-triangular groups.* Let $n = n_1 + \cdots + n_r$ (n_i integers $\geqslant 1$), and let us write the elements of G as matrices $g = (g_{ij})$ where g_{ij} is an $n_i \times n_j$ complex matrix; then the subgroup $P = P_{n_1,\dots,n_r}$ of matrices in G of the form

$$\begin{pmatrix} g_{11} & & * \\ & \ddots & \\ 0 & & g_{rr} \end{pmatrix}$$

is a parabolic subgroup, and every parabolic subgroup of G is conjugate to exactly one of the $P_{n_1,\dots,n_r}$. The homogeneous space G/P may be identified with the variety F of flags of type (d_j), $d_j = n_1 + \cdots + n_j$. Let us recall that a *flag of type* (d_j) is a chain $(L_j)_{1 \leqslant j \leqslant r}$ of subspaces with $L_1 \subset L_2 \subset \cdots \subset L_r$, and $\dim(L_j) = d_j$. G operates transitively on F; if $(e_i)_{1 \leqslant in}$ is the standard basis of $\mathbb{C}^n$ and L_j is the span of the e_i for $1 \leqslant i \leqslant d_j$, P is the stabilizer of (L_j); this gives the identification $G/P \cong F$. It is a well-known fact in classical geometry that F is a projective variety. For $r = 2$, $d = n_1$, F is just the Grassmannian of d-dimensional linear subspaces of $\mathbb{C}^n$. If $r = n$ and $d_j = j$ for all j, we speak of a *full flag*; P is then the subgroup of *all* upper-triangular matrices; it is a Borel subgroup and we shall usually write B for it.

The parabolic subgroups have natural decompositions known as *Levy–Langlands decompositions.* We shall not treat these in complete generality but discuss them briefly only for $G = SL(n, \mathbb{C})$. If $P = P_{n_1,\dots,n_r}$, we have the decomposition

$$P = M_1 N \quad \text{(semidirect product)}$$

where M_1 is the subgroup of all matrices of G of the form

$$\begin{pmatrix} g_{11} & & 0 \\ & \ddots & \\ 0 & & g_{rr} \end{pmatrix}$$

and N is the subgroup of all matrices of G of the form

$$\begin{pmatrix} 1 & & * \\ & \ddots & \\ 0 & & 1 \end{pmatrix}$$

It is clear that N is normal in P and that the map

$$\begin{pmatrix} g_{11} & & * \\ & \ddots & \\ 0 & & g_{rr} \end{pmatrix} \mapsto \begin{pmatrix} g_{11} & & 0 \\ & \ddots & \\ 0 & & g_{rr} \end{pmatrix}$$

is a morphism of P onto M_1 with kernel N. Thus the representations of M_1 may be viewed as representations of P that are trivial on N. The group N is unipotent, i.e., its matrices are unipotent, while M_1 is 'almost'

$$\prod_{1 \leqslant i \leqslant r} GL(n, \mathbb{C})$$

In fact M_1 is the subgroup of this group of elements (g_{ii}) with $\prod_i \det(g_{ii}) = 1$, and generates it together with $\mathbb{C}^\times \cdot 1$. If $P = B$, M_1 is the subgroup, written D, of diagonal matrices.

For the other complex classical groups there are analogous descriptions of parabolic subgroups and their decompositions. For instance, let V be an n-dimensional vector space over $\mathbb{C}$ and F a nondegenerate quadratic form over V. A subspace L of V is called *isotropic* if F is identically zero on L. An *isotropic flag* of type $(d_j)_{1 \leqslant j \leqslant m}$ is a chain $(L_j)_{1 \leqslant j \leqslant m}$ of isotropic subspaces such that $L_1 \subset \cdots \subset L_m$, $\dim(L_j) = d_j$. It is known that isotropic flags of a given type are acted upon transitively by $SO(F)$ and the corresponding stabilizers are precisely all the parabolic subgroups of $SO(F)$. One knows that the maximum dimension of an isotropic subspace is $m = [n/2]$; the isotropic flags $(L_j)_{1 \leqslant j \leqslant m}$ with $\dim(L_j) = j$ are called *full*, and their stabilizers are the Borel subgroups of $SO(F)$ ([Bo], pp. 3–19).

Let $G = SL(n, \mathbb{C})$ and B, the upper-triangular group. If χ is a character of D, we may view it as a character of B, and so construct the unitary representation

$$L_\chi = \operatorname{Ind}_B^G \chi$$

The family (L_χ) is called the *principal series of representations of* $SL(n, \mathbb{C})$. It was first introduced by Gel'fand and Naimark for the complex classical groups, although it is quite clear that they conceived of it in its full generality, for all complex semisimple groups.

To motivate this construction suppose that G is complex semisimple and B a Borel subgroup of it. In any finite-dimensional rational irreducible representation of G there is a unique one-dimensional subspace stabilized by B on which B acts through a rational character of D. It follows from this that the rational irreducible representations of G may be realized by letting G act by left translations on the spaces of everywhere regular rational functions f on G such that $f(gb) = \chi(b) f(g)$ for all $g \in G$, $b \in B$, χ being any rational character of D. If we drop the requirement that χ is rational and work with 'all' rather than 'rational' functions, we obtain the principal series of representations. This remark even suggests that it may be useful to extend the definition of the principal series and define the representations L_χ even when χ is only a *quasicharacter*, i.e., a continuous homomorphism of D into $\mathbb{C}^\times$. We shall see later on in this chapter that L_χ can then be viewed as a Hilbert-space representation, which is unitary for unitary χ.

The vindication of this definition arises from the beautiful theorem of Gel'fand and Naimark, which we shall prove later, that the L_χ are irreducible for all $\chi \in \hat{D}$. They proved moreover that the regular representation can be decomposed as a direct integral of the L_χ, $\chi \in \hat{D}$, and in fact obtained an explicit Plancherel formula for G. At the same time, their work exposed the remarkable nonclassical fact that not all irreducible unitary representations can be obtained by decomposing the regular representation; indeed, the L_χ are all infinite-dimensional, so that the trivial representation is certainly outside this description. Actually Gel'fand and Naimark constructed additional irreducible unitary representations, in essentially two ways, in both of which parabolic induction played a central role. The first method, leading to the so-called *degenerate series*, replaces the Borel subgroups by more general parabolic subgroups. If P is a parabolic subgroup and $P = M_1 N$ its Levi–Langlands decomposition, the map

$$\begin{pmatrix} g_{11} & & * \\ & \ddots & \\ 0 & & g_{rr} \end{pmatrix} \mapsto (\det(g_{11}), \ldots, \det(g_{rr}))$$

of M_1 into $(\mathbb{C}^\times)^r$, when composed with characters of $(\mathbb{C}^\times)^r$, gives one-dimensional characters χ of P; *the degenerate series of representations associated with* P is then the family $(L_{P,\chi})$ where

$$L_{P,\chi} = \operatorname{Ind}_P^G \chi$$

These are all irreducible. The case $P = G$ cannot be omitted; the corresponding degenerate series consists precisely of the trivial representation. The second method is, in a certain sense, one of *analytic unitary continuation*, and leads to the so-called *supplementary series of representations*. Its basis is the observation that, for the representations L_χ (or the $L_{P,\chi}$), there exist suitable *nonunitary* χ for which one can construct invariant positive-definite scalar products on G-stable dense subspaces of L_χ; the associated completions will then define irreducible unitary representations, not equivalent to any of the members of the principal series (L_χ). The significance of these additional constructions may already be seen from the result that, for $G = SL(2, \mathbb{C})$, an irreducible unitary representation (up to equivalence) belongs to either the principal series or the supplementary series, or else is just the trivial representation.

If π is one of these irreducible unitary representations, Gel'fand and Naimark associated to π a *character* θ_π which is a class function on G. It is characterized by the fact that it is locally summable on G and, for any $\alpha \in C_c^\infty(G)$,

$$\int_G \theta_\pi(x)\alpha(x)\,dx = \operatorname{tr}\left(\int_G \alpha(x)\pi(x)\,dx\right)$$

Unlike the character of finite-dimensional representations, θ_π will in general have singularities; nevertheless it will determine π up to equivalence. The explicit formulae for the θ_π play an essential role in the Plancherel formula.

For example let us look at the case $G = SL(2, \mathbb{C})$. Then B is the subgroup of matrices of the form $\begin{pmatrix} a & b \\ 0 & a^{-1} \end{pmatrix}$. G acts on the Riemann sphere $\mathbb{P}^1 = \mathbb{C} \cup (\infty)$ by Möbius transformations:

$$m(g): z \mapsto \frac{az+b}{cz+d} = m(g)(z), \quad g = \begin{pmatrix} a & b \\ c & d \end{pmatrix}$$

We work with the action $g, z \mapsto g[z]$ given by

$$g[z] = m((g^{-1})^t(z) = \frac{dz - c}{-bz + a}$$

For this action the stabilizer of 0 is B so that $G/B \cong \mathbb{P}^1$. We have a section defined on $\mathbb{C}$ given by

$$s(z) = \begin{pmatrix} 1 & 1 \\ -z & 1 \end{pmatrix} \quad s(z)[0] = z$$

The measure $\mathrm{d}x\,\mathrm{d}y = (\mathrm{i}/2)\,\mathrm{d}z\,\mathrm{d}\bar{z}$ is quasi-invariant under the action of G; since $z' = (az+b)/(cz+d)$ gives $\mathrm{d}z' = (cz+d)^{-2}\,\mathrm{d}z$, we have

$$z' = (az+b)/(cz+d), \quad \mathrm{d}x'\,\mathrm{d}y' = |cz+d|^{-4}\,\mathrm{d}x\,\mathrm{d}y$$

If χ is a character or a quasicharacter of D and we view it as defined on B, the representation L_χ acts on the Hilbert space $\mathscr{H}$ of all functions f on $\mathbb{C}$ with

$$\|f\|^2 = \int |f(z)|^2\,\mathrm{d}x\,\mathrm{d}y < \infty$$

For $g = \begin{pmatrix} a & b \\ c & d \end{pmatrix}$ the operator $L_\chi(g)$ is given by

$$(L_\chi(g)f)(z) = f(g^{-1}[z])|(\mathrm{d}g^{-1}[z]/\mathrm{d}z)|^{1/2}\chi(s(z)^{-1}gs(g^{-1}[z]))$$

(see Chapter 3, Sections 2–3). A simple calculation gives

$$s(z)^{-1}gs(g^{-1}[z]) = \begin{pmatrix} (bz+d)^{-1} & * \\ 0 & (bz+d) \end{pmatrix}$$

If we write $\chi_{m,\rho}$ for the quasicharacter

$$\begin{pmatrix} a & * \\ 0 & a^{-1} \end{pmatrix} \mapsto (a/|a|)^m |a|^\rho \quad (m \in \mathbb{Z}, \rho \in \mathbb{C})$$

then, for $\chi = \chi_{m,\rho}$, L_χ acts on $\mathscr{H}$ by the formula

$$(L_\chi(g)f)(z) = f\left(\frac{az+c}{bz+d}\right)|bz+d|^{m-\rho-2}(bz+d)^{-m}$$

If ρ is purely imaginary, L_χ is unitary and irreducible. The class function which represents the character of L_χ is given by

$$\theta_\chi(\operatorname{diag}(a, a^{-1})) = \frac{\chi(a) + \chi(a^{-1})}{|a - a^{-1}|^2}$$

($a \neq \pm 1$). Using these characters one can define the *Fourier transform* of a function $\alpha \in C_c^\infty(G)$ by

$$\hat{\alpha}(\chi) = \int_G \theta_\chi(x)\alpha(x)\,\mathrm{d}x$$

The Plancherel formula for G then takes the following explicit form: for a suitable (explicitly determined) normalization of $\mathrm{d}x$, we have, for all $\alpha \in C_c^\infty(G)$,

$$\alpha(1) = \sum_{m \in \mathbb{Z}} \int_{-\infty}^{\infty} \hat{\alpha}(\chi_{m,i\rho})(m^2 + \rho^2)\,\mathrm{d}\rho$$

Let me describe the supplementary series of representations. Let us take the representations L_χ where $\chi = \chi_{0,\rho}$, and ρ is *real*, and introduce the Hermitian form

$$(f_1 | f_2) = \iint |z_1 - z_2|^{-2+\rho} f_1(z_1)\overline{f_2(z_2)}\,\mathrm{d}x_1\,\mathrm{d}y_1\,\mathrm{d}x_2\,\mathrm{d}y_2$$

Gel'fand and Naimark proved that, if $0 < \rho < 2$, this form is *positive-definite*; and the action $f \mapsto L_\chi(g)f$,

$$(L_\chi(g)f)(z) = f\left(\frac{az + c}{bz + d}\right)|bz + d|^{-2-\rho}$$

defines an irreducible unitary representation in the Hilbert space corresponding to the above scalar product. (One can do this for $-2 < \rho < 0$ also, but the integral defining the scalar product has to be taken in the sense of regularization.) The character of this representation of the supplementary series is represented by the function θ_χ with

$$\theta_\chi(\operatorname{diag}(a, a^{-1})) = \frac{\chi(a) + \chi(a^{-1})}{|a - a^{-1}|^2}$$

This formula shows that it is not equivalent to any member of the unitary principal series.

If we let $\rho \to 2$ in the above formula for θ_χ we see that it converges to

$$\frac{|a|^2 + |a|^{-2}}{|a - a^{-1}|^2} = 1 + \frac{a^2|a|^{-2} + a^{-2}|a|^2}{|a - a^{-1}|^2}$$

The terms on the right side are respectively the characters of the trivial representation and the unitary principal series representation L_χ where $\chi = \chi_{2,0}$. This suggests that the representation $L_{\chi_{0,2}}$ has a composition series consisting of the trivial representations and the $L_{\chi_{2,0}}$. It is interesting to note that each of these is unitary but that $L_{\chi_{0,2}}$ is *not* unitarizable. As further evidence that it is worthwhile to consider *all* representations of semisimple groups, not just the unitary ones, and that the *full* series (L_χ) is already rich enough to yield all the irreducible representations by forming composition series, let me mention the fact (one can prove this easily; see [GGV]) that all finite-dimensional representations of $SL(2, \mathbb{C})$ may be obtained in this manner. If π_m is the holomorphic irreducible representation of highest weight m, the finite-dimensional irreducible representations are the $\pi_{n_1} \otimes \pi_{n_2}^{\text{conj}}$, realized in the space of polynomials in z and $\bar{z}$, of degree $\leqslant n_1$ in z and $\leqslant n_2$ in $\bar{z}$ with the action

$$\begin{pmatrix} a & b \\ c & d \end{pmatrix} : p(z, \bar{z}) \mapsto p\left(\frac{az + c}{bz + d}, \frac{\bar{a}\bar{z} + \bar{c}}{\bar{b}\bar{z} + \bar{d}}\right)(bz + d)^{n_1}(\bar{b}\bar{z} + \bar{d})^{n_2}$$

Comparing this with the principal-series representation $L_{\chi_{m,\rho}}$ and writing $|bz + d|^{m-\rho-2}$ as $[(bz + d)(\bar{b}\bar{z} + \bar{d})]^{(m-\rho-2)/2}$ we are led to study the composition series of $L_{\chi_{m,\rho}}$ with

$$n_2 = \frac{m - \rho - 2}{2} \qquad n_1 = \frac{-m - \rho - 2}{2}$$

which is equivalent to

$$m = n_2 - n_1, \rho = -(n_1 + n_2 + 2)$$

It can be shown by explicit calculation that the composition series of this L_χ is of length 2 and one of the two irreducible representations is $\pi_{n_1} \otimes \pi_{n_2}^{\text{conj}}$.

4.3 The Harish-Chandra perspective

The work of Bargmann in the case of $SL(2, \mathbb{R})$ made it clear that harmonic analysis for real groups is at a much deeper level than for complex groups. Let $G = SL(2, \mathbb{R})$ and let B be as before the upper-triangular group. Then the principal series of representations may be defined as the family

$$L_\chi = \operatorname{Ind}_B^G \chi$$

where χ runs over the group of characters of the diagonal subgroup viewed as characters of B:

$$\chi\left(\begin{pmatrix} a & b \\ 0 & a^{-1} \end{pmatrix}\right) = \operatorname{sgn}(a)^\varepsilon |a|^\rho \quad (\varepsilon = 0, 1, \rho \in i\mathbb{R})$$

As before we can extend this definition to quasicharacters χ $(\rho \in \mathbb{C})$. The L_χ are irreducible for all unitary χ with one exception $(\rho = 0,\ \varepsilon = 1)$; moreover, the supplementary series may be constructed just as in the case of $SL(2, \mathbb{C})$. But now *there are additional irreducible unitary representations for G, and these have the remarkable property that their matrix elements are square-integrable on G, and orthogonal to the matrix elements of the unitary principal series.* This means that these representations occur as *discrete direct summands* of the regular representation, and the principal series will no longer suffice to decompose the regular representation or to obtain an explicit Plancherel formula. These representations constitute the so-called *discrete series*; their definition, namely, as irreducible unitary representations with square-integrable matrix elements, makes sense for any second-countable locally compact group H; the corresponding subset of H will be denoted by $\hat{H}_{\mathrm{d}}$. It can be empty however.

The clue to the origin and significance of the discrete series lies in the formulae for their characters. To see this let me begin with some preparatory remarks. Recall that an element $x \in G$ is regular if its eigenvalues are distinct. The set G' of regular elements is open, dense, invariant, and $G \backslash G'$ is of measure zero, as in the case of complex groups. However, G' is now *not connected* and is the disjoint union of the *elliptic set* G'_{ell} and the *hyperbolic set* G'_{hyp}; here, $x \in G'_{\mathrm{ell}}$ (resp. $x \in G'_{\mathrm{hyp}}$) if and only if the eigenvalues of x are of absolute value 1 (resp. real). The hyperbolic conjugacy classes meet the diagonal group

$$L = \left\{ \begin{pmatrix} a & 0 \\ 0 & a^{-1} \end{pmatrix} \right\}$$

while the elliptic conjugacy classes meet the rotation group

$$K = \left\{ u_\theta = \begin{pmatrix} \cos\theta & \sin\theta \\ -\sin\theta & \cos\theta \end{pmatrix} \right\}$$

So, in order to specify a class function on G' it is necessary to determine it on $L \cap G'$ as well as $K \cap G'$. If now one computes the character of the principal series representation L_χ, one finds that

$$\theta_\chi(\operatorname{diag}(a, a^{-1})) = \frac{\chi(a) + \chi(a^{-1})}{|a - a^{-1}|} \quad (a \neq \pm 1)$$

$$\theta_\chi(u_\theta) = 0 \quad (\theta \not\equiv 0, \pi \ (\mathrm{mod}\ 2\pi))$$

The fact that the θ_χ are all trivial on the elliptic set is the real reason why the principal series fails to be decisive for harmonic analysis. On the other hand, concerning the discrete-series representations, we find the following: the discrete series is parametrized by the (regular) characters of K, the subgroup of rotations u_θ. Writing these characters as

$$\chi_n : u_\theta \mapsto \mathrm{e}^{in\theta}$$

there is a discrete-series representation associated to χ_n whenever $n \neq 0$, with a character Θ_n which is a locally integrable class function, given by the following formulae:

$$\Theta_n(u_\theta) = -\operatorname{sgn}(n)\frac{e^{in\theta}}{e^{i\theta} - e^{-i\theta}} \quad (\theta \not\equiv 0, \pi)$$

$$\Theta_n(\operatorname{diag}(a, a^{-1})) = \operatorname{sgn}(a)\frac{a^{-|n|}}{|a - a^{-1}|} \quad (|a| > 1)$$

These formulae, which were first obtained by Harish-Chandra [H8], strongly suggest that the discrete-series representations are associated to the torus K in the same way as the principal series is associated to L, and strengthen the analogy with the case of the compact groups. The completeness of the discrete and the unitary principal series now appears quite natural, and Harish-Chandra obtained it in [H8] as a corollary of an explicit Plancherel formula. Bargmann's proof of the completeness theorem was, as we mentioned earlier, based on the spectral theory of the Casimir operator; we shall take up this theme a little later.

These remarks on the discrete series for $SL(2, \mathbb{R})$ are a good starting point for a brief discussion of the general lines along which Harish-Chandra developed the harmonic analysis of real semisimple Lie groups. Essentially there are two fundamental principles:

(i) *restricted parabolic induction*, i.e., induction from parabolic subgroups which are associated to Cartan subgroups, using representations of the discrete series of the semisimple groups that occur in the Levy decomposition;

(ii) *independent construction* of the discrete-series representations.

A detailed amplification of these principles requires a familiarity with the theory of parabolic subgroups and Cartan groups of general real semisimple Lie groups. I shall therefore be sketchy, essentially restricting myself to the case of $SL(n, \mathbb{R})$.

We write $G = SL(n, \mathbb{R})$ and view it as the group of real points of $G_\mathbb{C} = SL(n, \mathbb{C})$; let θ be the involutive automorphism $x \mapsto (x^{-1})^t$ of $G_\mathbb{C}$, which stabilizes G and whose fixed points in G form the maximal compact subgroup $K = SO(n)$ of G. If A (resp. M) is the subgroup of diagonal matrices with diagonal entries > 0 (resp. ± 1), and N is the group of upper-triangular matrices with 1s on the diagonal, we have the *Iwasawa decomposition*

$$G = K \cdot A \cdot N$$

and the *Levy–Langlands decomposition*

$$B = M \cdot A \cdot N$$

of the group B of upper-triangular matrices of G. More generally, if

$n = n_1 + \cdots + n_r$ (n_i integers $\geqslant 1$), we have the subgroups $P_{n_1,\ldots,n_r}$ of matrices in G of the form

$$\begin{pmatrix} g_{11} & & * \\ & \ddots & \\ 0 & & g_{rr} \end{pmatrix} \quad (g_{ij} \text{ are } n_i \times n_j \text{ real matrices})$$

For $r = n$ and $n_i = 1$, $P_{n_1,\ldots,n_r} = B$. The $P_{n_1,\ldots,n_r}$ are the groups of real points of the corresponding subgroup of $G_{\mathbb{C}}$ which leads to the correct definition of the parabolic subgroups of G: they are the groups of real points of these parabolic subgroups of $G_{\mathbb{C}}$ which are *defined over* $\mathbb{R}$ (this is equivalent to saying that they are stable under complex conjugation). Any parabolic subgroup of G is conjugate in G to a unique $P_{n_1,\ldots,n_r}$. For $P = P_{n_1,\ldots,n_r}$ we have the Levy-Langlands decomposition

$$P = M_P \cdot A_P \cdot N_P$$

Here M_P, A_P, N_P are the groups of real matrices of the respective forms

$$M_P\colon \begin{pmatrix} g_{11} & & 0 \\ & \ddots & \\ 0 & & g_{rr} \end{pmatrix} \quad \det(g_{ii}) = \pm 1, \quad \prod_i \det(g_{ii}) = 1$$

$$A_P\colon \begin{pmatrix} a_1 \cdot 1 & & 0 \\ & \ddots & \\ 0 & & a_r \cdot 1 \end{pmatrix} \quad a_i > 0, \quad \prod_i a_i = 1$$

$$N_P\colon \begin{pmatrix} 1 & & * \\ & \ddots & \\ 0 & & 1 \end{pmatrix}$$

If η is an irreducible unitary representation of M_P and χ is a character of A_P we have a unitary representation

$$\eta \boxtimes \chi\colon man \mapsto \chi(a)\eta(m)$$

of P and so we can define

$$L_{P,\eta,\chi} = \mathrm{Ind}_P^G(\eta \boxtimes \chi)$$

We thus obtain a family of unitary representations associated to P. If $P = B$, this is just the *principal series* for G. As before these induced representations make sense for quasi-characters χ also.

Among the parabolic subgroups a special place is occupied by those which are associated, in the sense to be made clear now, to Cartan subgroups. A *Cartan subgroup* of G is by definition the group of real points of a maximal torus of $G_{\mathbb{C}}$ that is defined over $\mathbb{R}$. They are precisely the centralizers in G of regular elements of G. Up to conjugacy in G they are the subgroups of G of the form

$$\begin{pmatrix} g_{11} & & 0 \\ & \ddots & \\ 0 & & g_{rr} \end{pmatrix}$$

where, for $1 \leqslant i \leqslant t, g_{ii}$ is a 2×2 real matrix of the form

$$\begin{pmatrix} a & b \\ -b & a \end{pmatrix} \quad (a^2 + b^2 \neq 0)$$

while for $t < i \leqslant r = t + s, g_{ii}$ is a real scalar $\neq 0$; so $n = 2t + s$, and $(\prod_{1 \leqslant i \leqslant t} \det(g_{ii})) \prod_{t < i \leqslant r} g_{ii} = 1$. The parabolic subgroups $P_{n_1,\ldots,n_r}$ where t of the n_is are equal to 2 and the rest equal to 1, are said to be *associated to this Cartan subgroup*; a standard choice will be $n_i = 2$ $(1 \leqslant i \leqslant t)$, $n_i = 1$ $(t < i \leqslant r)$. In this case the component of the identity M_P^0 of M_p is

$$\prod_{1 \leqslant i \leqslant t} G_i, \quad G_i = SL(2, \mathbb{R})$$

and M_P^0 is normal of finite index in M_P. *Thus M_P has discrete-series representations*, and we follow Harish-Chandra in defining the *principal series associated to the Cartan subgroup* as the family

$$(L_{P,\eta,\chi}) \quad (\eta \in (\hat{M}_P)_d, \ \chi \in \hat{A}_P)$$

It can be shown that this does not depend on the choice of P as long as it is associated to the given Cartan subgroup, and that (once P is fixed), for each η, the representation $L_{P,\eta,\chi}$ is irreducible for *general* χ.

The above definitions can be generalized in a more or less straightforward way to arbitrary real reductive groups in the Harish-Chandra class, in particular to the groups of real points of complex algebraic semisimple groups defined over $\mathbb{R}$. If P is a parabolic subgroup of such a group G, and

$$P = M_P \cdot A_P \cdot N_P$$

its *Levy–Langlands decomposition*, it turns out that *P is associated to a Cartan subgroup if and only if M_P has a compact Cartan subgroup*. The second great insight of Harish-Chandra was his realization *that this is also the precise condition that M_P admits a discrete series*.

Thus, from Harish-Chandra's perspective, there are two fundamental parts to the programme of L^2 harmonic analysis on a real semisimple group G.

(1) Construct the discrete series when G has a compact Cartan subgroup.
(2) Prove the completeness of the irreducible representations obtained by the method of restricted parabolic induction.

At this stage of these lectures we can only make some brief remarks concerning the first part. It can be shown, that, if G has compact Cartan subgroups at all, these form a single conjugacy class (among the $SL(n, \mathbb{R})$, only for $n = 2$ are there compact Cartan subgroups, while for $SO(1, m)$ the condition is that m be even). Let us suppose that U is a maximal compact subgroup of $G_{\mathbb{C}}$ such that $U \cap G = K$ is a maximal compact subgroup of G. If G has compact Cartan subgroups, we can arrange matters so that one of them, say T, is contained in K; it is then also a maximal torus of U (look at $G = SL(2, \mathbb{R})$, $T = K = (u_\theta)$, $U = SU(2)$, $G_{\mathbb{C}} = SL(2, \mathbb{C})$). There are now two Weyl groups associated to T: W, the geometric Weyl group, and W_G, the part of W coming from G. Let $\hat{T}$ be the character group of T. Then W operates on $\hat{T}$ and we denote by $\hat{T}'$ the subset of those characters which are not fixed by any element of W other than the identity (for $SL(2, \mathbb{R})$, $\hat{T} = (\chi_n)$, and $\hat{T}' = (\chi_n)_{n \neq 0}$). Then Harish-Chandra proved that the discrete series for G is parametrized by $\hat{T}'/W_G$; for any $\xi \in \hat{T}'$ the character of the corresponding representation is given on $T \cap G'$ by the formula (up to a sign factor)

$$\Theta_\xi(t) = \pm \left(\sum_{s \in W_G} \varepsilon(s) \xi^s(t) \right) \Big/ \Delta_T(t)$$

The reader should note that, unlike Weyl's formula, the Harish-Chandra formula involves summation only over W_G. Of course this does not completely specify Θ_ξ because no information about the restriction of the characters to nonelliptic conjugacy classes is given. Harish-Chandra strengthened the above formula by requiring a *boundedness condition at infinity on* G: for every Cartan subgroup L of G,

$$\sup_{a \in L \cap G'} |\Delta_L(a)| |\Theta_\xi(a)| < \infty$$

(For instance, this is true in $SL(2, \mathbb{R})$ for Θ_n because on the real diagonal subgroup the numerator in the formula for Θ_n $(\operatorname{diag}(a, a^{-1}))$ is $\pm a^{-|n|}$ when $|a| > 1$). He proved that the data consisting of the formula on T and the boundedness property on the other Cartan subgroups is sufficient to guarantee complete determination of the character. Two important comments must be added to this. The first is that the construction does not give anything when G does not have a compact Cartan subgroup; it is a crucial step in the Harish-Chandra theory that in that case there is no discrete series either. The second is a much more basic one, and has to do with what we mean by unique determination of Θ_ξ. For Harish-Chandra, characters of irreducible unitary representations are special instances of invariant distributions on the group which are eigendistributions for all the (bi-)invariant differential operators on the group; and underlying his entire theory is the remarkable regularity theorem he proved, namely, that *all such invariant distributions are locally summable class functions.* The uniqueness theorem for Θ_ξ simply asserts that there is one and only one invariant eigendistribution corresponding to the data specified above. Only after all the Θ_ξ have been constructed does one prove by harmonic analysis that the Θ_ξ are (up to sign), precisely the characters of the discrete series.

With this brief survey of the method of parabolic induction completed, I shall go back to the principal series for the rest of this chapter. We shall be mostly concerned with $SL(n, \mathbb{C})$ and $SL(n, \mathbb{R})$, and often only with $SL(2, \mathbb{R})$.

4.4 Character calculations

For exploring the properties of the principal series representations we need explicit 'models'. We begin by discussing the 'compact model' which will allow us to calculate the characters.

Haar measure calculations. Let H be a Lie group, K, B closed subgroups, and suppose we have a decomposition $H = K \cdot B$; this means the map $(k, b) \mapsto kb$ of $K \times B$ into H is an isomorphism of analytic manifolds. In computing Haar measures we shall generally ignore normalizing constants.

Lemma 1. *Let H be unimodular with Haar measure* $\mathrm{d}h$, *and let* $\mathrm{d}_l k$ (*resp.* $\mathrm{d}_r b$) *be a left* (*resp. right*) *Haar measure for K* (*resp. B*). *Then writing μ_B for the modular function of B,*

$$\mathrm{d}h = \mathrm{d}_l k\, \mathrm{d}_r b = \mu_B(b)\, \mathrm{d}_l k\, \mathrm{d}_l b \quad (h = kb)$$

Proof. There is an analytic function $j > 0$ on $K \times B$ such that $\mathrm{d}h = j \mathrm{d}_l k\, \mathrm{d}_r b$. Since $\mathrm{d}h$ is left-invariant under K, it is immediate that $j(k, b)$ is independent of k; similarly, right invariance of $\mathrm{d}h$ under B implies $j(k, b)$ is independent of b also.

Lemma 2. *Suppose $B = A \cdot N$ where A and N are closed subgroups, both unimodular, with N normal in B. Then there is an analytic homomorphism δ of A into $\mathbb{R}_+^\times$ (positive reals) such that, for each $a \in A$,*

$$\mathrm{d}(ana^{-1}) = \delta(a)^2\, \mathrm{d}n$$

Moreover, $\mu_B(an) = \delta(a)^2$ *so that*

$$\mathrm{d}_l b = \mathrm{d}a\, \mathrm{d}n, \quad \mathrm{d}_r b = \delta(a)^2\, \mathrm{d}a\, \mathrm{d}n \quad (b = an)$$

Proof. For any $a \in A$, $n \mapsto ana^{-1}$ is an automorphism of N and so

$d(ana^{-1}) = c(a)\,dn$ for some constant $c(a) > 0$. It is trivial to check that $a \mapsto c(a)$ is an analytic homomorphism. We take $\delta = c^{1/2}$ to get the first statement. For proving the second statement, let $f \in C_c(B)$. Then, for $a_0 \in A$, $n_0 \in N$,

$$\iint f(a_0 an)\,da\,dn = \iint f(an)\,da\,dn$$

$$\iint f(n_0 an)\,da\,dn = \int da \int f(aa^{-1}n_0 an)\,dn = \iint f(an)\,da\,dn$$

which shows that $d_l b = da\,dn$. On the other hand,

$$\iint f(ann_0)\delta(a)^2\,da\,dn = \iint f(an)\delta(a)^2\,da\,dn$$

$$\begin{aligned}\iint f(ana_0)\delta(a)^2\,da\,dn &= \int \delta(a)^2\,da \int f(aa_0a_0^{-1}na_0)\,dn \\ &= \iint \delta(aa_0)^2 f(aa_0 n)\,da\,dn \\ &= \iint \delta(a)^2 f(an)\,da\,dn\end{aligned}$$

proving that $d_r b = \delta(a)^2\,da\,dn$.

Iwasawa decomposition. We shall take $G = SL(n, \mathbb{C})$ or $SL(n, \mathbb{R})$; K is accordingly $SU(n)$ or $SO(n)$, A is the subgroup of diagonal matrices with diagonal entries >0, and N is the group of upper-triangular matrices with 1's on the diagonal and entries from $\mathbb{C}$ or $\mathbb{R}$. $B = M \cdot A \cdot N$ is the upper-triangular group, with M as the diagonal subgroup of $SU(n)$ or $SO(n)$: compact in either case, a torus in $SL(n, \mathbb{C})$ and a finite group in $SL(n, \mathbb{R})$.

Suppose $x \in SL(n, \mathbb{C})$. Let $(e_j)_{1 \leqslant j \leqslant n}$ be the standard basis of $\mathbb{C}^n$. We regard $\mathbb{C}^n$ as a Hilbert space as usual. By the usual process of orthogonalization applied to the basis $(f_j)_{1 \leqslant j \leqslant n}$ of $\mathbb{C}^n$, $f_j = x \cdot e_j$, we determine, uniquely and inductively, constants $a_j > 0$ and an ON basis $(e'_j)_{1 \leqslant j \leqslant n}$ of $\mathbb{C}^n$ such that

$$\begin{aligned} f_1 &= a_1 e'_1 \\ f_j &= a_j e'_j + \sum_{1 \leqslant r < j} (f_j | e'_r) e'_r \quad (j > 1) \end{aligned}$$

An easy induction on j then shows that the a_j and the e'_r are analytic functions of x (in the real analytic structure of G). Let u be the unitary matrix such that $u \cdot e_j = e'_j$ for all j. Then

$$u^{-1}x \cdot e_j = a_j e_j + \sum_{1 \leqslant r < j} (f_j | e'_r) e_r$$

from which it is immediate that $x = uan$ where $a = \operatorname{diag}(a_1, \ldots, a_n)$ and n is the upper-triangular matrix with entries a_{ij},

$$a_{jj} = 1, \quad a_{rj} = a_j^{-1}(f_j | e'_r)$$

As $1 = \det(u)\ \det(a)$, we have $u \in SU(n)$, $a \in A$. This is the Iwasawa decomposition for $SL(n, \mathbb{C})$,

$$G = K \cdot A \cdot N$$

with the projections $x \mapsto u$, $x \mapsto a$, $x \mapsto n$ shown to be analytic functions. If x is real, so are u and n, and we have the decomposition for $SL(n, \mathbb{R})$. Write

$$x = k(x)a(x)n(x) \quad (k(x) \in K,\ a(x) \in A,\ n(x) \in N)$$

Let us now calculate $a(x)$. Write, for $1 \leqslant r \leqslant n$,

$$E_r = \mathbb{C}^n \wedge \cdots \wedge \mathbb{C}^n \quad (r \text{ factors})$$

and view it as a Hilbert space in the usual manner, so that $(e_{i_1} \wedge \cdots \wedge e_{i_r})_{1 \leqslant i_1 < \cdots < i_r \leqslant n}$ is an ON basis for E_r. It is then immediate that

$$a_1 = \| x \cdot e_1 \|, \quad a_1 \cdots a_r = \| (x \cdot e_1) \wedge \cdots \wedge (x \cdot e_r) \|$$

If $n = 2$ and $x = \begin{pmatrix} \alpha & \beta \\ \gamma & \delta \end{pmatrix}$, then

$$a(x) = (|\alpha|^2 + |\gamma^2|)^{1/2}$$

In the general case, if $x = (g_{ij})$,

$$a_1(x) \cdots a_r(x) = \left(\sum_{i_1 < \cdots < i_r} |g_{i_1, \ldots, i_r}|^2 \right)^{1/2}$$

where $g_{i_1, \ldots, i_r}$ is the determinant of the minor $(g_{i_p q})_{1 \leqslant p, q \leqslant r}$.

Lemma 3. (a) *If* $a \in A$, *then* $\mathrm{d}(ana^{-1}) = \delta(a)^2\, \mathrm{d}n$ *where*

$$\delta(\operatorname{diag}(a_1, \ldots, a_n)) = \begin{cases} a_1^{2(n-1)} a_2^{2(n-2)} \cdots a_{n-1}^2 & (\mathbb{C}) \\ a_1^{n-1} a_2^{n-2} \cdots a_{n-1} & (\mathbb{R}) \end{cases}$$

(b) $\mathrm{d}x = \delta(a)^2\, \mathrm{d}u\, \mathrm{d}a\, \mathrm{d}n \quad (x = uan)$

Proof. (a) Let $C = A \cdot N$ so that $G = K \cdot C$. In view of Lemmas 1 and 2 it is a question of calculating δ. Let us use the matrix entries a_{ij} $(i < j)$ as coordinates on N. Then $a_{ij}(ana^{-1}) = (a_i/a_j)a_{ij}(n)$ and hence

$$\delta(\operatorname{diag}(a_1, \ldots, a_n))^2 = \begin{cases} \prod_{i<j} (a_i/a_j)^2 & (\mathbb{C}) \\ \prod_{i<j} (a_i/a_j) & (\mathbb{R}) \end{cases}$$

where the parentheses correspond to the cases $G = SL(n, \mathbb{C})$ and $G = SL(n, \mathbb{R})$.

Since $a_1 \cdots a_n = 1$, we multiply by $(a_1 \cdots a_n)^m$, $m = 2(n-1)$ in the complex case and $m = (n-1)$ in the real case, to get the required formulae.

The action of G on K. Let us begin in the general setting of Lemma 1 and view K as an H-space by identifying H/B with K. We write $x, u \mapsto x[u]$ for this action. Fix $x \in H$, write $x = k(x)b(x)$, $k(x) \in K$, $b(x) \in B$. Then, for $x \in H$, $u \in K$,

$$xu = x[u] \cdot b(xu), \quad x[u] = k(xu)$$

We wish to calculate the Radon–Nikodym derivatives $\mathrm{d}(x^{-1}[u])/\mathrm{d}u$. If we replace H by G and B by $A \cdot N$, we are in the framework of the Iwasawa decomposition.

Lemma 4. *For any $x \in H$,*

$$\mathrm{d}(x^{-1}[u]) = \mu_B(b(x^{-1}u))^{-1}\,\mathrm{d}u$$

In other words, for any continuous function f on K,

$$\int_K f(u)\,\mathrm{d}u = \int_K f(x^{-1}[u])\,\mu_B(b(x^{-1}u))^{-1}\,\mathrm{d}u$$

In particular, for $G = K \cdot A \cdot N$, we have

$$\mathrm{d}(x^{-1}[u]) = \delta(a(x^{-1}u))^{-2}\,\mathrm{d}u$$

$$\int_K f(u)\,\mathrm{d}u = \int_K f(x^{-1}[u])\delta(a(x^{-1}u))^{-2}\,\mathrm{d}u$$

Proof. This is a simple consequence of the left invariance of $\mathrm{d}x$ and the formula $\mathrm{d}x = \mu_B(b)\,\mathrm{d}u\,\mathrm{d}_l b$. Select $E \in C_c(B)$ such that

$$\int_B E(b)\mu_B(b)\,\mathrm{d}_l b = 1$$

and, for $f \in C(K)$, write $g \in C_c(H)$ for the function $ub \mapsto f(u)E(b)$. Then

$$\int_K f(u)\,\mathrm{d}u = \int_H g(y)\,\mathrm{d}y = \int_H g(x^{-1}y)\,\mathrm{d}y = \int \mu_B(b)g(x^{-1}ub)\,\mathrm{d}u\,\mathrm{d}_l b.$$

But

$$x^{-1}ub = k(x^{-1}u)b(x^{-1}u)b$$

so that

$$g(x^{-1}ub) = f(x^{-1}[u])E(b(x^{-1}u)b)$$

This implies that

$$\int \mu_B(b)g(x^{-1}ub)\,\mathrm{d}u\,\mathrm{d}_l b = \int f(x^{-1}[u])E(b(x^{-1}u)b)\mu_B(b)\,\mathrm{d}u\,\mathrm{d}_l b$$

$$= \int f(x^{-1}[u]\mu_B(b(x^{-1}u))^{-1}\,du \int E(b)\mu_B(b)\,d_l b$$

$$= \int f(x^{-1}[u]\mu_B(b(x^{-1}u))^{-1}du$$

which proves the lemma.

The function a as a cocycle: compact model for principal series. The following identities are easy to check.

$$k(xy) = k(xk(y)) = x[k(y)]$$
$$a(xy) = a(xk(y))a(y)$$
$$n(xy) = (a(y)^{-1}n(xk(y))a(y))n(y)$$

Let us put

$$a(x:u) = a(xu) \quad (x \in G,\ u \in K)$$

Then

(i) $a(1:u) = 1$,
(ii) $a(x_1x_2:u) = a(x_1:x_2[u])a(x_2:u)$,
(iii) $a(x:x^{-1}[u]) = a(x^{-1}u)^{-1}$.

The identities (i) and (ii) show that $a(\cdot:\cdot)$ is an analytic A-valued cocycle, and hence, for any quasicharacter $\eta(A \to \mathbb{C}^\times)$, $\eta(a(\cdot:\cdot))$ is an analytic $\mathbb{C}^\times$-valued cocycle. We may thus define a representation of G in $\mathscr{H} = L^2(K)$ by setting

$$(\pi_\eta(x)f)(u) = (\eta\delta)(a(x^{-1}u))^{-1}f(x^{-1}[u])$$

The cocycle property of a implies that $x \mapsto \pi_\eta(x)$ is multiplicative. Further,

$$\|\pi_\eta(x)f\|^2 = \int_K |\eta(a(x^{-1}u)|^{-2}\delta(a(x^{-1}u))^{-2}f(x^{-1}[u])\,du$$

so that, if η is unitary, we conclude from Lemma 4 that all the $\pi_\eta(x)$ are unitary and that in fact, for any quasicharacter η,

$$\|\pi_\eta(x)f\|^2 \leqslant \sup_{u\in K}|\eta(a(x^{-1}u)|^{-2}\|f\|^2$$

Similarly,

$$\|\pi_\eta(x)f - f\|^2 \leqslant \sup_{u\in K}|\eta(a(x^{-1}u)^{-1} - \delta(a(x^{-1}u))|^2\|f\|^2$$

The continuity properties needed to assert that π_η is a representation follow from these estimates and the continuity on $G \times K$ of a $(\cdot:\cdot)$.

To get the principal series we must cut down these representations further.

Lemma 5. *The restriction of π_η to K is the left regular representation of K. Moreover, for any $m \in M$, right translation by m commutes with all the $\pi_\eta(x)$.*

Proof. If $x = v \in K$, $a(v^{-1}u) = 1$, and $v^{-1}[u] = v^{-1}u$, so that $\pi_\eta(v)f(u) = f(v^{-1}u)$. If $m \in M$ and $(r(m)f)(u) = f(um)$, the relation $\pi_\eta(x)r(m) = r(m)\pi_\eta(x)$ follows from the formula for $\pi_\eta(x)$ if we note the relations

$$a(ym) = a(y), \quad k(ym) = k(y)m \quad (y \in G)$$

that are implied by the fact that m normalizes N and commutes with A.

For any character ξ of M let

$$\mathcal{H}_\xi = \{ f \in \mathcal{H} \mid r(m)f = \xi(m)^{-1}f \quad \text{for all } m \in M \}$$

where $\mathcal{H} = L^2(K)$. By Lemma 5, $\mathcal{H}_\xi$ is stable under all the $\pi_\eta(x)$. We put $\chi = (\xi, \eta)$ and

$$\pi_\chi(x) = \pi_\eta(x)|_{\mathcal{H}_\xi}$$

Theorem 6. *For fixed $\chi = (\xi, \eta)$, π_χ is a representation of G in $\mathcal{H}_\xi$. If χ (i.e., η) is unitary, π_χ is unitary and is equivalent to the principal series representation $L_\chi = \mathrm{Ind}_B^G \chi$.*

Proof. We assume that $\chi = (\xi, \eta)$ is unitary and realize the induced representation L_χ in the usual way in the Hilbert space of (equivalence classes of) functions g on G such that

(i) $g(yb) = \chi(b)^{-1}g(y)$ $(y \in G, b \in B)$
(ii) $\int_K |g(u)|^2 \, du = \| g \|^2 < \infty$.

We note that (i) is equivalent to

$$g(yman) = \xi(m)^{-1}\eta(a)^{-1}g(y)$$

for $y \in G$, $m \in M$, $a \in A$, $n \in N$. In particular, $|g(u)|$ depends only on the coset uM, so that the integral in (ii) is really over K/M, which is canonically identified with G/B, via the maps

$$uM \mapsto uB, \quad yB \mapsto k(y)M$$

The action is given by (cf. Chapter 3)

$$(L_\chi(x)g)(y) = \delta(a(x^{-1}k(y)))^{-1}g(x^{-1}y)$$

The restriction map $g \mapsto g|_K$ takes this Hilbert space isomorphically on to $\mathcal{H}_\xi$; for $f \in \mathcal{H}_\xi$, the corresponding g is given by $g(uan) = \eta(a)^{-1}f(u)$ $(u \in K, a \in A, n \in N)$. A simple verification shows that under this isomorphism $L_\chi(x)$ goes over to $\pi_\eta(x)$.

We shall not distinguish between L_χ and π_χ. The reader should note that the parameter ξ, varying over $\hat{M}$, is discrete; and that for all quasicharacters $\chi = (\xi, \eta)$, where η is any quasicharacter of A, the representations L_χ act on the *fixed* Hilbert space $\mathscr{H}_\xi$.

Characters for the L_χ. We shall now use the compact model $\mathscr{H}_\xi$ to prove that the operators of the representation π_χ are integral operators with smooth kernel and hence of trace class.

***Lemma* 7.** *Let η be any quasicharacter of A. Then, for any $\alpha \in C_c(G)$, $\pi_\eta(\alpha)$ is an integral operator in $L^2(K)$ with kernel*

$$M_\alpha(u:v:\eta) = \iint (\eta\delta)(a)\alpha(uanv^{-1})\,\mathrm{d}a\,\mathrm{d}n \quad (u, v \in K)$$

M_α is continuous on $K \times K$, of class C^∞ if α is C^∞. In particular, $\pi_\eta(\alpha)$ is of Hilbert–Schmidt class for all continuous α, and of trace class for all α in $C_c^\infty(G)$; in the latter case ($\mathrm{d}x = \delta(a)^2\,\mathrm{d}k\,\mathrm{d}a\,\mathrm{d}n$)

$$T_\eta(\alpha) := \operatorname{tr} \pi_\eta(\alpha) = \int_K \mathrm{d}u \left(\int_{A \times N} (\eta\delta)(a)\alpha(uanu^{-1})\,\mathrm{d}a\,\mathrm{d}n \right)$$

Proof. It is enough to check the first statement; the second is just the fact that the trace is

$$\int_K M_\alpha(u:u)\,\mathrm{d}u$$

For $g, h \in C(K)$, $\alpha \in C_c(G)$,

$$(\pi_\eta(\alpha)g \mid h) = \int_K \left(\int_G \alpha(x) g(x^{-1}[u])(\eta\delta)(a(x^{-1}u)^{-1})\,\mathrm{d}x \right) \overline{h(u)}\,\mathrm{d}u$$

Hence

$$\begin{aligned}
(\pi_\eta(\alpha)g)(u) &= \int_G \alpha(x) g(k(x^{-1}u)(\eta\delta)(a(x^{-1}u))^{-1}\,\mathrm{d}x \\
&= \int_G \alpha(uy^{-1}) g(k(y))(\eta\delta)(a(y))^{-1}\,\mathrm{d}y \\
&= \int_G \alpha(un^{-1}a^{-1}v^{-1}) g(v)(\eta\delta)(a)^{-1}\delta(a)^2\,\mathrm{d}v\,\mathrm{d}a\,\mathrm{d}n \\
&= \int_G \alpha(unav^{-1}) g(v)\eta(a)\delta(a)^{-1}\,\mathrm{d}v\,\mathrm{d}a\,\mathrm{d}n \\
&= \int_G \alpha(uanv^{-1}) g(v)(\eta\delta)(a)\,\mathrm{d}v\,\mathrm{d}a\,\mathrm{d}n
\end{aligned}$$

on writing na as $a \cdot (a^{-1}na)$ and going over to $n' = a^{-1}na$.

We are actually interested in the operators $\pi_\chi(\alpha) = \pi_\eta(\alpha)|_{\mathscr{H}_\xi}$. We shall identify this operator with $\pi_\eta(\alpha)Q_\xi$ ($= Q_\xi \pi_\eta(\alpha)$) where Q_ξ is the orthogonal projection $\mathscr{H} \to \mathscr{H}_\xi$.

Theorem 8. *Fix a quasicharacter $\chi = (\xi, \eta)$. Then, for $\alpha \in C_c(G)$, $\pi_\eta(\alpha)Q_\xi$ is an integral operator with kernel $R_\alpha(u{:}v{:}\chi)$ where*

$$R_\alpha(u{:}v{:}\chi) = \int \alpha(umanv^{-1})\xi(m)(\eta\delta)(a)\,\mathrm{d}m\,\mathrm{d}a\,\mathrm{d}n$$

In particular, for any $\alpha \in C_c^\infty(G)$, $\pi_\chi(\alpha)$ is of trace class; and if

$$T_\chi(\alpha) = \mathrm{tr}\,\pi_\chi(\alpha)$$

then T_χ is the invariant complex measure given by

$$T_\chi(\alpha) = \int_K \mathrm{d}u \int \alpha(umanu^{-1})\xi(m)(\eta\delta)(a)\,\mathrm{d}m\,\mathrm{d}a\,\mathrm{d}n$$

for $\alpha \in C_c(G)$; and for $G = SL(n, \mathbb{R})$, T_χ is supported in $\mathrm{Cl}((MA)^G)$, the closed invariant set of elements whose eigenvalues are all real ($\mathrm{d}x = \delta(a)^2\,\mathrm{d}k\,\mathrm{d}a\,\mathrm{d}n$, $\int_M \mathrm{d}m = 1$).

Proof. The projection Q_ξ is given by (as $\int_M \mathrm{d}m = 1$)

$$(Q_\xi f)(u) = \int_M f(um)\xi(m)\,\mathrm{d}m \quad (u \in K)$$

and hence $\pi_\eta(\alpha)Q_\xi$ is an integral operator with kernel

$$R_\alpha(u{:}v{:}\chi) = \int_M M_\alpha(u{:}vm^{-1}{:}\eta)\xi(m)\,\mathrm{d}m$$

$$= \int \alpha(uanmv^{-1})(\eta\delta)(a)\xi(m)\,\mathrm{d}m\,\mathrm{d}a\,\mathrm{d}n$$

which reduces easily to the required expression since m commutes with A and normalizes N with $d(m^{-1}nm) = \mathrm{d}n$. The formula for $T_\chi(\alpha)$ is immediate also. If E is a compact subset of G, there is a compact subset F of $K \times M \times A \times N$ such that, for $\alpha \in C_c(E)$, the function $u, m, a, n \mapsto \alpha(umanu^{-1})$ has support in F. Hence, for some constant $c = c(E{:}\chi) > 0$, we have

$$|T_\chi(\alpha)| \leqslant c \sup|\alpha|,$$

which proves that T_χ is a measure. To see that it is invariant, we note that, if $x \in G$ and $\alpha^x(y) = \alpha(x^{-1}yx)$, then $\pi(\alpha^x) = \pi(x)\pi(\alpha)\pi(x)^{-1}$ ($\pi = \pi_\chi$) so that $T_\chi(\alpha^x) = T_\chi(\alpha)$. For the assertion regarding $\mathrm{supp}(T_\chi)$, we must show that if $R \subset G$ is the set of matrices with real eigenvalues and α is 0 on R, then $T_\chi(\alpha) = 0$. But the matrix $umanu^{-1}$ has the same characteristic polynomial as *man* which in turn, because of its upper-triangular nature, has the same characteristic polynomial as *ma*. So, $umanu^{-1} \in R$, proving that $T_\chi(\alpha) = 0$.

Remark. The argument for proving that $\mathrm{supp}(T_\chi)$ is contained in $\mathrm{Cl}((MA)^G)$ can be refined to prove an even more striking result: T_χ is the Fourier transform of the orbital integrals associated to the Cartan subgroup $H = M \cdot A$. We have already encountered this result in the case of compact groups. In the present case, it offers the first explanation as to why one should view the principal series (L_χ) as associated to H. When there is only one conjugacy class of Cartan subgroups, for instance when G is complex, it leads to the Plancherel formula exactly as for compact G.

Let now $G = SL(2, \mathbb{R})$. Then $M = \{1, \gamma\}$, $\gamma = -\begin{pmatrix} 1 & 0 \\ 0 & 1 \end{pmatrix}$ and so M is the centre of G. $\hat{M} \cong \{0, 1\}$ where $\varepsilon(=0$ or $1)$ represents the character that takes γ to $(-1)^\varepsilon$. K is the group of rotations u_θ,

$$u_\theta = \begin{pmatrix} \cos\theta & \sin\theta \\ -\sin\theta & \cos\theta \end{pmatrix}$$

while A consists of the matrices

$$a_t = \begin{pmatrix} e^t & 0 \\ 0 & e^{-t} \end{pmatrix} \quad (t \in \mathbb{R})$$

$\mathscr{H}_\varepsilon$ is the subspace of $L^2(K)$ of functions f such that

$$f(u_{-\theta}) = (-1)^\varepsilon f(u_\theta)$$

so that $\mathscr{H}_0$ (resp. $\mathscr{H}_1$) is the space of symmetric (resp. skew-symmetric) functions. The quasicharacters of A are

$$\eta_\lambda : a_t \mapsto e^{\lambda t} \quad (\lambda \in \mathbb{C})$$

The principal-series representations may therefore be denoted by $\pi_{\varepsilon,\lambda}$ ($\varepsilon = 0, 1$, $\lambda \in \mathbb{C}$) with ε corresponding to the character taking γ to $(-1)^\varepsilon$. Similarly we write π_λ and T_λ for the representation π_{η_λ} and its characters; $T_{\varepsilon,\lambda} = \text{character } (\pi_{\varepsilon,\lambda})$.

Theorem 9. *Let $G = SL(2, \mathbb{R})$, and let $l(\gamma)$ denote left translation by γ. Then the character $T_{\varepsilon,\lambda}$ of $\pi_{\varepsilon,\lambda}$ is given by*

$$T_{\varepsilon,\lambda} = \tfrac{1}{2}(T_\lambda + (-1)^\varepsilon l(\gamma) T_\lambda)$$

Proof. For $m \in M$, left and right translations by m coincide. Hence the projection $Q_\varepsilon(\mathscr{H} \to \mathscr{H}_\varepsilon)$ is given by

$$Q_\varepsilon = \tfrac{1}{2}(1 + (-1)^\varepsilon \pi_\lambda(\gamma))$$

and so, for any $\alpha \in C_c^\infty(G)$,

$$\pi_\lambda(\alpha)Q_\varepsilon = \tfrac{1}{2}(\pi_\lambda(\alpha) + (-1)^\varepsilon \pi_\lambda(l(\gamma)\alpha))$$

This leads to the required formula for $T_{\varepsilon,\lambda}$.

Principal series characters as Fourier transforms of orbital integrals. Let

$$H = M \cdot A$$

be the group of diagonal matrices. We shall now prove the result alluded to earlier, exhibiting T_χ as the Fourier transform of the orbital integrals on H. This requires a little preparation. The first step is to define the orbital integrals which are integrals of an arbitrary function on G on the conjugacy classes meeting H. However, unlike the compact case, we need more care in doing this, because the conjugacy classes are not compact, and so the invariant measures on them have to be normalized properly.

For the Haar measure $\mathrm{d}x$ on G we take $\mathrm{d}x = \delta(a)^2\,\mathrm{d}u\,\mathrm{d}a\,\mathrm{d}n$ $(x = uan)$. Since $\mathrm{d}(ana^{-1}) = \delta(a)^2\,\mathrm{d}n$, we have

$$\mathrm{d}x = \mathrm{d}x\,\mathrm{d}n\,\mathrm{d}a \quad (x = una)$$

We thus have a unique invariant measure $\mathrm{d}\bar{x}$ on G/A such that $\mathrm{d}x = \mathrm{d}\bar{x}\,\mathrm{d}a$; thus, for any $g \in G_C(G/A)$,

$$\int_{G/A} g\,\mathrm{d}\bar{x} = \int_{K\times N} g(unA)\,\mathrm{d}u\,\mathrm{d}n$$

Since $H \cong M \times A$, we define $\mathrm{d}h = \mathrm{d}m\,\mathrm{d}a$ $(h = ma)$ on H where $\int_M \mathrm{d}m = 1$ and define $\mathrm{d}\dot{x}$ on G/H by $\mathrm{d}x = \mathrm{d}\dot{x}\,\mathrm{d}h$; if $\dot{g} \in C_c(G/H)$ and $\bar{g}$ is its lift on G/A via the map $G/A \to G/H$ $(\bar{g} \in C_c(G/A))$, then

$$\int_{G/H} \dot{g}\,\mathrm{d}\dot{x} = \int_{G/A} \bar{g}\,\mathrm{d}\bar{x}$$

We now define

$$\Delta(h) = \prod_{i<j} (h_i - h_j) \quad (h = \mathrm{diag}(h_1, \ldots, h_n) \in H)$$

and

$$|\Delta|_F(h) = \begin{cases} |\Delta(h)|^2 & (F = \mathbb{C}) \\ |\Delta(h)| & (F = \mathbb{R}) \end{cases}$$

Finally let

$$\delta(h) = \delta(a) \quad (h = ma \in H,\ m \in M,\ a \in A)$$

Since M is compact, $d(mnm^{-1}) = \mathrm{d}n$ for $m \in M$ and so $d(hnh^{-1}) = \delta(h)^2\,\mathrm{d}n$ for $h \in H$. To define the orbital integral we need the following.

Lemma 10. *Let H' be the set of regular elements of H and $h \in H'$. Then the* *conjugacy class h^G is closed in G, and the map $xH \mapsto h^x = xhx^{-1}$ is a G-equivariant homeomorphism of G/H onto h^G.*

Proof. If F is any field and g is an element of $SL(n, F)$ with *distinct eigenvalues all of which are in F*, the conjugacy class of g in $SL(n, F)$ consists precisely of all matrices having the same characteristic polynomial as g. For $F = \mathbb{R}$ or $\mathbb{C}$ this shows at once that h^G is closed in G, even algebraic. Moreover, as the centralizer of h in G is precisely H, the map $xH \mapsto h^x$ is a well-defined continuous G-equivariant bijection of G/H with h^G. Since both G/H and h^G are second-countable locally compact G-spaces, an easy application of the Baire category theorem shows that this map is also open (see [V1], Theorem 5.11, p. 164).

It follows from this lemma that, if $f \in C_c(G)$, the function

$$xH \mapsto f(xhx^{-1})$$

is an element of $C_c(G/H)$. We now define

$$\begin{aligned} F_{f,H}(h) &= |\Delta(h)|_F \int_{G/H} f(xhx^{-1})\, \mathrm{d}\dot{x} \\ &= |\Delta(h)|_F \int_{G/A} f(xhx^{-1})\, \mathrm{d}\bar{x} \end{aligned}$$

So, $F_{f,H}$ is a well-defined function on H'. We shall speak of it as the *orbital integral of f associated to H.*

This definition requires several comments. We have used a specific Haar measure on G, which we shall call *standard* sometimes. It is also clear that we cannot take h to be singular. For instance, if $h = \operatorname{diag}(h_1, \ldots, h_n)$ and $h_1 = h_2$, the function $f(xhx^{-1})$ is constant on the cosets of the subgroup S of matrices of the form

$$\begin{pmatrix} A_{11} & & 0 \\ & a_{33} & \\ & & \ddots \\ 0 & & & a_{nn} \end{pmatrix} \quad (A_{11} \text{ a } 2 \times 2 \text{ matrix})$$

and, as S/H is not compact the function $xH \mapsto f(xhx^{-1})$ *will not have compact support in G/H*. The reader should compare this situation with the one in the compact case where the orbital integral is defined and smooth everywhere for trivial reasons.

Lemma 11. *Fix $h \in H'$. Then the map $n \mapsto h^{-1}nhn^{-1}$ is an analytic diffeomorphism of N with itself; and*

$$\mathrm{d}(h^{-1}nhn^{-1}) = \delta(h)^{-1} |\Delta(h)|_F \, \mathrm{d}n$$

Proof. Let $f(n) = h^{-1}nhn^{-1}$. It is easy to see that f is one-one: $f(n) = f(n')$ implies that $n'^{-1}n$ centralizes h, hence lies in $H \cap N = (1)$. We shall next prove that f is surjective. Let $n' \in N$ be given and let us search for $n \in N$ such that $n'n = h^{-1}nh$. Writing (a'_{ij}) and (a_{ij}) for n' and n respectively we find that the equations for a_{ij} ($a_{jj} = a'_{jj} = 1$, $a_{ij} = a'_{ij} = 0$ for $i > j$) reduce to

$$(h_j h_i^{-1} - 1)a_{ij} = a'_{ij} + \sum_{i<l<j} a'_{il} a_{lj}$$

If a_{lj} are already determined for $i < l \leqslant j$, this can be solved uniquely for a_{ij} since $h_j h_i^{-1} - 1 \neq 0$. So starting with $a_{nn} = 1$ all the a_{ij} can be determined in succession. In fact the same inductive argument proves that

$$a_{ij} = (h_j h_i^{-1} - 1)^{-1} a'_{ij} + b_{ij}$$

where b_{ij} is a *polynomial* in all the a'_{rs} with $i \leqslant r < s \leqslant j$ with the exception of a'_{ij}. Hence f^{-1} is *actually a polynomial map.* Let us introduce the lexicographic ordering $\ll$ on the set of pairs (i,j) with $i < j$ by defining $(i,j) \ll (i',j')$ to be when $i > i'$ or $i = i'$ and $j < j'$ (note the reversing of the inequality). Then b_{ij} above depends only on the a'_{rs} with $(r,s) \ll (i,j)$. The system of equations of the differentials

$$\mathrm{d}a_{ij} = (h_j h_i^{-1} - 1)^{-1} \mathrm{d}a'_{ij} + \mathrm{d}b_{ij}$$

is then *triangular* with respect to $\ll$, from which we get the result that, if $n' = h^{-1}nhn^{-1}$, then the volume elements $\mathrm{d}n'$ and $\mathrm{d}n$ are related by $\mathrm{d}n' = j(h)\,\mathrm{d}n$ where

$$j(h) = \begin{cases} \prod_{i<j} |h_j h_i^{-1} - 1|^2 & (\mathbb{C}) \\ \prod_{i<j} |h_j h_i^{-1} - 1| & (\mathbb{R}) \end{cases}$$

A simple computation now shows, in view of Lemma 3, that

$$j(h) = |\Delta(h)|_F \delta(h)^{-1}$$

This proves the lemma.

Proposition 12. *For any $f \in C_c(G)$, $h \in H'$,*

$$F_{f,H}(h) = \delta(h) \iint f(uhnu^{-1})\,\mathrm{d}u\,\mathrm{d}n$$

In particular, $F_{f,H}$ extends to an element of $C_c(H)$, which is of class C^∞ if $f \in C_c^\infty(G)$. This extension is symmetric with respect to the action of W, the group of permutations of $\{1, \ldots, n\}$.

Proof. Let $f \in C_c(G)$, $h \in H'$. Then

$$F_{f,H}(h) = |\Delta(h)|_F \int_{G/A} f(xhx^{-1})\,\mathrm{d}\bar{x}$$

$$= |\Delta(h)|_F \iint_{K\times N} f(unhn^{-1}u^{-1})\,du\,dn$$

$$= |\Delta(h)|_F \iint_{K\times N} f(uh(h^{-1}nhn^{-1})u^{-1})\,du\,dn$$

$$= \delta(h) \iint_{K\times N} f(uhnu^{-1})\,du\,dn$$

by Lemma 11. The function $u, h, n \mapsto f(uhnu^{-1})$ belongs to $C_c(K \times H \times N)$, from which it follows at once that the *above integral is defined for all h* and is an element of $C_c(H)$; if $f \in C_c^\infty(G)$, the same argument proves that the integral defines an element of $C_c^\infty(H)$.

It remains to verify the symmetry under W. Since $|\Delta|_F$ is W-invariant, it suffices to do this on H' for the function

$$I(h) = \int_{G/H} f(xhx^{-1})\,d\dot{x} \quad (h \in H')$$

Now the action of W is the one induced by the normalizer of H in G, and so we must prove that $I(whw^{-1}) = I(h)$ for any $w \in G$ normalizing H. But then

$$r_w : xH \mapsto xwH$$

is a well-defined analytic isomorphism of G/H with itself that preserves $d\dot{x}$ and changes $f(xhx^{-1})$ to $f(x(whw^{-1})x^{-1})$, so that the invariance of I under w follows immediately.

Remark. We shall write $F_{f,H}$ also for the extension guaranteed by the above proposition. The reader should observe the indirect way in which the orbital integral is defined at the singular points in H. The orbital integral can be introduced (when G is real) for the other Cartan subgroups also; however it will in general have jumps across the singular subvarieties in H. The Fourier transform of these other orbital integrals will thus have a more complex structure that will be reflected in the structure of the characters of the associated series of representations.

Since H is abelian we can define Fourier transforms of elements of $C_c^\infty(H)$ in the usual manner. Thus, for $g \in C_c^\infty(H)$

$$\hat{g}(\chi) = \int_{M\times A} g(ma)\xi(m)\eta(a)\,dm\,da$$

for $\chi = (\xi, \eta) \in \hat{M} \times \hat{A} \cong \hat{H}$. Actually $\hat{g}$ is defined for all quasicharacters $\chi = (\xi, \eta)$ by the same integral.

Theorem 13. *Let χ be any quasicharacter of H, $\chi = (\xi, \eta)$. Then for any $f \in C_c^\infty(G)$*

$$T_\chi(f) = \hat{F}_{f,H}(\chi)$$

This formula is independent of the choice of $\mathrm{d}x, \mathrm{d}h$, *and* $\mathrm{d}\dot{x}$ *provided* $\mathrm{d}x = \mathrm{d}\dot{x}\,\mathrm{d}h$, *the Fourier transform is defined with respect to* $\mathrm{d}h$, *and* $F_{f,H}$ *is defined by*

$$F_{f,H}(h) = ||\Delta(h)|_F \int_{G/H} f(xhx^{-1})\,\mathrm{d}\dot{x}$$

Proof. By Theorem 8 and Proposition 12,

$$\begin{aligned} T_\chi(f) &= \int \xi(m)\eta(a)\,\mathrm{d}m\,\mathrm{d}a\delta(a) \int f(umanu^{-1})\,\mathrm{d}u\,\mathrm{d}n \\ &= \int \chi(h)\,\mathrm{d}h\left(\int \delta(h) \int f(uhnu^{-1})\,\mathrm{d}u\,\mathrm{d}n\right) \\ &= \int \chi(h) F_{f,H}(h)\,\mathrm{d}h \end{aligned}$$

where $\mathrm{d}h = \mathrm{d}m\,\mathrm{d}a$ $(h = ma)$. To be sure we have always assumed $\int_M \mathrm{d}m = 1$. But this just had the consequence that

$$|\Delta(h)|_F \int_{G/H} f(xhx^{-1})\,\mathrm{d}\dot{x} = |\Delta(h)|_F \int_{G/A} f(xhx^{-1})\,\mathrm{d}\bar{x}$$

and so the final result remains true in general if we define $F_{f,H}$ by the integration on G/H rather than on G/A.

Explicit formula for T_χ on H. As a first application of Theorem 13 we shall prove that T_χ is actually defined by a locally integrable class-function vanishing outside $(H')^G$, and explicitly compute this function. In fact, the integral of any function of $(H')^G$ can be calculated by integrating the orbital integral of this function against a suitable density on H; and we can determine this formula via the identity $T_\chi(f) = \hat{F}_{f,H}(\chi)$.

For convenience in treating the real and complex cases simultaneously we shall use the abbreviation F-analytic to mean holomorphic when $F = \mathbb{C}$ and real-analytic when $F = \mathbb{R}$. If L is a unimodular Lie group over F with Lie algebra $\mathfrak{l}$ and ω_L is a nonzero exterior linear form on $\mathfrak{l}$ of maximal degree, ω_L may be viewed as a translation-invariant F-analytic exterior differential form on L of maximal degree; it defines an invariant volume form, hence a Haar measure, namely ω_L itself or $(i/2)^n \omega_L \wedge \omega_L^{\mathrm{conj}}$ $(n = \dim L)$ according as $F = \mathbb{R}$ or $\mathbb{C}$. We call ω_L a *Haar form.* We extend this terminology to homogeneous spaces for L also. Suppose L_1 is a closed unimodular subgroup and ω_{L_1} a Haar form on L_1. Then the homogeneous space L/L_1 admits a Haar form; if ω_{L/L_1} is one such, we say ω_L, ω_{L_1} and ω_{L/L_1} are *compatible* if the corresponding measures dL, dL_1 and $d(L/L_1)$ satisfy the relation $dL = d(L/L_1)\,dL_1$; the condition for this is that if $(X_i)_{1\leqslant i\leqslant l}$ is a basis for $\mathfrak{l}$ such that $(X_i)_{1\leqslant i\leqslant l_1}$ is a basis for $\mathfrak{l}_1$, and if $\bar{X}_i$ $(l_1 < i \leqslant l)$ is the image of X_i in the tangent space to L/L_1 at the coset L_1

$$\omega_L(X_1,\ldots,X_l) = \pm\,\omega_{L/L_1}(\bar{X}_{l_1+1},\ldots,\bar{X}_l)\omega_{L_1}(X_1,\ldots,X_{l_1})$$

We now consider the F-analytic map

$$\psi\colon xH, h \mapsto xhx^{-1}$$

of $G/H \times H'$ into G. Its image is the invariant set

$$G'_H = \{x \in G \mid \text{all eigenvalues of } x \text{ are distinct and lie in } F\}$$

We can calculate its differential exactly as in the case of $SU(n)$ that we treated in Section 3 of Chapter 2. Essentially that same argument gives the following lemma.

Lemma 14. *ψ has a bijective differential everywhere on $(G/H) \times H'$; in particular G'_H is open in G. If ω_G, ω_H and $\omega_{G/H}$ are compatible Haar forms on G, H and G/H respectively, then*

$$\psi^*(\omega_G) = \pm\Delta^2\omega_{G/H}\omega_H$$

Let W be the Weyl group, acting on H by permutations of diagonal entries; it is $N_G(H)/H, N_G(H)$ being the normalizer of H in G. For $w \in W$, we have seen that $xH \mapsto xw^{-1}H$ is a well-defined map of G/H with itself. We thus have an action

$$w, (xH, h) \mapsto (xw^{-1}H, whw^{-1})$$

of W on $(G/H) \times H'$. It is easy to check that the map ψ is constant precisely on the orbits of W. We thus have the identification

$$(G/H) \times H'/W \cong G'_H$$

In particular we see that a C^∞ (resp. analytic) function on H' extends to an invariant C^∞ (resp. analytic) function on G'_H if and only if it is W-invariant. We further have the following.

Lemma 15. *ψ is proper, hence a covering map; its degree is $|W| = n!$.*

Proof. The argument is a little more delicate than in the compact case. We now use the fact that the eigenvalues of the elements of a compact subset of G' are in compact subsets of $C^\times$ to come down to proving the following: if $x_n \in G$, $h_n \in H'$ and $h_n \to h_0$, $x_n h_n x_n^{-1} \to h_0$ as $n \to \infty$ for some

$h_0 \in H'$, then $x_n H \to H$; since $\psi(x_n H, h_n) = x_n h_n x_n^{-1} \to \psi(H, h_0)$ and since ψ is a local homeomorphism we can find $x'_n H \to H$, $h'_n \to h_0$ such that $\psi(x_n H, h_n) = \psi(x'_n H, h'_n)$ for all sufficiently large n, showing that $x'_n H = x_n w_n^{-1} H$, $h'_n = w_n h_n w_n^{-1}$ for some $w_n \in W$. As both h_n and $h'_n \to h_0$, w_n must be 1, hence $x_n H = x'_n H \to H$.

Lemmas 14 and 15 lead at once to the following integration formula.

Lemma 16 (Integration formula). *Let f be a Borel function on G'_H. Then f is integrable on G'_H if and only if $f \circ \psi$ is integrable on $(G/H) \times H'$; and if $\mathrm{d}x, \mathrm{d}h$ and $\mathrm{d}\dot{x}$ are compatible invariant measures on G, H and G/H, respectively,*

$$\int_{G'_H} f \,\mathrm{d}x = \frac{1}{n!} \iint_{(G/H) \times H'} f(xhx^{-1}) |\Delta(h)|_F^2 \,\mathrm{d}\dot{x} \,\mathrm{d}h$$

Let us write, for any Borel function f on G,

$$F_{f,H}(h) = |\Delta(h)|_F \int_{G/H} f(xhx^{-1}) \,\mathrm{d}\dot{x}$$

whenever this integral is absolutely convergent. We then have the following immediate consequence of the above lemma.

Lemma 17. *If f is a Borel function, f is integrable on G'_H if and only if $F_{|f|,H}(h)$ exists for almost all $h \in H'$ and $F_{|f|,H}$ is integrable with respect to $|\Delta|_F \,\mathrm{d}h$; and*

$$\int_{G'_H} f \,\mathrm{d}x = \frac{1}{n!} \int_{H'} |\Delta(h)|_F F_{f,H}(h) \,\mathrm{d}h$$

The character formula we are after can now be established very quickly. First we have

Theorem 18. *There is a unique invariant function, say D_H, which is 0 outside G'_H and equals $|\Delta|_F^{-1}$ on H'. It is real-analytic on G'_H and locally integrable on G.*

Proof. Since $|\Delta|_F^{-1}$ is invariant and real-analytic on H', the first statement is clear. Let D_H denote this invariant function. We must prove that D_H is integrable on any compact set of G. If f is any element of $C_c^\infty(G)$ which is $\geqslant 0$ everywhere and $\geqslant 1$ on the compact set in question, we must prove that $D_H \cdot f$ is integrable. Since everything in sight is $\geqslant 0$, Lemma 17 reduces this to proving that

$$\int_{H'} |\Delta(h)|_F F_{D_H f,H}(h)\,\mathrm{d}h < \infty$$

But

$$|\Delta(h)|_F F_{D_H f,H} = F_{f,H}(h)$$

and

$$\int_{H'} F_{f,H}(h)\,\mathrm{d}h < \infty$$

since we know $F_{f,H}$ extends to an element of $C_c(H)$.

Theorem 19. *Let χ be a quasicharacter of H. Then there is a unique invariant function θ_χ on G which is 0 outside G'_H and coincides with*

$$\left(\sum_{w\in W} \chi^w\right)\Big/ |\Delta|_F$$

on H'. θ_χ is real-analytic on G'_H, locally integrable on G and is precisely the character of the principal-series representation L_χ:

$$T_\chi(f) = \int_G \theta_\chi f\,\mathrm{d}x$$

Proof. The existence of θ_χ and its real-analyticity on G'_H are clear. The function $\sum_w \chi^w$ is smooth and W-invariant on H' and so there is a unique invariant function ϕ on G, vanishing outside G'_H and coinciding with $\sum_w \chi^w$ on H'. As $\theta_\chi = \phi D_H$, to prove that θ_χ is locally integrable on G it is enough to prove, in view of the local integrability of D_H, that ϕ is locally bounded on G. This would follow if we show that if R is a compact set in G there is a compact set $S \subset \mathbb{C}^\times$ such that the eigenvalues of all elements of R are in S. But this is immediate from the fact that the coefficients of the characteristic polynomial of the elements of R and their inverses are uniformly bounded. This said, let $f \in C_c^\infty(G)$. Then $\theta_\chi f$ is integrable on G, and so,

$$\begin{aligned}
\int_G \theta_\chi f\,\mathrm{d}x &= \frac{1}{n!}\int_H |\Delta|_F F_{\theta_\chi f,H}\,\mathrm{d}h \\
&= \frac{1}{n!}\int_H \left(\sum_w \chi^w\right) F_{f,H}\,\mathrm{d}h \\
&= \int F_{f,H}\chi\,\mathrm{d}h \\
&= T_\chi(f)
\end{aligned}$$

where we have used $F_{\theta_\chi f,H}(h) = \theta_\chi(h)F_{f,H}(h)$ as well as the W-invariance of $F_{f,H}$.

When $n = 2$, the diagonal matrices have diagonal entries a, a^{-1} ($a \in F^\times$), and so $W = \{1, s\}$ where s acts by $\mathrm{diag}\,(a, a^{-1}) \mapsto \mathrm{diag}\,(a^{-1}, a)$. Hence

$$\theta_\chi(\mathrm{diag}\,(a, a^{-1})) = \begin{cases} \dfrac{\chi(a) + \chi(a^{-1})}{|a - a^{-1}|^2} & (\mathbb{C}) \\[2ex] \dfrac{\chi(a) + \chi(a^{-1})}{|a - a^{-1}|} & (\mathbb{R}) \end{cases}$$

In the complex case the quasicharacters are of the form

$$\chi_{m,\rho}: \mathrm{diag}\,(a, a^{-1}) \mapsto (a/|a|)^m |a|^\rho \quad (m \in \mathbb{Z}, p \in \mathbb{C})$$

and in the real case

$$\chi_{\varepsilon,\rho}: \mathrm{diag}\,(a, a^{-1}) \mapsto \mathrm{sgn}\,(a)^\varepsilon |a|^\rho \quad (\varepsilon = 0 \text{ or } 1, \rho \in \mathbb{C})$$

These results suggest that it is possible to develop the theory of characters for semisimple groups in great depth and detail. If H is a Lie group and π is an infinite-dimensional representation of H in some Hilbert space $\mathscr{H}$, it is clear that the simple-minded definition of $\mathrm{tr}\,\pi(x)$ as $\sum_n (\pi(x)e_n | e_n)$, (e_n) an ON basis of $\mathscr{H}$, will not work; for instance, the terms of the series are all equal to 1 when $x = 1$. It is natural to try to overcome this difficulty by a *summation process.* A natural one, which in addition makes sense in the context of an arbitrary Lie group, is to try to define $\mathrm{tr}\,\pi(x)$ as a *regularization* of

$$\sum_n \int_H \alpha(x)(\pi(x)e_n | e_n)\,dx$$

Formally this comes down to considering the traces

$$\Theta_\pi(\alpha) = \mathrm{tr}\left(\int_G \alpha(x)\pi(x)\,dx\right) \quad (\alpha \in C_c^\infty(H))$$

It is then natural to say π *has a character* if Θ_π above is *well defined on $C_c^\infty(G)$ and determines a distribution in the sense of L. Schwartz.* Now the group H operates on itself by inner automorphisms, and this action induces actions on $C_c^\infty(H)$ and the space of distributions on H: $\alpha^x(h) = \alpha(x^{-1}hx)$, $T^x(\alpha) = T(\alpha^{x^{-1}})$ ($\alpha \in C_c^\infty(H)$, $x, h \in H$, T a distribution on H); T is called *invariant* if $T = T^x$ for all $x \in H$. Since

$$\pi(\alpha^x) = \pi(x)\pi(\alpha)\pi(x)^{-1}$$

it is immediate from the invariance of trace under similarity that Θ_π is an invariant distribution, and that Θ_π depends only on the equivalence class of π. It then makes sense to call a class function θ_π the *character of* π if

θ_π is locally integrable on H and

$$\Theta_\pi(\alpha) = \int_H \alpha(x)\theta_\pi(x)\,\mathrm{d}x \quad (\alpha \in C_c^\infty(H))$$

The measure $\mathrm{d}x$ on the right is the same as the one used in defining $\pi(\alpha)$, so that θ_π (if it exists) *is independent of any normalization of* $\mathrm{d}x$. All of these remarks hang in the air until one exhibits representations in infinite-dimensional spaces for which characters in this sense exist. This is exactly what Gel'fand and Naimark did for the group G and the representations L_χ.

The formula for θ_x shows that

$$\theta_\chi = \theta_{\chi'} \Leftrightarrow \chi' = \chi^w \quad \text{for some } w \in W$$

Hence, unless χ and χ' are in the same W-orbit, the representations L_χ and $L_{\chi'}$, cannot be equivalent. The converse, that, for χ, χ' in the same orbit of W in $\hat{H}$, the unitary representations L_χ and $L_{\chi'}$ are equivalent, was also proved by Gel'fand and Naimark. Thus the character of L_χ determines it up to equivalence. This was the first indication that the concept of the (distribution) character would play the same decisive role in the theory of infinite-dimensional representations that it does in the finite-dimensional theory.

The character formulae for the principal series suggest very strongly, if one keeps in mind the analogy with compact groups, that the principal-series representations are all irreducible and that the only equivalences among the L_χ are those between L_χ and L_{χ^s} ($s \in W$). This is true for complex groups while for real groups the L_χ may split occasionally when χ is special. We shall examine the question of irreducibility as well as the construction of the supplementary series later. However, the model constructed for the principal series of $SL(2, \mathbb{C})$ using the action on $\mathbb{P}^1$ – the so-called *affine model* – may be used to prove the irreducibility rather quickly. We have in fact a stronger result, as follows.

Theorem 20. *Let $G = SL(2, \mathbb{C})$. Then all the L_χ ($\chi \in \hat{D}$) are already irreducible when restricted to any Borel subgroup.*

Proof. It is enough to prove this for the 'opposite' Borel subgroup B^t. Let R be a bounded operator commuting with $L_\chi(b)$, $b \in B^t$. For $b_1 = \begin{pmatrix} 1 & 0 \\ u & 1 \end{pmatrix}$, $L_\chi(b_1)$ is just translation by u, $f(z) \mapsto f(z+u)$, while $b_2 = \begin{pmatrix} a & 0 \\ 0 & a^{-1} \end{pmatrix}$ gives $(L_\chi(b_2)f)(z) = f(a^2 z)a^m |a|^{-m+\rho+2}$. It is then natural to use *Fourier transform* theory, and so, for $f \in L^2(\mathbb{C}: \mathrm{d}x\,\mathrm{d}y)$ we introduce its L^2

Fourier transform

$$\hat{f}(w) = \frac{1}{2\pi}\int f(z)\exp(-\mathrm{i}\,\mathrm{Re}(\bar{z}w))\,\mathrm{d}x\,\mathrm{d}y$$

$$(L_\chi(b_1)f)\hat{}(w) = \exp(\mathrm{i}\,\mathrm{Re}(\bar{u}w))\hat{f}(w)$$
$$(L_\chi(b_2)f)\hat{}(w) = a^m|a|^{-m+\rho-2}\hat{f}(\bar{a}^{-2}w)$$

The operator R defines an operator $\hat{R}$ by $\hat{R}\hat{f} = (Rf)\hat{}$. Then it is immediate from the first of these formulae that $\hat{R}$ is multiplication by a bounded (Borel) function r; the second relation then shows that $r(w) = r(\bar{a}^{-2}w)$ for almost all w implying that r is a constant almost everywhere. This proves R is a scalar and completes the proof, and indeed, also the proof of the stronger assertion that, *for any quasicharacter* χ, *the commuting ring of the restriction of* L_χ *to* B (or B^t) *is just* $\mathbb{C}\cdot 1$.

The affine model arises when we work on the finite complex plane. This can be carried over to the general case by writing the flag manifold as a union of 'cells' and working on the 'biggest' cell which will be open in the flag manifold; the remaining cells will add up to a variety of lower dimension. We shall not discuss this theme here any more.

Notes and comments

The reader should consult [GN] for detailed discussions of the concepts of this chapter (parabolic subgroups, principal series, etc.) for the other classical groups. Here as elsewhere a monumental reference is [Kn].

Problems

1. Prove that the regular sets in $GL(n,\mathbb{C})$, $SL(n,\mathbb{C})$, $U(n)$, $SU(n)$ are all connected.
2. A subset of E of an analytic manifold (all manifolds are connected) M may be said to have dimension $\leqslant r$ if it is a finite union of sets E_i with the following property: there is an analytic manifold N_i of dimension r and an analytic map f_i of N_i into M such that $E_i \subset f_i(N_i)$; if r is the smallest integer with this property, E is said to have dimension exactly r. Prove that the singular set in $U(n)$ has dimension n^2-3.
3. For any integer $n \geqslant 2$ and $F = \mathbb{R}$ or $\mathbb{C}$, $Sp(n,F)$ denotes the subgroup of $GL(2n,F)$ of all x such that $x^tJx = J$ where

$$J = \begin{pmatrix} 0 & I \\ -I & 0 \end{pmatrix} \quad (n \times n \text{ blocks})$$

Prove that $Sp(n,F)$ has rank n. Prove also that $Sp(n,\mathbb{R})$ has Cartan subgroups that are compact as well as those that are isomorphic to $(\mathbb{R}^\times)^n$.
4. Describe the parabolic subgroups of $Sp(n,F)$ and their Levy–Langlands decompositions.

5

Representations of the Lie algebra

5.1 Preliminaries on Lie algebras and enveloping algebras

The theme in this chapter is the study of representations of semisimple Lie algebras and their relationship to representations of semisimple Lie groups. For finite-dimensional representations it is relatively simple to go back and forth between the Lie algebra ('infinitesimal') and the group ('global') pictures. However it is technically much more difficult to do this for infinite-dimensional representations. Harish-Chandra was the first to explore this theme systematically. In a remarkable series of papers [H1] [H2] [H3] he created the theory of infinite-dimensional representations of semisimple Lie algebras, and overcame all the difficulties which arise when one tries to 'lift' (or 'integrate') these representations to the group. His work showed that the infinitesimal view is as penetrating in the infinite-dimensional case as in the finite-dimensional one. For establishing the fundamental results on the structure of the infinite-dimensional representations of semisimple groups no other method can surpass the infinitesimal one in simplicity, elegance, and definitiveness.

We begin with some preliminary remarks on Lie groups and Lie algebras. Let G be any real Lie group with Lie algebra $\mathfrak{g}$. In all the calculations that we shall make it will be a closed subgroup of some $GL(n, \mathbb{R})$. It is a classical theorem of von Neumann (see [V2]) that any closed subgroup G of $GL(n, \mathbb{R})$ is a Lie group with Lie algebra $\mathfrak{g}$ given by

$$\mathfrak{g} = \{L \in \mathfrak{gl}(n, \mathbb{R}) \mid \exp tL \in G \quad \text{for all} \quad t \in \mathbb{R}\}$$

This is one of the basic results in the circle of ideas relating the topological aspects of Lie groups to the differentiable and analytic aspects. As a second example let me mention the fact that if $G_i (i = 1, 2)$ are real Lie groups, and $f(G_1 \to G_2)$ is a continuous homomorphism, then f is (real-)analytic. The most profound result of this type is of course the theorem that locally Euclidean topological groups are Lie groups, due to Gleason and Montgomery, answering a question originally posed as the fifth of twenty-three problems by Hilbert in his famous address in Paris in 1900.

If G is a Lie group we write $\mathfrak{g} = \mathrm{Lie}(G)$ for its Lie algebra and identify its elements with the *left-invariant* vector fields of G. Hence, for any $L \in \mathfrak{g}$ we have the first-order left-invariant differential operator $\partial(L)$ acting on

C^∞ functions f on G by

$$(\partial(L)f)(x) := f(x;L) = \left(\frac{d}{dt} f(x \exp tL)\right)_{t=0}$$

($x \in G$). Clearly it is natural to introduce the left-invariant differential operators of higher order also. It is a basic result that all such operators are analytic and that they are precisely the endomorphisms of $C^\infty(G)$ generated by the $\partial(L)$, $L \in \mathfrak{g}$. It is called the *universal enveloping algebra* of $\mathfrak{g}$ and denoted by $U(\mathfrak{g})$. We remark that the functions and differential operators are *complex-valued.* Thus, although $\mathfrak{g}$ is a Lie algebra over $\mathbb{R}$, $U(\mathfrak{g})$ is an algebra over $\mathbb{C}$ and contains $\mathfrak{g}_{\mathbb{C}}$, the complexification of $\mathfrak{g}$; and, for $X, Y \in \mathfrak{g}$, $\partial([X,Y]) = \partial(X)\partial(Y) - \partial(Y)\partial(X)$. The word *universal* refers to the following universal property: if A is any associative algebra over $\mathbb{C}$ with unit and $f(\mathfrak{g} \to A)$ an $\mathbb{R}$-linear map such that, for all $X, Y \in \mathfrak{g}$,

$$f([X,Y]) = f(X)f(Y) - f(Y)f(X),$$

then there is a unique homomorphism $U(\mathfrak{g}) \to A$ of complex associative algebras with units, also denoted by f, such that the diagram

$$\begin{array}{ccc} \mathfrak{g} & \hookrightarrow & U(\mathfrak{g}) \\ & {\scriptstyle f}\searrow \quad \swarrow{\scriptstyle f} & \\ & A & \end{array}$$

is commutative. Although $U(\mathfrak{g})$ has been introduced as an algebra of differential operators on G, we shall find it convenient to think of it as an abstract associative algebra containing $\mathfrak{g}$ with the above universal property, equipped with a distinguished isomorphism ∂ of it with the algebra of left-invariant differential operators on G. Thus, for $L_1, \ldots, L_r \in \mathfrak{g}$, we write $L_1 \cdots L_r$ for their product in $U(\mathfrak{g})$ and $\partial(L_1 \cdots L_r)$ for the differential operator $\partial(L_1)\partial(L_2)\cdots\partial(L_r)$:

$$(\partial(L_1 \cdots L_r)f)(x) = \left(\frac{\partial^r}{\partial t_1 \cdots \partial t_r} f(x \exp t_1 L_1 \cdots \exp t_r L_r)\right)_0$$

($x \in G$), the subscript 0 indicating that the derivative is to be taken at $t_1 = t_2 = \cdots = t_r = 0$. We shall also set up an *anti-isomorphism* of $U(\mathfrak{g})$, ∂_r, with the algebra of *right*-invariant differential operators on G:

$$(\partial_r(L)f)(x) := f(L;x) = \left(\frac{d}{dt} f(\exp tLx)\right)_{t=0}$$

$$\begin{aligned}(\partial_r(L_1 \cdots L_m)f)(x) &:= f(L_1 \cdots L_m; x) \\ &= \left(\left(\frac{\partial^m}{\partial t_1 \cdots \partial t_m f}\right)(\exp t_1 L_1 \cdots \exp t_m L_m x)\right)_0\end{aligned}$$

$$(\partial(a)f)(x) := f(x;a), \quad (\partial_r(a)f)(x) = f(a;x)$$

Thus, for $a, b \in U(\mathfrak{g})$

$$\partial(ab) = \partial(a)\partial(b), \quad \partial_r(ab) = \partial_r(b)\partial_r(a)$$

In Lie groups one goes from left to right by the adjoint representation Ad of G acting on $\mathfrak{g}$; we write $L^x = \mathrm{Ad}(x)L$, $L \in \mathfrak{g}$, $x \in G$. It extends to an action Ad of G on $U(\mathfrak{g})$, and we write a^x for $\mathrm{Ad}(x) \cdot a$, $a \in U(\mathfrak{g})$. If $G \subset GL(n, \mathbb{R})$ then $\mathfrak{g} \subset \mathfrak{gl}(n, \mathbb{R})$, and $L^x = xLx^{-1}$. Since

$$\exp tL \cdot x = x \cdot \exp t(L^{x^{-1}})$$

we have, for any $f \in C^\infty(G)$, $a \in U(\mathfrak{g})$, $x \in G$,

$$f(a; x) = f(x; a^{x^{-1}})$$

A most important role in harmonic analysis is played by the algebra of those differential operators which are invariant under both left and right translations. These are the $\partial(a)$ where $a \in U(\mathfrak{g})$ is such that $a^x = a$ for all $x \in G$. We denote it by $\mathfrak{z}$:

$$\mathfrak{z} = \{a \in U(\mathfrak{g}) \mid a^x = a \quad \text{for all} \quad x \in G\}$$

If G is connected, $\mathfrak{z}$ is the *centre* of $U(\mathfrak{g})$; to prove this we note that the representations Ad of G and ad of $\mathfrak{g}$, the latter acting on $U(\mathfrak{g})$ by

$$\mathrm{ad}\, X := a \mapsto Xa - aX \quad (X \in \mathfrak{g})$$

are related by

$$\mathrm{Ad}(\exp X) = \exp(\mathrm{ad}\, X)$$

Hence, for $a \in U(\mathfrak{g})$, $\mathrm{Ad}(\exp X)a = a$ for all $X \in \mathfrak{g}$ if and only if $\mathrm{ad}\, X(a) = 0$ for all $X \in \mathfrak{g}$. Since G is connected, it is generated by $\exp \mathfrak{g}$, while $U(\mathfrak{g})$ is generated by $\mathfrak{g}$. Hence,

$$\mathfrak{z} = \text{centre}\, U(\mathfrak{g}) = \{a \in U(\mathfrak{g}) \mid Xa = aX \quad \text{for all} \quad X \in \mathfrak{g}\}$$

for G connected. $\mathfrak{z}$ is of course abelian.

It may be worthwhile to make a few comments on the structure of $U(\mathfrak{g})$, which we shall denote by U for brevity, as an associative algebra. Let $U^{(r)}$ $(r \geqslant 0)$ be the linear span of elements of the form $Z_1 \cdots Z_s$ $(s \leqslant r, Z_j \in \mathfrak{g})$; then $U^{(r)}$ is finite-dimensional, $U^{(r)} \subset U^{(s)}$ for $r \leqslant s$, $U = \bigcup_r U^{(r)}$, and $U^{(r)}U^{(s)} \subset U^{(r+s)}$. So $(U^{(r)})_{r \geqslant 0}$ is a filtration for U and U is a *filtered algebra.* We define

$$S_r = U^{(r)}/U^{(r-1)}, \quad S = \bigoplus_{r \geqslant 0} S_r$$

$(U^{(-1)} = \varnothing$, and products with zero number of elements are interpreted as 1). The multiplication in U gives rise to one in S, converting S into a *graded algebra.* The key point now is that S *is commutative and is canonically isomorphic to the full symmetric algebra* $S(\mathfrak{g})$ *over* $\mathfrak{g}_{\mathbb{C}}$. Indeed, it follows from the commutation rules that if $Z_1, \ldots, Z_s \in \mathfrak{g}$, and $i_1, \ldots, i_s$ is any permutation of $(1, \ldots, s)$, $Z_1 \cdots Z_s \equiv Z_{i_1} \cdots Z_{i_s} \pmod{U^{(s-1)}}$, so that the commutativity of S is obvious. The map $\mathfrak{g} \to S$ now lifts to a homomorphism of the symmetric algebra $S(\mathfrak{g})$ over $\mathfrak{g}_\mathbb{C}$ onto S; the assertion is that this is an *isomorphism.* This is less obvious and is known as the Poincaré–Birkhoff–Witt theorem; it comes down to proving that, if $(X_1, \ldots, X_n)$ is a basis for $\mathfrak{g}$, the *monomials*

$$X_1^{r_1} \cdots X_n^{r_n}$$

form a basis for U. It is usual to identify S with $S(\mathfrak{g})$. For technical reasons it is convenient to work with some linear isomorphism $S \tilde{\to} U$ which inverts the canonical maps $U^{(r)} \to S_r$. Harish-Chandra generally worked with the *symmetrizer* map λ: if $Z_1, \ldots, Z_r \in \mathfrak{g}$,

$$\lambda(Z_1 \cdots Z_r) = (r!)^{-1} \sum z_{i_1} Z_{i_2} \cdots Z_{i_r}$$

where the products on the right are in $U(\mathfrak{g})$ and the sum is over all the permutations of $(1, \ldots, r)$, while the product on the left is in S and is commutative. Obviously, if $a, b \in S$ are homogeneous and of respective degrees r, s, then

$$\lambda(ab) \equiv \lambda(a)\lambda(b) \pmod{U^{(r+s-1)}}$$

The map λ is a powerful technical tool to reduce many questions involving U to questions involving S. The reader who is familiar with the general theory of differential operators will recognize that for any $a \in U$ of degree r, i.e., $a \in U^{(r)}$ but $\notin U^{(r-1)}$, its image in $S_r \subset S$ is just the *symbol* of $\partial(a)$. The extra structure present here allows us to go in the opposite direction, namely, from symbols to operators, in a *covariant* fashion. Here covariance refers to the obvious fact that λ commutes with the adjoint action of G on U and S.

As an illustration let us see how $\mathfrak{Z} =$ centre (U) can be determined once we know $I = S^G$, the algebra of G-invariant elements of S. The covariance of λ shows that $\lambda(I) = \mathfrak{Z}$ while

$$\lambda : I \to \mathfrak{Z}$$

is bijective. Moreover, if $a_1, \ldots, a_m$ are homogeneous generators of I, $\lambda(a_1), \ldots, \lambda(a_m)$ are generators of Z, and are algebraically independent if the a_i are so. Indeed, if $b = \lambda(a)$ where $a \in I$ is of degree r and $a = \sum c_r a^r$ where $r = (r_i)$, $a^r = \prod_i a_i^{r_i}$,

$$b - \sum_r c_r \lambda(a_1)^{r_1} \cdots \lambda(a_m)^{r_m}$$

is in $\mathfrak{Z}$ and has degree $< r$, so that induction on degree applies; the argument for algebraic independence is similar (see [V2], p. 184). If $\mathfrak{g}$ is semisimple, there is a famous theorem of Chevalley that says that I is a free polynomial algebra over l homogeneous generators where $l = \text{rank}(\mathfrak{g})$; so, in this case, $\mathfrak{Z}$ is also a free polynomial algebra over l generators [V2].

Take $G = SL(2, \mathbb{R})$. Then $\mathfrak{g} = \mathfrak{sl}(2, \mathbb{R})$, the Lie algebra of real matrices of order 2 and trace 0; we have the standard basis

$$H = \begin{pmatrix} 1 & 0 \\ 0 & -1 \end{pmatrix}, \quad X = \begin{pmatrix} 0 & 1 \\ 0 & 0 \end{pmatrix}, \quad Y = \begin{pmatrix} 0 & 0 \\ 1 & 0 \end{pmatrix}$$

with commutation rules

$$[H, X] = 2X, [H, Y] = -2Y, [X, Y] = H$$

The adjoint action is similarity and the invariants are generated by $H^2 + 4YX$. Hence $\mathfrak{Z}$ is generated by

$$H^2 + 2YX + 2XY$$

which is, up to a scalar, the *Casimir operator* of G. We shall usually work with some normalized form of it, such as

$$\begin{aligned} \Omega &= H^2 + 2YX + 2XY + 1 = (H + 1)^2 + 4YX \\ &= -(X - Y)^2 + (X + Y)^2 + H^2 + 1 \end{aligned}$$

If $G = SL(n, \mathbb{R})$, I is generated by $n - 1$ homogeneous elements, of respective degrees

$$2, 3, \ldots, n$$

The element in $\mathfrak{Z}$ corresponding to the quadratic invariant is essentially the Casimir operator.

We end these preliminary remarks with some comments on complex groups. If G is a complex Lie group, its Lie algebra $\mathfrak{g}$ is a complex Lie algebra. However, in representation theory we are interested in the *real-analytic group underlying* G, denoted by $G_{\mathbb{R}}$; the corresponding Lie algebra is $\mathfrak{g}_{\mathbb{R}}$, the real Lie algebra underlying $\mathfrak{g}$; and to obtain $U(\mathfrak{g}_{\mathbb{R}})$ we must first replace $\mathfrak{g}_{\mathbb{R}}$ by its complexification

$$\mathfrak{g}_{\mathbb{R}} \otimes_{\mathbb{R}} \mathbb{C} = \tilde{\mathfrak{g}}$$

and then go over to the universal enveloping algebra of $\tilde{\mathfrak{g}}$. We shall describe the pair $(\mathfrak{g}_{\mathbb{R}}, \tilde{\mathfrak{g}})$ in two special situations.

$\mathfrak{g}$ **arbitrary.** Let $\bar{\mathfrak{g}}$ be the Lie algebra *conjugate to* $\mathfrak{g}$: it has the same elements as $\mathfrak{g}$, but the product of $c \in \mathbb{C}$ and $X \in \bar{\mathfrak{g}}$ in $\bar{\mathfrak{g}}$ is $\bar{c}X$, the product of $\bar{c}$ and X in $\mathfrak{g}$. Let j be the diagonal map $Z \mapsto (Z, Z)$ of $\mathfrak{g}_{\mathbb{R}}$ into $\tilde{\mathfrak{g}} = \mathfrak{g} \times \bar{\mathfrak{g}}$. It is then quite easy to check that $(\mathfrak{g} \times \bar{\mathfrak{g}}, j)$ is a complexification of $\mathfrak{g}_{\mathbb{R}}$ (note that $\mathrm{i}(Z, Z) = (\mathrm{i}Z, -\mathrm{i}Z)$). The conjugation of $\tilde{\mathfrak{g}}$ that defines $j(\mathfrak{g}_{\mathbb{R}})$

is the map $(Z, Z') \mapsto (Z', Z)$. Clearly

$$U(\tilde{\mathfrak{g}}) \cong U \otimes \bar{U}$$

where $U = U(\mathfrak{g})$ and $\bar{U}$ is the algebra conjugate to U.

$\mathfrak{g}$ **semisimple.** From Weyl's theory we know $\mathfrak{g}$ has compact real forms. Let $\mathfrak{k}$ be one of those, and let θ be the conjugation that defines $\mathfrak{k}$. We now take $\tilde{\mathfrak{g}} = \mathfrak{g} \times \mathfrak{g}$ and j to be the map $Z \mapsto (\theta Z, Z)$ of $\mathfrak{g}$ into $\tilde{\mathfrak{g}}$. It is again easy to verify that $(\tilde{\mathfrak{g}}, j)$ is a complexification of $\mathfrak{g}_{\mathbb{R}}(\mathrm{i}\cdot(\theta Z, Z) = (-\theta(\mathrm{i}Z), \mathrm{i}Z))$. Since $j(Z) = (Z, Z)$ for $Z \in \mathfrak{k}$, and $\mathbb{C}\cdot\mathfrak{k} = \mathfrak{g}$, the complex span of $j(\mathfrak{k})$ is the *diagonal subalgebra* of $\tilde{\mathfrak{g}}$. Thus the complexification of $(\mathfrak{g}_{\mathbb{R}}, \mathfrak{k})$ is the pair ($\mathfrak{g} \times \mathfrak{g}$, diagonal subalgebra); the complexification of θ is the involution $(X, Y) \mapsto (Y, X)$ of $\mathfrak{g} \times \mathfrak{g}$.

5.2 Analytic vectors and the Harish-Chandra density theorem

Let us now consider representations of a Lie group and its Lie algebra. Let G be a Lie group with Lie algebra $\mathfrak{g}$. First of all we observe that, if π is a representation of $\mathfrak{g}$ in a complex vector space V (not necessarily finite-dimensional), the universal property of the pair ($\mathfrak{g}$, $U(\mathfrak{g})$) implies that there is a unique representation, also denoted by π, of the associative algebra $U(\mathfrak{g})$ in V, extending π; if $X_1, \ldots, X_r \in \mathfrak{g}$ then

$$\pi(X_1 \cdots X_r) = \pi(X_1) \cdots \pi(X_r)$$

We shall therefore make no distinction between representations of $\mathfrak{g}$ and $U(\mathfrak{g})$. Suppose G is a connected (real) Lie group with Lie algebra $\mathfrak{g}$ and π is a representation of G in a finite-dimensional vector space V over $\mathbb{C}$. Then π, regarded as a homomorphism of G into $\mathrm{GL}(V_{\mathbb{R}})$ ($V_{\mathbb{R}}$ is the real vector space underlying V) is analytic and so we can introduce the corresponding Lie algebra representation, also denoted by π, as follows; for $X \in \mathfrak{g}$,

$$\pi(X)v = \left(\frac{\mathrm{d}}{\mathrm{d}t}\right)_{t=0} \pi(\exp tX)v \quad (v \in V)$$

The $\pi(X)$ are endomorphisms of V, $X\,\pi(X)$ is $\mathbb{R}$-linear, and $\pi([X, Y]) = [\pi(X), \pi(Y)]$ for all $X, Y \in \mathfrak{g}$. So π extends uniquely to a representation of $\mathfrak{g}_{\mathbb{C}}$, the complexification of $\mathfrak{g}$, in V. To go back in the other direction, one must assume that G is simply connected; then, given a representation π of $\mathfrak{g}_{\mathbb{C}}$ in V, there is a unique representation, also denoted by π, of G in V, such that

$$\pi(\exp X) = \exp(\pi(X)) \quad (X \in \mathfrak{g})$$

It is necessary to work with greater care when dealing with infinite-dimensional representations. Let V now be a *Banach space* and π a representation of G in V. It is then not reasonable to expect that the functions $g \mapsto \pi(g)v$, for $v \in V$, are all differentiable, so that the operators $\pi(X)$ will in general not be defined everywhere on V. Let us call a vector $v \in V$ *differentiable* or C^∞ (for π) if the function

$$g \mapsto \pi(g)v$$

from G to V is of class C^∞. It is known that this is equivalent to requiring that for any $v^* \in V^*$, the Banach space dual to V, the complex-valued function

$$g \mapsto (\pi(g)v, v^*)$$

is of class C^∞. Let V^∞ be the space of differentiable vectors. Since

$$(\pi(g)\pi(x)v, v^*) = (\pi(gx)v, v^*)$$

it is clear that V^∞ is stable under G. We now define, as in the finite-dimensional case,

$$\pi(X)v = \left(\frac{\mathrm{d}}{\mathrm{d}t}\pi(\exp tX)v\right)_{t=0} \qquad (X \in \mathfrak{g}, v \in V^\infty)$$

If we write $F(g{:}v, v^*)$ for $(\pi(g)v, v^*)$, then

$$(\pi(X)v, v^*) = F(1; X{:}v, v^*)$$
$$(\pi(g)\pi(X)v, v^*) = F(g; X{:}v, v^*)$$

so that $\pi(X)v \in V^\infty$. Thus $\pi(X)$ is an endomorphism of V, and it is easy to show that $X \mapsto \pi(X)$ is a representation of $\mathfrak{g}$ in V^∞.

As the simplest example, let $G = \mathbb{R}$ and let π be the regular representation in $L^2(\mathbb{R})$. If $v \in C_c^\infty(\mathbb{R})$, the function $(\pi(t)v|w)(w \in L^2(\mathbb{R}))$ is nothing but the convolution integral

$$\int_{-\infty}^{\infty} v(s+t)w(t)\,\mathrm{d}t$$

This clearly is C^∞ in s. If $X = \mathrm{d}/\mathrm{d}t$ is the basis of $\mathfrak{g}$, then

$$\pi(X)v = \mathrm{d}v/\mathrm{d}t, \quad \pi(X^n)v = \mathrm{d}^n v/\mathrm{d}t^n$$

It is clear from this example that one can expect the $\pi(a)$ $(a \in U(\mathfrak{g}))$ only to be densely defined at best.

Let us now return to the general case. Although V^∞ may not be all of V, one may still expect it to be *dense* in V; this is true in the above example. This was first proved by Gårding [Ga].

Lemma 1 (Gårding). *If $f \in C_c^\infty(G)$, the vectors*

$$\pi(f)v = \int_G f(x)\pi(x)v\,\mathrm{d}x \quad (v \in V)$$

are in V^∞. Their linear span, V_0, the so-called Gårding subspace of V, is dense in V.

Proof. We have

$$\pi(y)\pi(f)v = \int_G f(y^{-1}x)\pi(x)v\,\mathrm{d}x$$

Since f is C^∞ and has compact support, this integral may be differentiated under the integral sign, i.e., it is of class C^∞ in y. We prove the density by finding a sequence (f_n) such that $\pi(f_n)v \to v$ as $n \to \infty$ for each $v \in V$. Take (f_n) such that

(i) $f_n \in C_c^\infty(G)$,
(ii) $\mathrm{supp}(f_n) \searrow (1)$, i.e., if W is any neighbourhood of 1, $\mathrm{supp}(f_n) \subset W$ for all sufficiently large n,
(iii) $\int_G f_n \,\mathrm{d}x = 1$ for all n and, for some constant $C > 0$ and all n, $\int_G |f_n| \,\mathrm{d}x \leqslant C$.

If then W is a neighbourhood of 1 and n is sufficiently large, so that $\mathrm{supp}(f_n) \subset W$,

$$\begin{aligned}\|\pi(f_n)v - v\| &= \left\| \int f_n(x)(\pi(x)v - v)\,\mathrm{d}x \right\| \\ &\leqslant \int_W |f_n(x)|\,\|\pi(x)v - v\|\,\mathrm{d}x \\ &\leqslant C \sup_{x \in W} \|\pi(x)v - v\|\end{aligned}$$

Letting $n \to \infty$, and then shrinking W to 1, we see that $\lim_n \pi(f_n)v = v$.

We also have the following compatibility relations between $\pi(x)$ and $\pi(X)$:

$$\pi(x)\pi(X)\pi(x)^{-1} = \pi(X^x) \quad (x \in G, X \in \mathfrak{g})$$
$$\pi(x)\pi(a)\pi(x)^{-1} = \pi(a^x) \quad (a \in U(\mathfrak{g}))$$

Here $X^x = \mathrm{Ad}(x)\cdot X$, $a^x = \mathrm{Ad}(x)\cdot a$, and the relations are valid on V^∞.

We have therefore a canonical way of passing from a representation of G to a representation of $\mathfrak{g}_\mathrm{C}$. However, the reverse process is much more delicate; the example given above suggests that the recovery of the group representation from the Lie algebra representation is analogous to the expression of a function in terms of its Taylor series. So one would expect to have an incisive correspondence only if one works with *analytic* functions. This suggests the following definition. A vector $v \in V$ is said to be *analytic* (for π) if the functions

$$g \mapsto \pi(g)v$$

are analytic on G. We must recall here that a function from a real-analytic manifold M into V, say $f(m \mapsto f(m))$, is *analytic* if for each point of M, in local coordinates $x_1, \ldots, x_n$ around that point, f can be represented as

$$f(m) = \sum x_1(m)^{r_1} \cdots x_n(m)^{r_n} f_{r_1, \ldots, r_n}$$

where $f_{r_1, \ldots, r_n} \in V$ and the series is convergent, i.e.,

$$\| f_{r_1, \ldots, r_n} \| \leqslant C^{r_1 + \cdots + r_n}$$

for some $C > 0$. It is not difficult to verify that it is enough to require the analyticity of the scalar-valued functions

$$g \mapsto \pi(gv, v^*) \quad (v^* \in V^*)$$

We write V^{an} for the space of analytic vectors in V. As before it is stable under both $\pi(G)$ and $\pi(U)$ so that it carries a representation of U. The next theorem formalizes our comments on the transition from $\mathfrak{g}$ to G.

Theorem 2. *Let G be connected and π a representation of G in a Banach space V. Suppose $V' \subset V^{\mathrm{an}}$ is a $\pi(U)$-stable subspace. Then* $\mathrm{Cl}(V')$ *is $\pi(G)$-stable. In particular, if $v \in V^{\mathrm{an}}$, $\pi(U)v$ is stable under $\mathfrak{g}$ and* $\mathrm{Cl}(\pi(U)v)$ *is stable under G.*

Proof. It is enough to prove that, for any $v \in V'$, $\pi(x)v \in \mathrm{Cl}(V')$ for all $x \in G$. Suppose, for some $x_0 \in G$, $\pi(x_0)v \notin \mathrm{Cl}(V')$. By the Hahn–Banach theorem there is a $v^* \in V^*$ such that $v^* = 0$ on $\mathrm{Cl}(V')$ but $v^*(\pi(x_0)v) \neq 0$. Let $f(x) = (\pi(x)v, v^*)$. Then, for any $a \in U(\mathfrak{g})$, $f(1; a) = (\pi(a)v, v^*) = 0$. In other words all derivatives of f vanish at 1. But f is analytic on G and G is connected. Hence $f = 0$, a contradiction, since $f(x_0) \neq 0$.

The analogue of Garding's theorem for analytic vectors is quite deep. It was first proved by Harish-Chandra [H2] for a large class of Lie groups that include all connected real semisimple Lie groups with finite centre. The Lie groups G that we shall consider are assumed to have a decomposition

$$G = K \cdot A \cdot N \tag{I}$$

where K, A, N are closed connected subgroups with K compact, A abelian, N nilpotent, N being normalized by A. The *decomposition property* (I) means that the map $(k, a, n) \mapsto kan$ is an isomorphism of $K \times A \times N$ with G (as analytic manifolds). If π is a representation of G in a Banach space V, and $v \in V$, we say v is *K-finite* if the vectors $\pi(k)v$ $(k \in K)$ span a finite-dimensional space. All connected semisimple groups with finite centre have the decomposition property (I) namely, the Iwasawa decomposition.

Theorem 3 (Harish-Chandra density theorem). *Let G be a connected Lie group with decomposition property* (I). *Let π be a representation of G in a Banach space V and let V′ be the subspace of K-finite analytic vectors. Then V′ is dense in V.*

I shall not prove this theorem here but only sketch the argument needed to reduce it to the case when $K = (1)$. Later on in this chapter I shall give an indication of the proof when $K = (1)$.

Let us begin with an arbitrary Lie group G and a compact subgroup $K \subset G$. If π is a representation of G in a Banach space V, and V^0 is the subspace of K-finite vectors, we can decompose the action of K on V^0. Let $\hat{K}$ be the unitary dual of K, and for any $\xi \in \hat{K}$ let us say that a vector $v \in V$ *transforms according to* ξ if we can write $v = v_1 + \cdots + v_m$ where each v_j lies in a subspace of V which is stable under K, and on which K acts irreducibly according to ξ. It is obvious that, if V_ξ is the space of all such vectors, it is a linear subspace of V^0 and

$$V^0 = \sum_{\xi \in \hat{K}} V_\xi \quad \text{(algebraic sum)}$$

The V_ξ are actually closed in V and one can even construct projections $V \to V_\xi$. Let χ_ξ be the character of ξ and let ($\int_K dk = 1$)

$$E_\xi = d(\xi) \int_K \chi_\xi(k^{-1})\pi(k)\,dk$$

where $d(\xi)$ is the degree (dimension) of ξ. It is an easy exercise to show that E_ξ is a projection operator and its range is V_ξ:

$$E_\xi^2 = E_\xi, \quad E_\xi V = V_\xi$$

If $v \in V$ and $E_\xi v = v_\xi$, we may think of the formal series

$$\sum_\xi v_\xi$$

as the *Fourier series* of v with respect to K. As in classical Fourier series theory one should not expect convergence of this series always; if v is, however, differentiable for K, one can prove that this series as well as its formal derivatives converges. We do not need this result now. We can also describe the vectors in V^0 in another way. Let $\mathscr{A}(K)$ be the space of all matrix elements of finite-dimensional representations of K. Then the vectors in V^0 are precisely those of the form

$$\int_K f(k)\pi(k)v\,dk \quad (v \in V, f \in \mathscr{A}(K))$$

The function in $\mathscr{A}(K)$ are all analytic on K. The reduction of the density theorem to the case when $K = (1)$ is accomplished by the following proposition.

Proposition 4. *Assume that $G = K \cdot B$ where B is a closed subgroup and the map $(k, b) \mapsto kb$ of $K \times B$ into G is an analytic isomorphism. Then, for any vector $v \in V$ which is analytic for B (i.e., for $\pi|_B$) and any analytic function f on K,*

$$v' = \int_K f(k)\pi(k)v \, \mathrm{d}k$$

is an analytic vector for G, i.e., for π itself.

To prove the proposition we first establish the following lemma.

Lemma 5. *Let X, Y be analytic manifolds; v, an analytic map of $X \times Y$ into a Banach space V; and π a Borel function on X with values in the space of bounded operators in V such that $\|\pi(x)\|$ is bounded on compact subsets of X. If m is a complex Borel measure on X with compact support, and if*

$$v'(y) = \int_X \pi(x)v(x:y) \, \mathrm{d}m(x),$$

then v' is an analytic map of Y into V.

Proof (of lemma). The lemma is local in Y and so we may take Y to be a neighbourhood of the origin in $\mathbb{R}^q$ with coordinates y_j $(1 \leqslant j \leqslant q)$. Moreover, since $\operatorname{supp}(m)$ is compact, we can come down to the following set up: X is the neighbourhood of the origin in some $\mathbb{R}^p$, with coordinates x_i, $|x_i| < a < 1$, $1 \leqslant i \leqslant p$;

$$v(x:y) = \sum_{r,s} x^r y^s v_{r,s}$$

where $x = (x_1, \ldots, x_p)$, $y = (y_1, \ldots, y_q)$, $r = (r_1, \ldots, r_p)$, $s = (s_1, \ldots, s_q)$, $x^r = x_1^{r_1} \cdots x_p^{r_p}$, $y^s = y_1^{s_1} \cdots y_q^{s_q}$, and $v_{r,s}$ are vectors in V such that, for some constant $C > 0$,

$$\|v_{r,s}\| \leqslant C^{|r| + |s|}$$

for all r, s with $|r| = r_1 + \cdots + r_p$, $|s| = s_1 + \cdots + s_q$; and $|x_i| < \dfrac{a}{2C}$ for $x \in \operatorname{supp}(m)$. Then

$$\pi(x)v(x:y) = \sum_{r,s} x^r y^s \pi(x) v_{r,s}$$

since m has compact support and $\|\pi(x)\|$ is bounded on compact subsets of X, we can integrate with respect to m and obtain

$$v'(y) = \sum_{r,s} y^s v'_{r,s}$$

where

$$v'_{r,s} = \int_X x^r \pi(x) v_{r,s} \, \mathrm{d}m(x)$$

It is clear that, for some constant $C_1 > 0$,

$$\sum_r \| v'_{r,s} \| \leqslant C_1 \cdot C^{|s|} \quad \text{(for all } r, s)$$

This proves that v' is analytic.

Proof of Proposition 4. The existence of the decomposition $G = K \cdot B$ allows us to use the integral calculations of Chapter 4 (cf. Lemma 4). We have, for $x \in G$,

$$\begin{aligned} \pi(x)v' &= \int_K f(k)\pi(xk)v \, dk \\ &= \int_K f(k)\pi(x[k])\pi(b(xk))v \, dk \\ &= \int_K f(x^{-1}[u])\mu_B(b(x^{-1}u))^{-1}\pi(u)\pi(b(x \cdot x^{-1}[u]))v \cdot du \\ &= \int_K \pi(u)v(x:u) \, du \end{aligned}$$

where

$$v(x:u) = f(x^{-1}[u])\mu_B(b(x^{-1}u))^{-1}\pi(b(x \cdot x^{-1}[u]))v$$

Since $b \mapsto \pi(b)v$ is an analytic map of B into V, v is an analytic map of $G \times K$ into V. Hence the analyticity of $x \mapsto \pi(x)v'$ is immediate from Lemma 5.

5.3 Representations of the Lie algebra. Finiteness and density theorems

We shall now use the Harish-Chandra density theorem for analytic vectors to set up a correspondence between representations of the group and the Lie algebra. We shall then follow it up with a discussion of the representation theory of the Lie algebra.

Lemma 1. *Let G be a Lie group, K a compact subgroup. Suppose π is a representation of G in a Banach space V, and V^0 the space of K-finite analytic vectors. Then V^0 is stable under K as well as $\mathfrak{g}$. In particular,*

$$V^0 = \sum_{\xi \in \hat{K}} V^0_\xi \quad \text{(algebraic direct sum)}$$

Proof. It is not in general true that V^0 is stable under G, so that the invariance under $\mathfrak{g}$ is remarkable. We must show that if $F \subset V^0$ is a K-stable subspace $\pi(\mathfrak{g})[F] \subset V^0$ also. Now K acts on F (via π) and on $\mathfrak{g}$ (via the adjoint representation), hence on $\mathfrak{g} \otimes F$. Let f be the linear map $\mathfrak{g} \otimes F \to V$ such that $f(X \otimes v) = \pi(X)v$ for $X \in \mathfrak{g}$, $v \in V$. Then

$$\begin{aligned} f(k \cdot (X \otimes v)) &= f((\mathrm{Ad}(k) \cdot X \otimes \pi(k)v)) \\ &= \pi(X^k)\pi(k)v \\ &= \pi(k)\pi(X)v \\ &= \pi(k) f(X \otimes v) \end{aligned}$$

So f commutes with the action of K, showing that $f(\mathfrak{g} \otimes F) = \pi(\mathfrak{g})[F]$ is finite-dimensional and stable under $\pi(K)$. Since analyticity is preserved under the action of $\mathfrak{g}$, we see that $\pi(\mathfrak{g})[F] \subset V^0$.

This lemma tells us what category of representations of the Lie algebra should be singled out for our study. Let V be a complex vector space and π a representation of $\mathfrak{g}_{\mathbb{C}}$ on V. We say that V is a $(\mathfrak{g}, K)$-*module* if there is a representation π of K in V such that

(i) (compatibility) $\pi(X^k) = \pi(k)\pi(X)\pi(k)^{-1}$ for all $X \in \mathfrak{g}$, $k \in K$,
(ii) (K-finiteness) $V = \sum_{\xi \in \hat{K}} V_\xi$ where the sum is algebraic and direct, and V_ξ is the linear space of all vectors in V that transform according to ξ.

The class of all $(\mathfrak{g}, K)$-modules forms a category. *It is the basic object of study in the theory of representations of semisimple Lie algebras.*

Although K appears in the definition of a $(\mathfrak{g}, K)$-module, one should regard the latter as an algebraic object. We could have replaced K by its Lie algebra $\mathfrak{k}$ in (ii) of the above definition; this is what Harish-Chandra did in his original papers. The more general modules that are obtained are then called $(\mathfrak{g}, \mathfrak{k})$-*modules*. If V is a $(\mathfrak{g}, \mathfrak{k}$-module, we have $V = \sum_{\xi \in \hat{\mathfrak{k}}} V_\xi$ where $\hat{\mathfrak{k}}$ is the set of equivalence classes of finite-dimensional irreducible representations of $\mathfrak{k}$; and V will be a $(\mathfrak{g}, K)$-module if and only if only those V_ξ are $\neq 0$ for which ξ lifts to a representation of K. *The restriction to* $(\mathfrak{g}, K)$*-modules is necessary in order that the module in question come from G and not from a covering group of G.*

From now on we shall be dealing only with connected semisimple Lie groups with finite centre. For any such group G, the symbols K, $\mathfrak{g}$, $U = U(\mathfrak{g})$, A, N have the same meaning as always. Although we shall be in this general setting, for those proofs which we give and the explicit calculations which we make we shall only work with $SL(2, F)$ ($F = \mathbb{R}$ or $\mathbb{C}$), and occasionally in $SL(n, F)$.

For $(\mathfrak{g}, \mathfrak{k})$-modules we have all the usual notions of representation theory, and everything is completely algebraic. For instance invariant subspace means invariance under $\mathfrak{g}$ (and K, for a $(\mathfrak{g}, K)$-module), and irreducibility means nonexistence of nonzero proper invariant subspaces. We often write $X \cdot v$ or $k \cdot v$ instead of $\pi(X)v$ or $\pi(k)v$.

A $\mathfrak{g}$-module V is said *to have an infinitesimal character* if the elements of the algebra $\mathfrak{Z}$ act as scalars on V, $\mathfrak{Z}$ being the centre of $U(\mathfrak{g})$. In this case there is a homomorphism χ, called the *infinitesimal character* of V, such that

$$z \cdot v = \chi(z) \cdot v \quad (z \in \mathfrak{Z}, v \in V)$$

There is also the more general notion of $\mathfrak{Z}$-finiteness. A $\mathfrak{g}$-module V is $\mathfrak{Z}$-finite if there is an ideal $J \subset \mathfrak{Z}$ such that

$$J \cdot V = 0, \quad \dim(\mathfrak{Z}/J) < \infty$$

The fundamental results in the theory of $(\mathfrak{g},\mathfrak{k})$-modules are the finiteness and density theorems for them which were discovered by Harish-Chandra. Let V be a $(\mathfrak{g},\mathfrak{k})$-module and for any $\xi \in \hat{\mathfrak{k}}$ let V_ξ be the subspace of V vectors that transform according to ξ.

***Lemma* 6.** *Let* $\mathfrak{Q} = U^{\mathfrak{k}}$ *be the centralizer of* $\mathfrak{k}$ *in* $U = U(\mathfrak{g})$. *Then each* V_ξ *is stable under* $\mathfrak{Q}$, *in particular under* $\mathfrak{Z}$.

Proof. Let $L \subset V_\xi$ be finite-dimensional and stable under $\mathfrak{k}$ and let $u \in \mathfrak{Q}$. If $M = u \cdot L$, the map $v \mapsto u \cdot v$, from L to M, commutes with $\mathfrak{k}$. Hence the $\mathfrak{k}$-module M is a quotient of the $\mathfrak{k}$-module L. Since L is a multiple of ξ, M is also a multiple of ξ. So $M \subset V_\xi$.

***Theorem* 7.** *Let* V *be a* $(\mathfrak{g},\mathfrak{k})$*-module and assume that it is finitely generated. Then each* V_ξ *is a finite* $\mathfrak{Z}$*-module.*

The V_ξ need not be free $\mathfrak{Z}$-modules. Nevertheless, in view of Theorem 7, it makes sense to speak of the minimum number of generators of V as a $\mathfrak{Z}$-module. We denote this number by $n(V:\xi)$.

***Theorem* 8.** *There is a constant* $c > 0$, *depending only on* $\mathfrak{g}$ *and having the following property. Let* V *be any finitely generated* $(\mathfrak{g},\mathfrak{k})$ *module generated by* $v_1, \ldots, v_m$ *where* $v_j \in V_{\xi_j}$, $1 \leqslant j \leqslant m$, *and write* $d = \max d(\xi_j)$. *Then*

$$n(V,\xi) \leqslant c \cdot m d d(\xi)^2$$

for all $\xi \in \hat{\mathfrak{k}}$. *In particular, given any finitely generated* $\mathfrak{Z}$*-finite* $(\mathfrak{g},\mathfrak{k})$*-module* V, *there is a constant* $c = c(V) > 0$ *such that for all* $\xi \in \hat{\mathfrak{k}}$

$$\dim(V_\xi) \leqslant c d(\xi)^2$$

From this theorem we can derive important consequences for irreducible modules. We need the following result, first noticed by Dixmier [Di]. It is an infinite-dimensional variant of Schur's lemma.

Lemma 9. *Let A be an associative algebra over $\mathbb{C}$ acting irreducibly on a complex vector space V. Let A' be the algebra of endomorphisms of V commuting with A. Then*

(i) *if V has at most countable dimension, $A' = \mathbb{C}\cdot 1$,*
(ii) *if A has at most countable dimension, then V has at most countable dimension, so that in this case $A' = \mathbb{C}\cdot 1$ for all (irreducible) V.*

Proof. If $a' \in A'$, the null space and range of a' are both A-stable. It then follows from the irreducibility of V that, if $a' \neq 0$, a' is invertible and lies in A'. Thus A' is a division algebra. If V were finite-dimensional the proof would end now since there are no finite-dimensional division algebras over $\mathbb{C}$ other than $\mathbb{C}$ itself. In the infinite-dimensional case one needs a more refined argument. We note first that the previous argument can still be made to imply that, if $a' \in A'$ and $a' \notin \mathbb{C}\cdot 1, a'$ is transcendental over $\mathbb{C}$. For, if a' is algebraic, it satisfies a polynomial equation over $\mathbb{C}$ and hence has an eigenvalue, say λ; $(a' - \lambda \cdot 1)$ would then not be invertible, a contradiction. Hence the field $\mathbb{C}(a')$ is isomorphic to the field of rational functions over $\mathbb{C}$, and so is *not* countable-dimensional. We shall now obtain the contradiction by showing that A' has at most countable dimension. Let $v \in V$ be nonzero. Since $a'v \neq 0$ unless $a' = 0$, $a' \mapsto a'v$ is an imbedding of A' in V, proving the countable dimensionality of A'. For proving (ii) we choose a nonzero v in V and note that $a \mapsto a\cdot v$ is a map of A onto V by irreducibility of V since its range is A-stable. As A is of countable dimension, the same is true of V.

Since $U(\mathfrak{g})$ is of countable dimension, it follows from this lemma that irreducible $\mathfrak{g}$-modules have infinitesimal characters, since, in any $\mathfrak{g}$-module, the action of $\mathfrak{z}$ commutes with that of $U(\mathfrak{g})$. Thus, if V is an irreducible $(\mathfrak{g}, \mathfrak{k})$-module, $\mathfrak{z}$ acts via scalars on each V_ξ and hence we obtain from Theorems 7 and 8 the following.

Theorem 10. *Let V be an irreducible $(\mathfrak{g}, \mathfrak{k})$-module. Then*

$$\dim(V_\xi) < \infty \quad (\text{for all } \xi \in \hat{\mathfrak{k}})$$

Moreover, given any $\xi_0 \in \hat{\mathfrak{k}}$, we can find a constant $c = c(\mathfrak{g}, \xi_0) > 0$ such that, for any irreducible $(\mathfrak{g}, \mathfrak{k})$-module V for which $V_{\xi_0} \neq 0$,

$$\dim(V_\xi) \leqslant cd(\xi)^2 \qquad (\xi \in \hat{\mathfrak{k}})$$

These results clearly suggest that it would be natural and interesting to consider $(\mathfrak{g}, \mathfrak{k})$-modules V such that $\dim(V_\xi) < \infty$ for all $\xi \in \hat{\mathfrak{k}}$. They are known as *Harish-Chandra modules for* $\mathfrak{g}$. In a completely analogous manner, if V is a Banach space and π is a representation of G in V such that $\dim(V_\xi) < \infty$ for all $\xi \in \hat{K}$, we call V a *Harish-Chandra module for G.* Obviously, if V is a Harish-Chandra G-module, V^0 (the space of K-finite analytic vectors) is a *Harish-Chandra* $(\mathfrak{g}, K)$*-module.* The principal series representations defined in Chapter 4 are Harish-Chandra modules; in fact we saw that on K these representations are subrepresentations of the regular representation of K, so that, for any such module V,

$$\dim(V_\xi) \leqslant d(\xi)^2 \quad (\xi \in \hat{K})$$

Thus the category of Harish-Chandra modules is quite large.

We shall not prove these results here, nor the density theorem that we shall describe presently. However, it is clear that for proving these theorems it is necessary to have some algebraic method of constructing all finitely generated $(\mathfrak{g}, \mathfrak{k})$-modules. I shall now indicate one such method which is actually quite simple. Let $U = U(\mathfrak{g})$ be as usual. Since $\mathfrak{k} \subset \mathfrak{g}$ is a subalgebra, we can identify $U(\mathfrak{k})$ with the subalgebra generated by $\mathfrak{k}$ (and 1) in U; in fact, by the universal property of $U(\mathfrak{k})$, we have a natural homomorphism of $U(\mathfrak{k})$ onto the above mentioned subalgebra; and this map is an isomorphism by the Poincaré–Birkhoff–Witt theorem. Let now W be a finite-dimensional irreducible $\mathfrak{k}$-(or $U(\mathfrak{k})$-)module. If we take a vector $w \in W$, $w \neq 0$, and define

$$M = M(w) = \{a \in U(\mathfrak{k}) \mid a \cdot w = 0\},$$

then M is a *left ideal* in $U(\mathfrak{k})$, and the $U(\mathfrak{k})$-module $U(\mathfrak{k})/M$ is naturally isomorphic to W; indeed, if $a \mapsto \bar{a}$ is the natural map of $U(\mathfrak{k})$ onto $U(\mathfrak{k})/M$, the isomorphism in question takes $\bar{a}$ to $a \cdot w$. The advantage of passing from W to M is that $U(\mathfrak{g})$ can now operate on M. Let

$$\bar{M} = U(\mathfrak{g}) \cdot M$$

Obviously $\bar{M}$ is a left ideal in $U(\mathfrak{g})$. We now have the following.

Lemma 11. *$\bar{M}$ is proper, i.e., $1 \notin \bar{M}$. Moreover, under the natural action of $U(\mathfrak{g})$, $U(\mathfrak{g})/\bar{M}$ is a $(\mathfrak{g}, \mathfrak{k})$-module, and there is a unique imbedding of W as a sub-$\mathfrak{k}$-module of $U(\mathfrak{g})/\bar{M}$ which takes w to $\bar{1}$, the image of 1 in $U(\mathfrak{g})/\bar{M}$.*

Proof. The proof that M is proper is an easy consequence of the Poincaré–Birkhoff–Witt theorem. Select a basis $(Y_1, \ldots, Y_p, X_1, \ldots, X_k)$ for $\mathfrak{g}$ so that $(X_1, \ldots, X_k)$ is a basis for $\mathfrak{k}$. Let A be the linear span of the monomials $Y_1^{r_1} \cdots Y_p^{r_p}$ in the Y's; the corresponding linear span associated to the Xs is just $U(\mathfrak{k})$ itself. The map $(a, b) \mapsto ab$ of $Y \times U(\mathfrak{k})$ into $\mathfrak{g}$ then extends uniquely to a linear isomorphism of $Y \otimes U(\mathfrak{k})$ with $U(\mathfrak{g})$ that sends $a \otimes b$ to ab; in particular, $U(\mathfrak{g}) = Y \cdot U(\mathfrak{k})$. Now, if M' is a subspace of $U(\mathfrak{k})$ containing 1 and complementary to M in $U(\mathfrak{k})$, we have

$$Y \otimes U(\mathfrak{k}) = (Y \otimes M) \oplus (Y \otimes M')$$

so that

$$U(\mathfrak{g}) = Y \cdot M \oplus Y \cdot M'$$

On the other hand, $U(\mathfrak{k}) \cdot M = M$, and so,

$$U(\mathfrak{g})M = Y \cdot U(\mathfrak{k}) \cdot M = Y \cdot M$$

Since $1 \in Y \cdot M'$, we are done.

Let $\bar{U} = U(\mathfrak{g})/\bar{M}$ and $a \mapsto \bar{a}$ the natural map of $U(\mathfrak{g})$ onto $\bar{U}$. If we write $a \cdot \bar{b} = \overline{(ab)}$ for $a, b \in U(\mathfrak{g})$, we see that $\bar{U}$ gets converted to a $U(\mathfrak{g})$-module that is generated by $\bar{1}$. The map $u \cdot w \mapsto u \cdot \bar{1}$ then gives rise to a $\mathfrak{k}$-module homomorphism of W onto $U(\mathfrak{k}) \cdot \bar{1}$. But $U(\mathfrak{k}) \cdot \bar{1} \neq 0$ and W is irreducible, so that this map is a $\mathfrak{k}$-module imbedding and is obviously the only one that takes w to $\bar{1}$. It remains to show that $\bar{U}$ is a $(\mathfrak{g}, \mathfrak{k})$-module. If $\bar{U}^0 = \sum_{\xi \in \hat{\mathfrak{k}}} \bar{U}_\xi$, the same argument as in Lemma 1 proves that U^0 is stable under $\mathfrak{g}$. As $\bar{1} \in \bar{U}^0$, we must have $\bar{U}^0 = \bar{U}$.

We write j for the $\mathfrak{k}$-module map $u \cdot w \mapsto u \cdot \bar{1}$ so that

$$j: W \hookrightarrow \bar{U}$$

The next proposition shows that $(\bar{U}, j)$ has the appropriate universal property.

Proposition 12. *Suppose V is a $(\mathfrak{g}, \mathfrak{k})$-module with the following property: there is a $\mathfrak{k}$-module imbedding*

$$j_V: W \hookrightarrow V$$

such that $V = U(\mathfrak{g}) \cdot j_V(W)$. Then V is a quotient module of $\bar{U}$; more precisely, there is a $\mathfrak{g}$-module homomorphism f of U onto V such that the diagram

$$\begin{array}{ccc} W & \hookrightarrow & \bar{U} \\ & {\scriptstyle j_V} \searrow \quad \swarrow {\scriptstyle f} & \\ & V & \end{array}$$

is commutative.

Proof. This is immediate if we define f by the rule

$$f(a \cdot \bar{1}) = a \cdot j_V(w)$$

In fact, since W is irreducible, $j_V(W) = U(\mathfrak{k}) \cdot j_V(w)$ and so $V = U(\mathfrak{g}) \cdot j_V(w)$ and the only thing to verify is that f is well defined, since it is trivial that f commutes with the action of $U(\mathfrak{g})$. But, as $M \cdot j_V(w) = 0$,

$$a \cdot \bar{1} = 0 \Rightarrow a \in U(\mathfrak{g}) \cdot M \Rightarrow a \cdot j_V(w) = 0$$

Since every finitely generated $(\mathfrak{g}, \mathfrak{k})$-module is a sum of submodules such as V above, any such is a quotient of a finite direct sum of modules of the form $\bar{U}$. Harish-Chandra proved the finiteness theorems stated above through a deep study of the $\mathfrak{k}$-module structure of the universal modules $\bar{U}$ constructed above.

We now come to the density theorem. For stating it we need a little preparation. We say a class $\xi \in \hat{\mathfrak{k}}$ is of *finite type* if it occurs in the reduction of a finite-dimensional irreducible representation of $\mathfrak{g}$ restricted to $\mathfrak{k}$. If G is a connected semisimple group with finite centre and K its maximal compact subgroup which corresponds to $\mathfrak{k}$, and if G has a finite-dimensional representation which is faithful on K, it is easy to show that any $\xi \in \hat{K} \subset \hat{\mathfrak{k}}$ is of finite type. If $G = SL(n, \mathbb{R})$ and $K = SO(n, \mathbb{R})$, the spin representation defines an element of $\hat{\mathfrak{k}}$ which is not of finite type. If $\mathfrak{g}$ is a *complex* semisimple Lie algebra, and G is a simply connected group with Lie algebra $\mathfrak{g}$, the (maximal compact) subgroup K corresponding to $\mathfrak{k}$ is also simply connected, and every member of $\hat{\mathfrak{k}}$ is of finite type.

Fix W, w as above and let $\mathcal{M}$ be the category of all pairs (V, j_V) as in Proposition 12 and let $\mathcal{M}_f$ be the subcategory of $\mathcal{M}$ of all (V, j_V) with V finite dimensional and irreducible. For $(V, j_V) \in \mathcal{M}$ we write

$$I_{V,j_V} = \{a \mid a \in U(\mathfrak{g}),\ a \cdot j_V(w) = 0\}$$

Recall the definition of M as the left ideal in $U(\mathfrak{k})$ which annihilates w. Finally let $\xi \in \hat{\mathfrak{k}}$ be the class of the representation defined on W.

Theorem 13. *Suppose ξ is of finite type. Then*

$$U(\mathfrak{g}) \cdot M = \bigcap_{(V, j_V) \in \mathcal{M}_f} I_{V,j_V}$$

In particular, let $a \in U(\mathfrak{g})$ be such that a is zero on V_ξ whenever V is an irreducible finite-dimensional $\mathfrak{g}$*-module with $V_\xi \neq 0$; then a is zero on V_ξ for any* $(\mathfrak{g}, \mathfrak{k})$*-module V with $V_\xi \neq 0$.*

The second assertion makes it clear why we call this a *density theorem.* To deduce it from the first we need only show that if $a \in U(\mathfrak{g})$ has the property described above and $(V, j_V) \in \mathcal{M}$ then $a \cdot j_V(w) = 0$, for $w \in W$ is completely arbitrary. But $a \in I_{V', j_{V'}}$) whenever $(V', j_{V'}) \in \mathcal{M}_f$, hence by the first assertion $a \in U(\mathfrak{g}) \cdot M$, thus $a \cdot j_V(w) = 0$. This theorem suggests that *the structure of the finite-dimensional representations of* $\mathfrak{g}$ *and their restrictions to* $\mathfrak{k}$ *determine very substantially the structure of arbitrary* $(\mathfrak{g}, \mathfrak{k})$*-modules.*

Harish-Chandra supplemented these results with a criterion for equivalence of irreducible $(\mathfrak{g}, \mathfrak{k})$-modules. For this purpose let us introduce

$\mathfrak{Q}$, the algebra of all elements in $U(\mathfrak{g})$ which commute with $U(\mathfrak{k})$, and let

$$\mathscr{A} = U(\mathfrak{k})\cdot\mathfrak{Q}.$$

$\mathscr{A}$ is an algebra, and if V is a $(\mathfrak{g},\mathfrak{k})$-module each V_ξ is stable under $\mathscr{A}$ and so may be viewed as an $\mathscr{A}$-module. Suppose now V is a Harish-Chandra module and $\xi\in\hat{\mathfrak{k}}$ is such that $V_\xi \neq 0$. By the finiteness theorem, V_ξ is finite-dimensional and it is clear that we can write (noncanonically) V_ξ in the form $W\otimes W'$ where $U(\mathfrak{k})$ acts only on the first factor and $\mathfrak{Q}$ only on the second, $W\cong\xi$ under $U(\mathfrak{k})$. In general there is no need that V_ξ be irreducible under $\mathscr{A}$; this can happen if and only if W' is irreducible under $\mathfrak{Q}$. The equivalence class of the representation of $\mathfrak{Q}$ in W' is uniquely determined; let us write it as $\chi(V:\xi)$. Harish-Chandra's theorem is then the following.

Theorem 14. *Let V be an irreducible $(\mathfrak{g},\mathfrak{k})$-module and let $\xi\in\hat{\mathfrak{k}}$ be such that $V_\xi\neq 0$. Then V_ξ is an irreducible $\mathscr{A}$-module, so that $\chi(V:\xi)$ is an irreducible class of $\mathfrak{Q}$. Suppose V' is another irreducible $(\mathfrak{g},\mathfrak{k})$-module. Then V and V' are equivalent if and only if $V'_\xi\neq 0$ and $\chi(V:\xi)=\chi(V':\xi)$.*

I shall conclude this summary of the theory of $(\mathfrak{g},\mathfrak{k})$-modules with a statement of the celebrated *subquotient theorem* of Harish-Chandra. If π is a representation of G in a Banach space V, and if V_1, V_2 are closed invariant subspaces for π with $V_2\subset V_1$, we have a representation of G on V_1/V_2; it is called a *subquotient* of π. It is obvious that an analogous notion can be introduced for $(\mathfrak{g},\mathfrak{k})$-modules also.

Theorem 15. *Let W be an irreducible $(\mathfrak{g},K)$-module. Then there is an irreducible representation π of G on a Hilbert space $\mathscr{H}$ with the following properties:*

(i) *π is a subquotient of a principal series representation of G;*
(ii) *$\mathscr{H}$ is a Harish-Chandra module for G and the associated $(\mathfrak{g},K)$-module $\mathscr{H}^0$ is equivalent to W.*

Within the framework of these notes it is not possible to prove these results in detail. At the end of this chapter I shall make a few comments on the methods of proof.

5.4 Some consequences: finiteness of multiplicity, $(\mathfrak{g},K)$-modules, existence of characters

I shall now take up some of the consequences of the theory described so far.

Correspondence between Harish-Chandra modules for G and $\mathfrak{g}$. Fix a representation π of G in a Banach space V and suppose that $\dim(V_\xi)<\infty$

for all $\xi \in \hat{K}$, i.e., V is a Harish-Chandra G-module. We claim that

$$V^0 = \sum_{\xi \in \hat{K}} V_\xi \quad \text{(algebraic sum)}$$

For, V^0 is dense in V, so that, as the projection $E_\xi(V \to V_\xi)$ is continuous,

$$V_\xi = \mathrm{Cl}(E_\xi V^0) = \mathrm{Cl}(V_\xi^0) = V_\xi^0$$

in view of the finiteness of $\dim(V_\xi^0)$. We now consider the collection $\mathfrak{L}(V)$ (resp. $\mathfrak{L}(V^0)$) of all closed G-stable subspace of V (resp. $\mathfrak{g}$-stable subspaces of V^0), both partially ordered by inclusion.

Theorem 16. *For $W \in \mathfrak{L}(V)$, $W^0 = W \cap V^0$. The map $W \mapsto W^0$ is an order-preserving bijection of $\mathfrak{L}(V)$ with $\mathfrak{L}(V^0)$, whose inverse is the bijection $Z^0 \mapsto \mathrm{Cl}(Z^0)$ of $\mathfrak{L}(V^0)$ with $\mathfrak{L}(V)$. In particular, V always has irreducible subquotients; and V itself is irreducible if and only if V^0 is irreducible.*

Proof. We have seen that, for $Z^0 \in \mathfrak{L}(V^0)$, $\mathrm{Cl}(Z^0) \in \mathfrak{L}(V)$. If $\bar{Z} = \mathrm{Cl}(Z^0)$, Z_ξ^0 is dense in $\bar{Z}_\xi$ for all $\xi \in \hat{K}$, hence $Z_\xi^0 = \bar{Z}_\xi$ for all ξ, due to the finite dimensionality of Z_ξ^0. Hence $Z^0 = \bar{Z}^0$ and the first assertion is proved. The isomorphism of $\mathfrak{L}(V)$ and $\mathfrak{L}(V^0)$ implies the last statement. To construct an irreducible subquotient it is enough to do it for V^0. We can clearly choose $Z^0 \in \mathfrak{L}(V^0)$ and $\xi_0 \in \hat{K}$ so that $Z_{\xi_0}^0 \neq 0$, $\dim(Z_{\xi_0}^0) \leqslant \dim(W_\xi^0)$ for all $W^0 \in \mathfrak{L}(V^0)$ and all ξ with $W_\xi^0 \neq 0$. Replacing Z^0 by the $\mathfrak{g}$-module generated by $Z_{\xi_0}^0$ will not change the defining property so that we may as well assume that $Z^0 = U(\mathfrak{g})Z_{\xi_0}^0$. Write $Z' = \sum_{\xi \neq \xi_0} Z_\xi^0$. By the minimality of $\dim(Z_{\xi_0}^0)$, if $L^0 \in \mathfrak{L}(V^0)$ is contained in Z^0, then either L^0 contains $Z_{\xi_0}^0$, in which case it coincides with Z^0, or else $L^0 \cap Z_{\xi_0}^0 = 0$, in which case $L^0 \subset Z'$. Hence all the L^0 that are *properly contained* in Z^0 are inside Z', showing that there is a *largest such*, namely *the span of all such.* If M^0 is this span, Z^0/M^0 is an irreducible $(\mathfrak{g}, K)$-module.

Unitary representations. Recall that a unitary representation π of G in a Hilbert space H is a *factor representation* if $R(\pi) \cap R'(\pi) = \mathbb{C} \cdot 1$ where $R(\pi)$ is the von Neumann algebra generated by the $\pi(x)$ $(x \in G)$, and $R'(\pi)$ is the algebra of operators in H commuting with $R(\pi)$. Another way of saying this is that, if A is a bounded operator fixed under conjugation by all the $\pi(x)$ as well as all the unitary operators of H that commute with all the $\pi(x)$, then A is a scalar; this is because of the well known fact that $R'(\pi)$ is generated as a von Neumann algebra by the unitary operators in it. If π is a multiple (finite or countable) of an irreducible unitary representation, π is a factor representation. These are the type I representations.

To apply the preceding theory to unitary representations it is necessary to look a little more closely into the structure of the Lie-algebra representation associated to a unitary representation, say π, on a Hilbert space H. We shall now define the notion of *adjoints* of elements of $U(\mathfrak{g})$. More precisely we introduce the map

$$a \mapsto a^\dagger$$

of $U(\mathfrak{g})$ into itself with the following properties:

(i) it is $\mathbb{R}$-linear;
(ii) $(ca)^\dagger = \bar{c}a^\dagger \quad (c \in \mathbb{C}, a \in U(\mathfrak{g}))$;
(iii) $(ab)^\dagger = b^\dagger a^\dagger$;
(iv) $a^{\dagger\dagger} = a$;
(v) $X^\dagger = -X \quad (X \in \mathfrak{g})$.

These are precisely the formal properties characteristic of the adjoint operation. In (v) note that $\mathfrak{g}$ is the real Lie algebra corresponding to G. The uniqueness of † is obvious. For existence we define an *anti-automorphism* $a \mapsto a'$ (formal transpose) of $U(\mathfrak{g})$ such that $X' = -X$ for $X \in \mathfrak{g}$; this is done using the universal property of $(\mathfrak{g}, U(\mathfrak{g}))$ mentioned earlier. On the other hand, the conjugation of $\mathfrak{g}_\mathbb{C}$ that defines $\mathfrak{g}$ extends to a conjugation $a \mapsto \bar{a}$ of $U(\mathfrak{g})$. We take $a^\dagger = (\bar{a})'$. Observe that $\mathfrak{Z}$ is stable under †. Returning to π, we get from the relations $(\pi(x)u|v) = (u|\pi(x^{-1})v)$ $(x \in G, u, v \in H)$ the identities

$$(\pi(a)u|v) = (u|\pi(a^\dagger)v) \quad (a \in U(\mathfrak{g}), u, v \in H^\infty)$$

In particular, $\pi(a)$, viewed as an operator defined on the dense domain H^∞, has a closure $[\pi(a)]$, whose adjoint is an extension of $\pi(a^\dagger)$ defined on H^∞; and $\pi(a)$ is a symmetric operator on H^∞ if $a = a^\dagger$. The key point is now contained in the two lemmas below.

Lemma 17. *Suppose $a \in \mathfrak{Z}$ is self-adjoint with respect to †, i.e., $a = a^\dagger$. Then $\pi(a)$ is essentially self-adjoint on H^∞, and its spectral projections are in $R(\pi) \cap R'(\pi)$. In particular, if π is a factor representation, the elements of $\mathfrak{Z}$ act as scalars on H^∞.*

Proof. We must show that the equations

$$[\pi(a)]^\dagger \cdot u = \pm iu$$

have only the solution $u = 0$. It is obvious that such u form closed subspaces L ; let $L_\pm \neq 0$. Since $\pi(x)\pi(a)\pi(x)^{-1} = \pi(a)$ on H^∞ for all $x \in G$, $\pi(x)[\pi(a)]^\dagger \pi(x)^{-1} = [\pi(a)]^\dagger$ for all $x \in G$; it follows that $L_\pm$ is G-stable. So $L_\pm^\infty = L_\pm \cap H^\infty$ is nonzero, showing that our equation has nonzero solutions in H^∞. This is impossible since H^∞ is the domain of our symmetric operator. If U is a unitary operator commuting with all $\pi(x)$, $U \cdot H^{\infty} = H^{\infty}$ and $U\pi(x)U^{-1} = \pi(x)$ on H^{∞}, giving, on differentiation, $U\pi(a)U^{-1} = \pi(a)$. So U centralizes $[\pi(a)]$. Since $\pi(x)\pi(a)\pi(x)^{-1} = \pi(a)$ on H^{∞}, $[\pi(a)]$ commutes with all $\pi(x)$. So the spectral projections of $[\pi(a)]$ lie in $R(\pi) \cap R'(\pi)$. If now $R(\pi) \cap R'(\pi) = \mathbb{C} \cdot 1$, the spectral projections are 0 or 1, so that $[\pi(a)]$ is a scalar.

Lemma 18. *Let π be a unitary factor representation of G in H and $0 \neq u \in H^0$. Let $H(u)$ be the smallest closed G-stable subspace containing u. Then $H(u)$ is a Harish-Chandra module for G.*

Proof. Since $U(\mathfrak{g}) \cdot u = Z^0$ is contained in $H(u)^0$ and $U(\mathfrak{g})$ stable, $\mathrm{Cl}(Z^0)$ is G-stable. Hence $\mathrm{Cl}(Z^0) = H(u)$. By the previous lemma Z^0 has an infinitesimal character; and it is finitely (singly) generated. So $\dim(Z^0_\xi) < \infty$ for all $\xi \in \hat{K}$. As $\mathrm{Cl}(Z^0_\xi) = H(u)_\xi$, we see that $\dim(H(u))_\xi < \infty$ for all ξ.

From this and the earlier work we see that $H(u)$, and hence H itself, has irreducible subquotients, which are *subrepresentations* in the unitary case. If H is irreducible, $H(u) = H$. Thus we get the following.

Theorem 19. *All unitary Harish-Chandra modules for G are direct sums of irreducible modules. All unitary factor representations of G are of type I. If π is an irreducible unitary representation of G in a Hilbert space H, all the H_ξ are finite-dimensional and there is a constant $c = c(\pi) > 0$ such that*

$$\dim(H_\xi) \leqslant c\,d(\xi)^2 \quad (\forall \xi \in \hat{K})$$

Unitary $(\mathfrak{g}, \mathfrak{k})$*-modules.* The infinitesimal analogue of a unitary representation is a *unitary* $(\mathfrak{g}, \mathfrak{k})$*-module* which is a $(\mathfrak{g}, \mathfrak{k})$-module V on which there is a positive-definite scalar product $(\cdot|\cdot)$ such that

$$(a \cdot u | v) = (u | a^{\dagger} \cdot v) \quad (u, v \in V, a \in U(\mathfrak{g}))$$

If the Hilbert space H carries a unitary representation for G, H^0 is a unitary $(\mathfrak{g}, K)$-module. Now the subquotient theorem shows that every Harish-Chandra $(\mathfrak{g}, K)$-module is equivalent to the module H^0 of a Hilbert-space Harish-Chandra module H for G. For unitary modules Harish-Chandra refined this to the following

Theorem 20. *Let V be a unitary irreducible Harish-Chandra $(\mathfrak{g}, K)$-module. Then there is an irreducible unitary representation π of G in a Hilbert space H such that H^0 is equivalent to V. The representation π is unique up to unitary equivalence. If π_1 and π_2 are two irreducible unitary representations of G in Hilbert spaces H_1 and H_2, then π_1 and π_2 are unitary equivalent if and only if H_1^0 and H_2^0 are equivalent as $(\mathfrak{g}, K)$ modules.*

The last statement is quite easy to verify. Suppose $T(H_1^0 \to H_2^0)$ is a $(\mathfrak{g}, K)$-morphism. Then T is bijective and maps $H_{1,\xi}^0$ onto $H_{2,\xi}^0$ for all ξ. If T_ξ is the restriction of T to $H_{1,\xi}^0$ we define $T^\dagger(H_2^0 \to H_1^0)$ by saying that $T^\dagger$ coincides with the adjoint $(T_\xi)^\dagger$ on $H_{2,\xi}^0$; since $\dim(H_{i,\xi}^0) < \infty$ for $i = 1, 2$, there is no difficulty in defining $(T_\xi)^\dagger$. Then it is clear that

$$(Tu_1 | u_2) = (u_1 | T^\dagger u_2) \quad (u_i \in H_i^0)$$

and that $T^\dagger(H_2^0 \to H_1^0)$ is also a $(\mathfrak{g}, K)$-map. This implies that $T^\dagger T$ commutes with $U(\mathfrak{g})$ and so must be a scalar (either by Lemma 9 or more simply by observing that $T^\dagger T$ must have eigenvalues because it stabilizes the $H_{1,\xi}^0$). Replacing T by a suitable multiple of it we may thus assume that $T^\dagger T$ is the identity, and hence that T is the restriction to H_1^0 of a unitary isomorphism T of H_1 with H_2. The functions $(\pi_1(x)u | v)$ and $(T^{-1}\pi_2(x)Tu | v)$ for $u, v \in H^0$ are analytic and have the same derivatives at $x = 1$. Hence they are identical, proving that $\pi_1 = T^{-1}\pi_2 T$.

The early work of Harish-Chandra, of which I have been able to give only a bare outline, made it very clear that the 'correct' object dual to G is the *set of all equivalence classes of irreducible* $(\mathfrak{g}, K)$*-modules.* Let us write $\overline{G}$ for this. Then we have a natural map from $\hat{G}$ to $\overline{G}$ that takes the equivalence class of an irreducible unitary representation to the equivalence class of the $(\mathfrak{g}, K)$-module defined by its K-finite vectors. The results on the unitary representations show that this map is a bijection of $\hat{G}$ with the subset of $\overline{G}$ of classes of unitary irreducible $(\mathfrak{g}, K)$-modules. We identify $\hat{G}$ with its image in $\overline{G}$ so that

$\hat{G} \subset \overline{G}$

$\hat{G} =$ set of equivalence classes of irreducible unitary $(\mathfrak{g}, K)$-modules.

We have thus reached an algebraic description of the fundamental objects of study in representation theory.

Let us now consider the category of all Banach-space representations of G which are finitely generated Harish-Chandra modules for G. Two such, say V_1, V_2, are said to be *infinitesimally equivalent* if the $(\mathfrak{g}, K)$-modules V_1^0 and V_2^0 are equivalent. It is then clear that we may view $\overline{G}$ as the set of infinitesimal equivalence classes of irreducible Banach-space Harish-Chandra modules for G.

Finally this work also suggested that the category of finitely generated Harish-Chandra modules is worth studying in depth. Recent work due to Casselman, Jacquet, Wallach, Schmid and others has revealed many beautiful properties of these modules and underscored their fundamental importance in representation theory. For instance Casselman proved that any such module is equivalent to the module of K-finite vectors of a Hilbert-space Harish-Chandra G-module. For a deeper study of this and most other issues of representation theory the advanced reader should refer to Wallach's forthcoming book [Wa1] as well as [Kn].

Matrix elements and their differential equations. Let π be a representation of G in a Banach space V, $v \in V$, $v^* \in V^*$. Then the function f_{v,v^*}: $x \mapsto (\pi(x)v, v^*)$ is called a *matrix element* of π. It is always a continuous function on G, and is C^∞ (resp. analytic) if $v \in V^\infty$ (resp. v is analytic). The structure of the representation of $U(\mathfrak{g})$ on V^∞ is then reflected in the behaviour of the matrix elements under the differential operators from $U(\mathfrak{g})$, in view of the formulae

$$\partial(a) f_{v,v^*} = f_{\pi(a)v,v^*} \quad (a \in U(\mathfrak{g}))$$

If now π has an infinitesimal character – for instance if π is irreducible unitary – then

$$\partial(z) f_{v,v^*} = \chi_\pi(z) f_{v,v^*} \quad (z \in \mathfrak{Z})$$

where χ_π is the infinitesimal character of π. The matrix elements of such π are thus *eigenfunctions* on the group for the bi-invariant differential operators from $\mathfrak{Z}$ for the eigenvalues $\chi_\pi(z)$; this reveals the analytic significance of the infinitesimal character. This link between representation theory and analysis on G first emerged in Bargmann's work and eventually became one of the most fundamental themes in Harish-Chandra's theory.

If $V = H$, a Hilbert space, it is more usual to replace H^* by H itself and define

$$f_{\pi:u,v}(x) = (\pi(x)v \mid u)$$

It may be assumed without loss of generality that $\pi(k)$ is unitary for all $k \in K$. We shall always suppose this. If u and v are K-finite, $f_{\pi:u,v}$ transforms according to finite-dimensional representations of K under translations by elements of K. To describe these transformation laws succinctly we go over to the functions

$$f_{\pi:\xi,\eta}(x) = E_\xi \pi(x) E_\eta \quad (\xi, \eta \in \hat{K})$$

where E_ξ are the projections $H \to H_\xi$ and $f_{\pi:\xi,\eta}$ is viewed as a map of G into $H_{\xi\eta} = \mathrm{Hom}_{\mathbb{C}}(H_\eta, H_\xi)$. These functions are most useful only when $\dim(H_\xi) < \infty$ for all ξ. Then they are analytic; and if (e_i) (resp. (e'_j)) is an orthonormal basis of H_ξ (resp. H_η), then, with respect to these bases, $f_{\pi:\xi,\eta}$ is represented by the matrix $(f_{\pi:e_i,e'_j})$. Now $H_{\xi\eta}$ is a bimodule, for $H_{\xi\xi} = \mathrm{End}(H_\xi)$ from the left and $H_{\eta\eta} = \mathrm{End}(H_\eta)$ from the right. If we write $\pi_\xi(k) = \pi(k)|_{H_\xi} (k \in K)$, we then have

$$f_{\pi:\xi,\eta}(k_1 x k_2) = \pi_\xi(k_1) f_{\pi:\xi,\eta}(x) \pi_\eta(k_2)$$

for $x \in G, k_1, k_2 \in K$. This is the fundamental transformation law for the

K-finite matrix elements. The $f_{\pi:\xi,\eta}$ are thus special cases of functions *spherical with respect to K-bimodules*, which are finite-dimensional vector spaces W on which K acts from both left and right, the functions being smooth maps $f(G\to W)$ such that

$$f(k_1 x k_2) = k_1 \cdot f(x) \cdot k_2$$

for $x \in G$, $k_1, k_2 \in K$. They were first introduced in this generality by Harish-Chandra. The terminology comes from the fact that, when $G = SO(3)$, $T =$ group of rotations around the z-axis, π, an irreducible representation of G and u, a unit vector fixed by $\pi(T)$, the function

$$f_{\pi:u,u}(x) = (\pi(x)u|u)$$

is bi-invariant under T, and, viewed as a function on the unit sphere, can be identified with a spherical harmonic.

Returning to the general case, suppose now that π is actually unitary. Then $(\pi(x)v|u) = (v|\pi(x^{-1})u)$ and hence

$$f_{\pi:u,v}(a;x;b) = f_{\pi:\pi(b)v,\pi(a^{\dagger})u}(x)$$

In particular, *all the derivatives of* $f_{\pi:u,v}$ *are uniformly bounded on G*. If π is not unitary, this will not be true but one can prove (with the proper definitions) that the $f_{\pi:u,v}$ and their derivatives grow at most exponentially.

We saw above that for irreducible π, or, more generally, for π with infinitesimal character, the $f_{\pi:u,v}$ are eigenfunctions for $\mathfrak{Z}$. The $f_{\pi:\xi,\eta}$ are actually eigenfunctions for a larger class of operators. In fact, let $t \in U(\mathfrak{k})$ and suppose that t belongs to the centre $\mathfrak{Z}_K$ of $U(\mathfrak{k})$. It is then clear that t acts as a scalar, say $c_\xi(t)$, on the representation space of ξ, and hence

$$f_{\pi:\xi,\eta}(x;t) = c_\eta(t) f_{\pi:\xi,\eta}(x)$$
$$f_{\pi:\xi,\eta}(t;x) = c_\xi(t) f_{\pi:\xi,\eta}(x)$$

So the $f_{\pi:\xi,\eta}$ are eigenfunctions for at least the algebra $\mathfrak{Z}\mathfrak{Z}_K$. Consider for instance a basis $(X_i)_{1\leqslant i\leqslant n}$ for $\mathfrak{g}$ such that

(i) $\langle X_i, X_j\rangle = 0 \quad (i \neq j)$, $\langle\cdot,\cdot\rangle$ being the Cartan–Killing form,
(ii) $(X_i)_{1\leqslant i\leqslant k}$ span $\mathfrak{k}$ and $\langle X_i, X_i\rangle = -1$, $1 \leqslant i \leqslant k$,
(iii) $\langle X_i, X_i\rangle = 1$, $k < i \leqslant n$.

It is not difficult to prove the existence of such a basis from the structure theory of $\mathfrak{g}$. If

$$\omega = -(X_1^2 + \cdots + X_k^2) + X_{k+1}^2 + \cdots + X_n^2$$
$$\omega_K = -(X_1^2 + \cdots + X_k^2)$$

then $\omega \in \mathfrak{Z}$, $\omega_K \in \mathfrak{Z}_K$; ω is the Casimir operator of G while ω_K is analogous to the Casimir operator of K (*analogous* because $\langle\cdot,\cdot\rangle$ may differ from the Killing form of $\mathfrak{k}$). If

$$\square = \omega - 2\omega_K = X_1^2 + \cdots + X_n^2$$

then $\square$ is an *elliptic* operator which is analytic. So the matrix elements are eigenfunctions of the 'Laplacian' $\square$.

As weak solutions of the equation $\square T = cT$ are analytic, by the general regularity theorem for elliptic operators, one may expect that there will be an alternative route to the main results (analytic vectors, finite multiplicity, and so on) from the point of view of differential equations on the group. This is the case in fact; for some discussion of this see [V3].

Characters. The finiteness theorems guarantee bounds on the multiplicities of the representations of K occurring in irreducible representations of G. These bounds lead immediately to the existence of the characters and the differential equations satisfied by them. The key estimates are contained in the following lemma.

Lemma 21. *Let $E = 1 + \omega_K, \omega_K$ being defined as above. For any $\xi \in \hat{K}$, let $c_\xi(E)$ be the scalar which represents E in the representations from the class ξ. We then have the following*:

(i) $c_\xi(E) \geqslant 1$ *and* $\sum_\xi c_\xi(E)^{-m} < \infty$ *if* $m \gg 0$;
(ii) *for some constants* $c > 0, r \geqslant 0, d(\xi) \leqslant c c_\xi(E)^r$ (for all $\xi \in \hat{K}$).

We shall not prove this here. If $G = SL(2, \mathbb{R})$, $K = (u_\theta)$, $E = 1 - \mathrm{d}^2/\mathrm{d}\theta^2$, and $\hat{K} = \mathbb{Z}$ where $n \in \mathbb{Z}$ corresponds to the representation $u_\theta \mapsto \mathrm{e}^{in\theta}$. So, $c_n(E) = 1 + n^2$, $d(n) = 1$, and the lemma is immediate. For $G = SL(2\text{-}\mathbb{C})$, we have $K = SU(2)$, and $\hat{K} = \{n | n \text{ integer } \geqslant 0\}$; then $d(n) = n + 1$. For $\mathfrak{k}$ we take the basis

$$X_1 = \begin{pmatrix} i & 0 \\ 0 & -i \end{pmatrix}, \quad X_2 = \begin{pmatrix} 0 & 1 \\ -1 & 0 \end{pmatrix}, \quad X_3 = \begin{pmatrix} 0 & i \\ i & 0 \end{pmatrix}$$

and let $\omega_K = -(X_1^2 + X_2^2 + X_3^2)$. Then $\mathfrak{k}_{\mathbb{C}} \simeq \mathfrak{sl}(2, \mathbb{C})$ with standard triple

$$H' = -iX_1, X' = (X_2 - iX_3)/2, \quad Y' = (X_2 + iX_3)/2$$

and

$$\omega_K = H'^2 + 2(X'Y' + Y'X')$$

It follows from this that ω_K acts as the scalar $n^2 + 2n$ on the space of the representations of dimension $n + 1$. Hence $c_n(E) = (n + 1)^2$. The lemma is clear in this case also. For the general case, we replace K by a finite cover and suppose that $K = K_1 \times T$ where K_1 is compact semisimple and T is a torus. The argument then combines the two types of arguments illustrated in the two special cases treated above.

Recall that a bounded operator $A(H \to H)$ is of *trace class* if, for every orthonormal basis (e_i) of H, the series $\sum_i (Ae_i|e_i)$ is absolutely convergent; the sum is then independent of the basis and is denoted by $\mathrm{tr}(A)$. One knows that in this case, for any bounded operators S, T, SAT is of trace class and

$$\mathrm{tr}(SAT) = \mathrm{tr}(ATS) = \mathrm{tr}(TSA)$$

We now introduce a very useful sufficient condition for an operator to be of trace class, which is easy to apply because it involves only a single orthonormal basis. A bounded operator $A(H \to H)$ is said to be *summable* with respect to an orthonormal basis (e_i) if $\sum_{i,j}|a_{ij}| < \infty$ where (a_{ij}) is the matrix of A relative to the basis (e_i), i.e., $a_{ij} = (Ae_j|e_i)$. It follows that, if S, T are bounded operators with respective matrices (s_{pq}), (t_{pq}), then $\sum_i |s_{ip}t_{qi}| \leqslant \|Se_p\| \, \|T^\dagger e_q\| \leqslant \|S\| \, \|T\|$, so that

$$\sum_i |(SATe_i|e_i)| \leqslant \sum_{i,p,q} |s_{ip}a_{pq}t_{qi}| \leqslant \|S\| \; \|T\| \sum_{p,q} |a_{pq}|$$

Replacing S by US and T by TU^{-1} where U is unitary we see that SAT is of trace class and that for *any orthonormal basis* (f_m),

$$\sum_m |(SATf_m|f_m) \leqslant \|S\| \; \|T\| \sum_{p,q} |a_{pq}|$$

If we take S to be a bounded operator with bounded inverse and take $T = S^{-1}$, we see that $\mathrm{tr}(SAS^{-1}) = \mathrm{tr}(A)$.

***Theorem* 22.** *Let H be a Hilbert space which is a Harish-Chandra G-module for a representation π, and suppose that, for some constants $c > 0$, $r \geqslant 0$,*

$$\dim(H_\xi) \leqslant cd(\xi)^r \quad (\xi \in \hat{K})$$

Then for any $f \in C_c^\infty(G)$, the operator

$$\pi(f) = \int_G f(x)\pi(x)\,\mathrm{d}x$$

is of trace class, and

$$\Theta_\pi : f \mapsto \mathrm{tr}(\pi(f))$$

is a distribution on G, invariant under the inner automorphisms of G. In particular, this is true if π is an irreducible representation.

***Proof*.** We assume that π is unitary when restricted to K. Let $n(\xi) = \dim(H_\xi)$ and let $(e_{\xi,j})$ $(1 \leqslant j \leqslant n(\xi))$ be an orthonormal basis of H_ξ. We shall show that the operators $\pi(f)$ are actually summable in this basis.

Write $f_{jk} = f_{\pi:e_{\xi,j},e_{\eta,k}}$. Then, with E as in Lemma 21,

$$f_{jk}(E^r; x; E^r) = c_\eta(E)^r c_\xi(E)^r f_{jk}(x)$$

and hence

$$\int_G f_{jk}(x) f(x)\,\mathrm{d}x = c_\eta(E)^{-r} c_\xi(E)^{-r} \int_G f_{jk}(E^r; x; E^r) f(x)\,\mathrm{d}x$$

$$= c(E)^{-r} c(E)^{-r} \int_G f_{jk}(x) f(E^r; x; E^r)\,\mathrm{d}x$$

The last step comes from the fact that if u, v are two smooth functions on G with v having compact support, and $a,b \in U(\mathfrak{g})$

$$\int_G u(a; x; b) v(x)\,\mathrm{d}x = \int_G u(x) v(a'; x; b')\,\mathrm{d}x$$

where $'$ is the formal transpose. Estimating the last integral by

$$|f_{jk}(x)| \leqslant \|\pi(x)\|$$

we have the estimate

$$|(\pi(f) e_{\eta,k} | e_{\xi,j})| \leqslant c_\xi(E)^{-r} c_\eta(E)^{-r} \int_G \|\pi(x)\| \, |f(E^r; x; E^r)|\,\mathrm{d}x$$

By choosing r sufficiently large we see that, for some constant $C > 0$,

$$\sum_{\xi,j:\eta,k} |(\pi(f) e_{\eta,k} | e_{\xi,j})| \leqslant C \int_G \|\pi(x)\| \, |\Delta f(x)|\,\mathrm{d}x$$

where Δ is the (two-sided) differential operator defined by $(\Delta f)(x) = f(E^r; x; E^r)$. The right side is a continuous seminorm on $C_c^\infty(G)$ and so Theorem 22 is clear.

Let us recall how distributions are to be differentiated. If $f \in C_c^\infty(G)$ and $T \in C^\infty(G)$, we have (for $a \in U(\mathfrak{g})$)

$$\int_G T(x; a) f(x)\,\mathrm{d}x = \int_G T(x) f(x; a')\,\mathrm{d}x$$

and so, for arbitrary distributions T we define the derivative $\partial(a)T$ by

$$(\partial(a)T)(f) = T(a'f) \quad (f \in C_c^\infty(G))$$

The operator $T \mapsto \partial(a)T$ is continuous in the topology of pointwise convergence. Let us now apply these remarks to the characters. We have

$$\Theta_\pi(f) = \sum_i \int (\pi(x) e_i | e_i) f(x)\,\mathrm{d}x$$

where (e_i) is some orthonormal basis of K-finite vectors, the series converging absolutely. If π has an infinitesimal character, the matrix elements $f_{\pi:e_i,e_i}$ satisfy the differential equations

$$\partial(z)f_{\pi:e_i,e_i} = \chi_\pi(z)f_{\pi:e_i,e_i} \quad (z \in \mathfrak{Z})$$

Hence, rewriting the earlier equation as

$$\Theta_\pi = \sum_i f_{\pi:e_i,e_i}$$

in the sense of distributions in the topology of pointwise convergence, we see that

$$\partial(z)\Theta_\pi = \chi_\pi(z)\Theta_\pi \quad (z \in \mathfrak{Z})$$

We describe this by saying that Θ_π is *an eigendistribution for $\mathfrak{Z}$ for the eigenvalue χ_π*.

Theorem 23. *The character of an irreducible Harish-Chandra G-module in a Hilbert space is an invariant eigendistribution on the group associated to its infinitesimal character (as its eigenvalue).*

We shall conclude this brief treatment of characters with an explanation of how the character of a representation determines it up to infinitesimal equivalence. Let π be as above. Then, for any $g \in C_c^\infty(G)$,

$$\Theta_\pi(g) = \sum_{\xi \in \hat{K}} \operatorname{tr}(E_\xi \pi(g) E_\xi)$$

while

$$\operatorname{tr}(E_\xi \pi(g) E_\xi) = \int_G \operatorname{tr}(f_{\pi:\xi,\xi}(x))g(x)\,\mathrm{d}x$$

where

$$f_{\pi:\xi,\xi}(x) = E_\xi \pi(x) E_\xi$$

Fix $\xi \in \hat{K}$. Then

$$E_\xi \pi(g) E_\xi = d(\xi)^2 \iint g(x)\chi_\xi(k^{-1})\chi_\xi(k'^{-1})\pi(kxk')\,\mathrm{d}k\,\mathrm{d}k'$$

If we now set

$$g_\xi(x) = d(\xi)^2 \iint g(k^{-1}xk'^{-1})\chi_\xi(k^{-1})\chi_{\xi'}(k'^{-1})\,\mathrm{d}k\,\mathrm{d}k'$$

then $g_\xi \in C_c^\infty(G)$ also and

$$E_\xi \pi(g) E_\xi = \pi(g_\xi)$$

so that

$$\operatorname{tr}(E_\xi \pi(g) E_\xi) = \Theta_\pi(g_\xi)$$

This means that

$$\Theta_\pi(g_\xi) = \int_G \operatorname{tr}(f_{\pi:\xi,\xi})g\,\mathrm{d}x$$

Since g is arbitrary, we see that Θ_π determines all the functions $\mathrm{tr}(f_{\pi:\xi,\xi})$. Suppose now $\mathscr{A} = U(\mathfrak{k})\mathfrak{Q}$ and $a \in \mathscr{A}$. Then $\pi(a)$ stabilizes H_ξ and we know that

$$f_{\pi:\xi,\xi}(1; a) = E_\xi \pi(a) E_\xi$$

Hence

$$\mathrm{tr}(f_{\pi:\xi,\xi}(1; a)) = \mathrm{tr}(E_\xi \pi(a) E_\xi)$$

This shows that Θ_π determines the trace of the representation of $\mathscr{A}$ on H_ξ, and hence the equivalence class of the representation itself. By Theorem 14 this determines the infinitesimal equivalence class of π if π is irreducible. Thus we get the following

Theorem 24. *Let π be any Harish-Chandra module for G in a Hilbert space. Let π be irreducible and let Θ_π be its character. Then π is determined up to infinitesimal equivalence by Θ_π. If π is unitary, it is determined up to unitary equivalence by Θ_π.*

5.5 Some explicit calculations

We shall now illustrate some of the aspects of the general theory with explicit calculations when $G = SL(2, F)$, $F = \mathbb{R}$ or $\mathbb{C}$.

$SL(2, \mathbb{R})$

Finiteness theorem and classification of irreducible modules. We have $G = SL(2, \mathbb{R})$, $\mathfrak{g} = \mathfrak{sl}(2, \mathbb{R})$ and the standard basis

$$H = \begin{pmatrix} 1 & 0 \\ 0 & -1 \end{pmatrix}, \quad X = \begin{pmatrix} 0 & 1 \\ 0 & 0 \end{pmatrix}, \quad Y = \begin{pmatrix} 0 & 0 \\ 1 & 0 \end{pmatrix}$$

The important subgroups K, A, N are now given by

$$K = (u_\theta), \quad u_\theta = \exp\theta(X - Y) = \begin{pmatrix} \cos\theta & \sin\theta \\ -\sin\theta & \cos\theta \end{pmatrix}$$

$$A = (a_t), \quad a_t = \exp tH = \begin{pmatrix} e^t & 0 \\ 0 & e^{-t} \end{pmatrix} \quad (t \in \mathbb{R})$$

$$N = (n_s), \quad n_s = \exp sX = \begin{pmatrix} 1 & s \\ 0 & 1 \end{pmatrix} \quad (s \in \mathbb{R})$$

The element $-i(X - Y)$ has eigenvalues ± 1 and so gives rise to the standard basis of $\mathfrak{g}_\mathbb{C} = \mathfrak{sl}(2, \mathbb{C})$:

$$H' = -i(X - Y), \quad X' = (H + i(X + Y))/2, \quad Y' = (H - i(X + Y))/2$$

Clearly $H \mapsto H'$, $X \mapsto X'$, $Y \mapsto Y'$ is an automorphism of $\mathfrak{g}_\mathbb{C}$. As $\mathfrak{k} = \mathbb{R}\cdot(X - Y)$, a $(\mathfrak{g}, \mathfrak{k})$-module is a $\mathfrak{g}$-module V on which H' acts semisimply, i.e., diagonal in some basis of V; in this case, for $v \in V$,

$$u_\theta \cdot v = e^{in\theta} v \Leftrightarrow H' \cdot v = +nv$$

and so V is a $(\mathfrak{g}, K)$-module if and only if the eigenvalues of H' are all integers. Let

$$s(H') = s(H' : V) = \text{set of eigenvalues of } H'$$

It is usual to call the eigenvalues of H' *weights* (of V) and the eigenspaces *weight spaces*. Clearly $s(H' : V)$ is an important invariant of V. For $\xi \in \mathbb{C}$, V_ξ denotes the eigenspace corresponding to ξ if $\xi \in s(H')$, and 0 for $\xi \notin s(H')$. As usual,

$$X' \cdot V_\xi \subset V_{\xi+2}, \quad Y' \cdot V_\xi \subset V_{\xi-2}$$

The centre $\mathfrak{Z}$ of $U(\mathfrak{g}) = U$ is generated by

$$\Omega = (H+1)^2 + 4YX = (H'+1)^2 + 4Y'X'$$

For any $c \in \mathbb{C}$ we write χ_c for the homomorphism of $\mathfrak{Z}$ to $\mathbb{C}$ such that $\chi_c(\Omega) = c$:

$$\chi_c : \mathfrak{Z} \to \mathbb{C}, \quad \chi_c(\Omega) = c$$

The proof of the finiteness theorem is quite simple in the present case and contains the basic idea of the general argument. In view of Proposition 12, we start with a one-dimensional module W on which H' acts as $h \cdot 1$, $h \in \mathbb{C}$, and consider $V = U/(U \cdot (H' - h'))$, we must prove that the V_ξ are finite $\mathfrak{Z}$-modules. Now the linear span of $X'^m Y'^n$ $(m, n \geqslant 0)$ is clearly a linear complement to $U \cdot (H' - h)$ in U and so we may suppose that V is this linear span. Let π be the representation of $\mathfrak{g}$ in V. To describe the action of H', note that $H'X' = X'(H'+2)$, $H'Y' = Y'(H'-2)$, hence $H'X'^m Y'^n = X'^m Y'^n (H' + 2(m-n))$, and hence

$$\pi(H') \cdot X'^m Y'^n = (h + 2(m-n)) X'^m Y'^n$$

In other words,

$$V = \bigoplus_{q \in \mathbb{Z}} V_{h+2q}, \quad V_{h+2q} = \bigoplus_{m-n=q} (\mathbb{C} \cdot X'^m Y'^n)$$

But if $\omega = \Omega/4$

$$\omega X'^m Y'^n = X'^m Y'^n \omega \equiv X'^{m+1} Y'^{n+1}$$

where the congruence is modulo terms in U of degree $< m+n+2$. Hence

$$\pi(\omega) \cdot X'^m Y'^n \equiv X'^{m+1} Y'^{n+1} \left(\operatorname{mod} \sum_{\substack{r+s<m+n+2 \\ r-s=m-n}} \mathbb{C} \cdot X'^r Y'^s \right)$$

It is easy to show from this that for all $q \geqslant 0$

$$V_{h+2q} = \pi(\mathfrak{Z}) \cdot X'^q, \quad V_{h-2q} = \pi(\mathfrak{Z}) Y'^q$$

By the finiteness theorem, the V_ξ are finite-dimensional in any irreducible $(\mathfrak{g},\mathfrak{k})$-module V. Actually they are *one-dimensional*. This is clear from Theorem 14; for the centralizer of $\mathfrak{k}_{\mathbb{C}}=\mathbb{C}\cdot H'$ in U is easily seen to be generated by H' and $Y'X'$; and V_ξ, being irreducible under these two *commuting* endomorphisms, has to be of dimension 1. It is also easy to give an independent proof.

Lemma 25. *Let V be any $(\mathfrak{g},\mathfrak{k})$-module and $0\neq v\in V_\xi$. Let $M(v)$ be the cyclic module generated by v. Then the following are equivalent:*

(i) $M(v)\cap V_\xi=\mathbb{C}\cdot v$;

(ii) *V is an eigenvector for $Y'X'$.*

If these conditions are satisfied,

$$M(v)=\mathbb{C}\cdot v\oplus\sum_{n\geqslant 1}(\mathbb{C}\cdot X'^n v)\oplus\sum_{n\geqslant 1}(\mathbb{C}\cdot Y'^n v)$$

Proof. Since $Y'X'\cdot V_\xi\subset V_\xi$, (i)$\Rightarrow$(ii) immediately. We now show that, if $Y'X'v=av$,

$$M'=\mathbb{C}\cdot v+\sum_{n\geqslant 1}(\mathbb{C}\cdot X'^n v)+\sum_{n\geqslant 1}(\mathbb{C}\cdot Y'^n v)$$

is $\mathfrak{g}$-stable; this will prove that $M'=M(v)$ and hence (i) also. Note that the sums above are all direct because $X'^n v\in V_{\xi+2n}$, $Y'^n v\in V_{\xi-2n}$. The relations

$$X'Y'^n=Y'^n X'+nY'^{n-1}(H'-n+1)$$
$$Y'X'^n=X'^n Y'-nX'^{n-1}(H'+n-1)$$

imply that (when $Y'X'v=av$) for $n\geqslant 1$

$$X'Y'^n v=(a+n(\xi-n+1))Y'^{n-1}v$$
$$Y'X'^n v=(a-(n-1)(\xi+n))X'^{n-1}V$$

from which it is immediate that M' is $\mathfrak{g}$-stable.

Corollary 26. *Under the conditions* (i) *and* (ii), *$M(v)$ has the infinitesimal character χ_c, $c=(\xi+1)^2+4a$, a being the eigenvalue of $Y'X'$ on V;* $\dim(M(v))_\xi\leqslant 1\,(\forall\xi)$.

Proof. We have $\Omega v=cv$ and hence $\Omega av=a\Omega v=cav$, $a\in U$.

For any $(\mathfrak{g},\mathfrak{k})$-module V let us consider the following two properties:

(a) $\dim(V_\xi)\leqslant 1$ for all ξ;

(b) for some $\xi\in\mathbb{C}$, $s(H':V)\subset\xi+2\mathbb{Z}$.

The key to the classification is then the following lemma.

Lemma 27. *If V is irreducible, it has the properties* (a) *and* (b); *and then $s(H'\!:V) = \xi + 2\mathbb{Z}$ if and only if X' and Y' act injectively. Conversely, if V has properties* (a) *and* (b) *and both X' and Y' act injectively, they act bijectively, $s(H'\!:V) = \xi + 2\mathbb{Z}$ and V is irreducible.*

Proof. The first statement follows from the preceding lemma since we can generate V by an element of V which is an eigenvector for $Y'X'$. If X' is not injective, its kernel, being H'-stable, will contain a nonzero weight vector v. If $v \in V_\xi$, the lemma above shows that $s(H'\!:V) \subset \{\xi, \xi - 2, \ldots\} \neq \xi + 2\mathbb{Z}$. Similarly for noninjective Y'. If X' and Y' are both injective, $X'^k(V\xi) = V_{\xi+2k}$, $Y'^k(V_\xi) = V_{\xi-2k}$ and hence $s(H'\!:V) = \xi + 2\mathbb{Z}$. If we only knew of V_ξ that it had properties (a) and (b), this argument would still prove that $s(H'\!:V) = \xi + 2\mathbb{Z}$; and then it is immediate that X' and Y' are both bijective and V is irreducible.

To classify the irreducible $(\mathfrak{g}, \mathfrak{k})$-modules, we look at modules V satisfying (a) and (b). If $s(H'\!:V)$ is a *full orbit* $\xi + 2\mathbb{Z}$, one can describe it quite explicitly and uniquely in terms of $s(H'\!:V)$ and the infinitesimal character χ_V of V. Otherwise, either X' or Y' must annihilate a vector and the module has a highest or lowest weight. We proceed to give the details, and as a first step introduce the modules of the various types.

(I) Modules $D(\mu\!:+)$ ($\mu \in \mathbb{C}$) with highest weight $\mu - 1$. $D(\mu\!:+)$ has a basis $(v_n)_{n \geqslant 0}$; and for $n \geqq 0$ ($v_{-1} = 0$)

$$\left.\begin{aligned} H'v_n &= (\mu - 1 - 2n)v_n,\ Y'v_n = v_{n+1} \\ X'v_n &= n(\mu - n)v_{n-1} \end{aligned}\right\} \quad (D(\mu\!:+))$$

We represent this by the picture

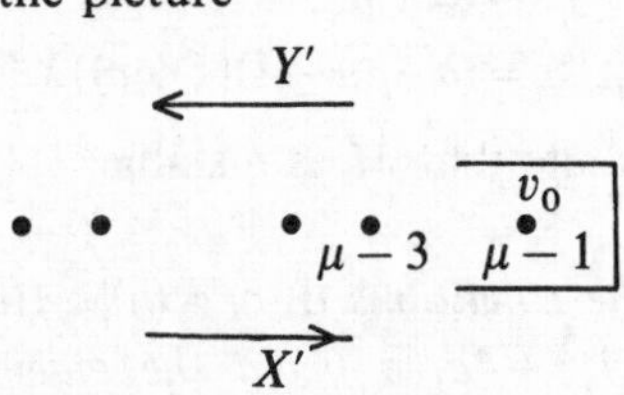

$$\begin{array}{c} \xleftarrow{\quad Y' \quad} \\ \bullet \quad \bullet \qquad \bullet \quad \bullet \quad \boxed{\overset{v_0}{\bullet}} \\ \qquad\qquad \mu - 3 \quad \mu - 1 \\ \xrightarrow[\quad X' \quad]{} \end{array}$$

The consistency of these definitions comes out of the relations involving $X'Y'^n$ and $Y'X'^n$ described earlier. Note that

$$s(H') = \{\ldots, \mu - 3, \mu - 1\}$$

$$\chi_{D(\mu:+)} = \chi_{\mu^2}$$

(II) Modules $D(\mu\!:-)$ ($\mu \in \mathbb{C}$) with lowest weight $\mu + 1$. $D(\mu\!:-)$ has a basis $(v_n)_{n \geqslant 0}$ and for $n \geqslant 0$ ($v_{-1} = 0$)

$$\left.\begin{aligned} H'v_n &= (\mu + 1 + 2n)v_n, X'v_n = v_{n+1} \\ Y'v_n &= -n(\mu + n)v_{n-1} \end{aligned}\right\} (D(\mu:-))$$

The picture is

$$\begin{array}{c} \xrightarrow{X'} \\ v_0 \\ \bullet \quad \bullet \quad \bullet\bullet \\ \mu+1 \quad \mu+3 \\ \xleftarrow{Y'} \end{array}$$

We have

$$s(H') = \{\mu + 1, \mu + 3, \ldots\}$$

$$\chi_{D(\mu:-)} = \chi_{\mu^2}$$

(III) Module $F(\mu)$ which is irreducible and of dimension μ, μ an integer $\geqslant 1$.

Let E be any orbit for the action of $\mathbb{Z}$ on $\mathbb{C}$ by translations with even integers: $E = \xi + 2\mathbb{Z}$ for some $\xi \in \mathbb{C}$.

(IV) Modules $P(c:E)$ where $c \in \mathbb{C}$, E as above, and $\pm c^{1/2} - 1 \notin E$. The space has basis $(v_n)_{n\in\mathbb{Z}}$. To define the module structure select $\xi \in E$, choose a by

$$c = (\xi + 1)^2 + 4a$$

and define

$$\left.\begin{aligned} &H'v_n = (\xi + 2n)v_n \quad (n \in \mathbb{Z}) \\ &X'v_n = v_{n+1}, Y'v_{n+1} = (a - n(\xi + n + 1))v_n \quad (n \geqslant 0) \\ &Y'v_n = v_{n-1}, X'v_{n-1} = (a - (n-1)(\xi + n))v_n \quad (n \leqslant 0) \end{aligned}\right\} (P(c:E))$$

The picture is

$$\begin{array}{c} \xleftarrow{Y'} \\ v_0 \\ \bullet \quad \bullet \quad \bullet \quad \bullet \quad \bullet \\ \xi - 2 \quad \xi \quad \xi + 2 \\ \xrightarrow{X'} \end{array}$$

The choice of constants is dictated as usual by the relations involving $X'Y''$ etc. It is trivial to check that we have a well-defined $(\mathfrak{g}, \mathfrak{k})$-module.

Lemma 28. *We have the following:*

(i) *$D(\mu:+)$ is irreducible if and only if $\mu \neq 1, 2, \ldots$;*
(ii) *$D(\mu:-)$ is irreducible if and only if $\mu \neq -1, -2, \ldots$;*

(iii) *$P(c{:}E)$ is always irreducible and does not depend (up to isomorphism) on the choice of $\xi \in E$; moreover $\chi_{P(c:E)} = \chi_c$.*

Proof. (i) If $\mu \neq 1, 2, \ldots, X'v_n$ is a nonzero multiple of v_{n-1} for all $n \geqslant 1$, so that $X'^n v_n$ is a nonzero multiple of v_0 for all $n \geqslant 1$. The irreducibility of $D(\mu{:}+)$ is immediate. Suppose on the other hand μ is an integer $\geqslant 1$. Then $X'v_\mu = 0$ and so

$$v_\mu, v_{\mu+1}, v_{\mu+2}, \ldots$$

span a $\mathfrak{g}$-stable subspace, showing that $D(\mu{:}+)$ is not irreducible in this case.

(ii) For $D(\mu{:}-)$ the argument is analogous, and is omitted.

(iii) We have $a - n(\xi + n + 1) = 0$ for some integer n if and only if $c - (\xi + 1)^2 = 4n(\xi + n + 1)$, i.e., if and only if $c = (\xi + 1 + 2n)^2$; so $\pm\sqrt{c} - 1 \notin E \Leftrightarrow a - n(\xi + n + 1) \neq 0$ for all $n \in \mathbb{Z}$. The formulae for $P(c{:}E)$ now show that X' and Y' act injectively and thence that $P(c{:}E)$ is irreducible. Since $Y'X'v_0 = av_0$, $\Omega v_0 = (4a + (\xi + 1)^2)v_0 = cv_0$. The module $P(c{:}E)$ is thus also generated by any $v_n = v'$ and so can be described in the same way, with $\xi' = \xi + 2n$ and $a' = a - n(\xi + n + 1)$; as

$$4a' = c - (\xi + 1)^2 - 4n(\xi + n + 1) = c - (\xi' + 1)^2$$

we are through.

Remark. When $\mu = 1, 2, \ldots, D(\mu{:}+)$ is reducible. It is easy to show that $\sum_{r \geqslant 0}(\mathbb{C}\cdot v_{\mu+r})$ is the only proper zero submodule, that it is isomorphic to $D(-\mu{:}+)$, and that the quotient is isomorphic to $F(\mu)$:

$$0 \to D(-\mu{:}+) \to D(\mu{:}+) \to F(\mu) \to 0$$

Similarly, if $\mu = -1, -2, \ldots$, $\sum_{r \geqslant 0}(\mathbb{C}\cdot v_{-\mu+r})$ is the unique proper nonzero submodule; it is isomorphic to $D(-\mu{:}-)$, and leads to the exact sequence

$$0 \to D(-\mu{:}-) \to D(\mu{:}-) \to F(-\mu) \to 0$$

Theorem 29. *The irreducible $(\mathfrak{g}, \mathfrak{k})$-modules are precisely the $D(\mu{:}+)$ $(\mu \neq 1, 2, \ldots)$, the $D(\mu{:}-)$ $(\mu \neq -1, -2, \ldots)$, the $F(\mu)(\mu = 1, 2, \ldots)$, and the $P(c{:}E)$ $(\pm\sqrt{c} - 1 \notin E)$; also no two of these are equivalent, and*

$$\chi_{F(\mu)} = \chi_{D(\mu:\pm)} = \chi_{\mu^2}, \quad \chi_{P(c:E)} = \chi_c$$

The irreducible $(\mathfrak{g}, K)$-modules are the $D(\mu{:}+)(\mu = 0, -1, \ldots)$, the $D(\mu{:}-)(\mu = 0, 1, \ldots)$, the $F(\mu)(\mu = 1, 2, \ldots)$, $P(\mu^2{:}2\mathbb{Z})$ $(\mu \neq$ odd integer$)$ and $P(\mu^2{:}1 + 2\mathbb{Z})$ $(\mu \neq$ even integer$)$.

Proof. Once the classification of $(\mathfrak{g}, \mathfrak{k})$-modules is settled, the $(\mathfrak{g}, K)$-

modules are singled out by requiring that $s(H') \subset \mathbb{Z}$; and there are two orbits for $2\mathbb{Z}$ inside $\mathbb{Z}$, namely, $2\mathbb{Z}$ and $1+2\mathbb{Z}$. That there are no equivalences in the given list is trivial to see. Suppose V is an irreducible $(\mathfrak{g}, \mathfrak{k})$-module. If X' and Y' act injectively, $s(H') = \xi + 2\mathbb{Z}$ by Lemma 27, $Y'X'$ is the scalar a on V_ξ since $\dim(V_\xi) = 1$, and $a \neq n(\xi + n + 1)$ for any $n \in \mathbb{Z}$. So $V \cong P(c:E)$ where $E = \xi + 2\mathbb{Z}$, $c = (\xi+1)^2 + 4a$. Suppose one of X' and Y' has nonzero kernel, but not both. If X' has nontrivial kernel, we can find $0 \neq v \in V_{\mu-1}$ such that $X'v = 0$ while $Y'^n v \neq 0$ for any $n \geqslant 0$ and $V = U \cdot v$. It is then immediate that $V \cong D(\mu:+)$. Similarly, if Y' but not X' has nonzero kernel, $V \cong D(\mu:-)$ for some μ. If both kernels are nonzero, let $0 \neq v \in V_{\mu-1}$ be such that $X'v = 0$; then $Y'^n v = 0$ for some $n \geqslant 1$ as otherwise Y' would be injective. So V is spanned by the $Y'^m v$, $0 \leqslant m < n$, hence finite-dimensional. This proves the theorem.

Because there are only two orbits $2\mathbb{Z}$ and $1+2\mathbb{Z}$ in $\mathbb{Z}$, we can speak of the *parity* of an irreducible $(\mathfrak{g}, K)$-module. The classification shows that for a given infinitesimal character χ_{μ^2}, if $\mu \notin \mathbb{Z}$, $P(\mu^2:E)$, $E = 2\mathbb{Z}$ or $1+2\mathbb{Z}$ are the two irreducible $(\mathfrak{g}, K)$-modules with infinitesimal character χ_{μ^2}, while for $\mu \in \mathbb{Z}$ there will be four: namely

$$D(-|\mu|:+), \quad D(|\mu|:-), \quad F(|\mu|), \quad P(\mu^2:E_\mu)$$

where E_μ is the set of odd or even integers according as μ is odd or even.

Splitting of principal series. We shall now analyze, using the infinitesimal method, the principal series representations. Recall the 'projections' $k(G \to K)$, $a(G \to A)$, $n(G \to N)$ associated to the Iwasawa decomposition. Let us define $t(x)$, $s(x) \in \mathbb{R}$, $\varphi(x) \in \mathbb{R}/2\pi\mathbb{Z}$, for any $x \in G$, by

$$k(x) = u_{\varphi(x)}, \quad a(x) = \exp(t(x)H), \quad n(x) = \exp(s(x)X)$$

The principal series representations $\pi_{\varepsilon,\lambda}$ ($\varepsilon = 0$ or 1, $\lambda \in \mathbb{C}$) act on the Hilbert space $\mathscr{H}_\varepsilon \subset L^2(K)$ of all functions f with $f(u_{-\theta}) = (-1)^\varepsilon f(u_\theta)$, by

$$(\pi_{\varepsilon,\lambda}(x)f)(u_\theta) = e^{-(\lambda+1)t(x^{-1}u_\theta)} f(x^{-1}[u_\theta])$$

If $f \in C^\infty(K)_\varepsilon$, the right side is C^∞ on $G \times K$, and so $f \in \mathscr{H}_\varepsilon^\infty$; similarly, if f is analytic on K, the right side is analytic on $G \times K$ and so f is an analytic vector (Lemma 5). Thus

$$\mathscr{H}_\varepsilon^0 = \mathscr{F}_\varepsilon := \text{space of finite Fourier series of parity } \varepsilon$$

Let

$$e_n(u_\theta) = \mathrm{e}^{-in\theta} \quad (\pi_{\varepsilon,\lambda}(u_\varphi)e_n = \mathrm{e}^{in\varphi}e_n)$$

The sub-$\mathfrak{g}$-modules of $\mathscr{H}_\varepsilon^0$ and the closed G-stable subspaces of $\mathscr{H}_\varepsilon$ are in natural correspondence: to $W \subset \mathscr{H}_\varepsilon^0$ we associate its closure $\mathrm{Cl}(W) \subset$

$\mathscr{H}_\varepsilon$, and to $\bar{W}\subset\mathscr{H}_\varepsilon$ we associate $\bar{W}\cap\mathscr{H}_\varepsilon^0=\bar{W}^0$. We have

$$\mathrm{Cl}(W)^0=W,\quad \mathrm{Cl}(\bar{W}^0)=\bar{W}$$

The main step is to calculate the operators $\pi_{\varepsilon,\lambda}(Z)$, $Z\in\mathfrak{g}_{\mathbb{C}}$.

Lemma 30. *For $L\in\mathfrak{g}$, let $x_r=\exp rL(r\in\mathbb{R})$ and for any $u_\theta\in K$ let $\dot{t}_L$, $\dot{s}_L$, $\dot{\varphi}_L$ be the derivatives at $r=0$ of $t(x_r^{-1}u)$, $s(x_r^{-1}u)$, $\varphi(x_r^{-1}u)$ (φ is uniquely determined for small r by requiring its value to be θ for $r=0$). Then, for*

$$L=\begin{pmatrix} a & b\\ c & -a\end{pmatrix}$$

$$\dot{t}_L=((b+c)/2)\sin 2\theta-a\cos 2\theta$$
$$\dot{s}_L=-2a\sin 2\theta-(b+c)\cos 2\theta$$
$$\dot{\varphi}_L=a\sin 2\theta-b\sin^2\theta+c\cos^2\theta$$

Proof. We have

$$\exp(-rL)\begin{pmatrix}\cos\theta & \sin\theta\\ -\sin\theta & \cos\theta\end{pmatrix}=u_\varphi a_t n_s$$

So, differentiating, we get

$$-\begin{pmatrix} a & b\\ c & -a\end{pmatrix}\begin{pmatrix}\cos\theta & \sin\theta\\ -\sin\theta & \cos\theta\end{pmatrix}=\dot{\varphi}_L\begin{pmatrix}-\sin\theta & \cos\theta\\ -\cos\theta & -\sin\theta\end{pmatrix}$$
$$+\begin{pmatrix}\cos\theta & \sin\theta\\ -\sin\theta & \cos\theta\end{pmatrix}\begin{pmatrix}\dot{t}_L & 0\\ 0 & -\dot{t}_L\end{pmatrix}$$
$$+\begin{pmatrix}\cos\theta & \sin\theta\\ -\sin\theta & \cos\theta\end{pmatrix}\begin{pmatrix}0 & \dot{s}_L\\ 0 & 0\end{pmatrix}$$

These are linear equations in $\dot{\varphi}_L$, $\dot{t}_L$, $\dot{s}_L$ and their solution gives the lemma.

Proposition 31. *For L as above, $\pi_{\varepsilon,\lambda}(L)$ is the differential operator*

$(a\sin 2\theta-b\sin^2\theta+c\cos^2\theta)\mathrm{d}/\mathrm{d}\theta-(\lambda+1)((b+c)/2)\sin 2\theta-a\cos 2\theta)$

In particular we have

(i) $\pi_{\varepsilon,\lambda}(H')=\mathrm{i}\,\mathrm{d}/\mathrm{d}\theta$, $\pi_{\varepsilon,\lambda}(X')=\frac{1}{2}\mathrm{i}\mathrm{e}^{-2\mathrm{i}\theta}\mathrm{d}/\mathrm{d}\theta+\frac{1}{2}(\lambda+1)\mathrm{e}^{-2\mathrm{i}\theta}$,
$\pi_{\varepsilon,\lambda}(Y')=-\frac{1}{2}\mathrm{i}\mathrm{e}^{2\mathrm{i}\theta}\mathrm{d}/\mathrm{d}\theta+\frac{1}{2}(\lambda+1)\mathrm{e}^{2\mathrm{i}\theta}$,

(ii) $\pi_{\varepsilon,\lambda}(\Omega)=\lambda^2$,

(iii) $\pi_{\varepsilon,\lambda}(H')e_n=ne_n$, $\pi_{\varepsilon,\lambda}(X')e_n=\frac{1}{2}(n+1+\lambda)e_{n+2}$,
$\pi_{\varepsilon,\lambda}(Y')e_n=\frac{1}{2}(-n+1+\lambda)e_{n-2}$.

Proof. For $f\in C^\infty(K)_\varepsilon$, writing $\pi=\pi_{\varepsilon,\lambda}$,

$$(\pi(L)f)(u_\theta)=(\mathrm{d}/\mathrm{d}r)_{r=0}\mathrm{e}^{-(\lambda+1)t(x_r^{-1}u_\theta)}f(u_{\varphi(x_r^{-1}u_\theta)})$$

So,

$$\pi(L)f=\dot{\varphi}_L(\mathrm{d}f/\mathrm{d}\theta)-(\lambda+1)\dot{t}_Lf$$

The rest is straightforward calculation which uses the expressions for H', X', Y' in terms of H, X, Y.

On $\mathscr{F}_\varepsilon$ we thus have a representation $\pi_{\varepsilon,\lambda}$ of $\mathfrak{g}$ with weights as the set of integers $\varepsilon + 2\mathbb{Z}$, all weight multiplicities 1, and infinitesimal character χ_{λ^2}. The decomposition theory is now quite simple. We just state the final result. If $n_1, n_2, \ldots$ are integers, we write $\langle e_{n_1}, e_{n_2}, \ldots \rangle$ for the linear span of the $e_{n_1}, e_{n_2}, \ldots$. Since the e_n are analytic vectors, the closures in $L^2(K)$ of the $\mathfrak{g}$-stable subspaces are G-stable and so the theorem describes the decomposition of the $\pi_{\varepsilon,\lambda}$ also. From now on we use the same symbol for a sub-$\mathfrak{g}$-module of $\mathscr{F}_\varepsilon$ as well as the G-module defined on its closure.

Theorem 32. *We have the following*:

(i) if $\lambda \notin \mathbb{Z}$, $\pi_{\varepsilon,\lambda}$ *is irreducible and equivalent to* $P(\lambda^2 : \varepsilon + 2\mathbb{Z})$;

(ii) if $\lambda \in \mathbb{Z}$, $\pi_{\varepsilon,\lambda}$ *is again irreducible if* ε *and* λ *have the same parity, and is equivalent to* $P(\lambda^2 : \varepsilon + 2\mathbb{Z})$;

(iii) *let* $\lambda \in \mathbb{Z}$ *and suppose* ε *and* λ *have opposite parities; then*
$\pi_{\varepsilon,\lambda}(X')e_{-\lambda-1} = 0, \quad \pi_{\varepsilon,\lambda}(Y')e_{\lambda+1} = 0,$
$\langle \ldots, e_{-\lambda-3}, e_{-\lambda-1} \rangle \cong D(-\lambda : +), \langle e_{\lambda+1}, e_{\lambda+3}, \ldots \rangle \cong D(\lambda : -).$

If $\lambda > 0$, *these two submodules are irreducible and linearly independent, and* $0 \to D(-\lambda : +) \oplus D(\lambda : -) \to \pi_{\varepsilon,\lambda} \to F(\lambda) \to 0$ *is exact. If* $\lambda < 0$,

$$D(-\lambda : +) \cap D(\lambda : -) \cong F(-\lambda)$$

and we have the exact sequence

$$0 \to F(-\lambda) \to \pi_{\varepsilon,\lambda} \to D(\lambda : +) \oplus D(-\lambda : -) \to 0$$

If $\lambda = 0$, ε *must be* 1, *and*

$$\pi_{1,0} \cong D(0 : +) \oplus D(0 : -)$$

The reader would have observed that the cases $\lambda > 0$ and $\lambda < 0$ are dual to each other. The pictures are

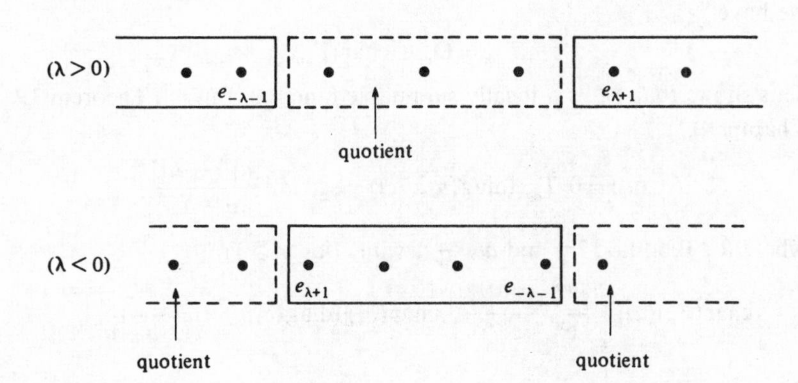

Corollary 33. *Every irreducible* ($\mathfrak{g}$, K)*-module is a submodule of some* $\pi_{\varepsilon,\lambda}$.

Since the characters of $\pi_{\varepsilon,\lambda}$ have been calculated already and those of the $F(\lambda)$ are known, we can obtain the characters of the modules $D(\lambda:+)\oplus D(-\lambda:-)$. We now define the modules $D_m(m\in\mathbb{Z}\backslash(0))$, $D_{0,\pm}$ by

$$D_m = D(m:-) \quad (m\geqslant 1)$$
$$D_m = D(m:+) \quad (m\leqslant -1)$$
$$D_{0,\pm} = D(0:\mp)$$

Thus D_m has lowest weight $m+1$ for $m\geqslant 1$ and highest weight $m-1$ for $m\leqslant -1$; $D_{0,+}$ has lowest weight 1 and $D_{0,-}$ has highest weight -1. Let

$$\Theta_m = \text{character of } D_m$$
$$\Theta_{0,\pm} = \text{character of } D_{0,+}$$

and put

$$\Theta_m^* = \Theta_m + \Theta_{-m} \quad (m\geqslant 1), \quad \Theta_0^* = \Theta_{0,+} + \Theta_{0,-}$$

Theorem 34. Θ_m^* $(m\geqslant 0)$ *are locally summable functions on G and are given on the elliptic and hyperbolic elements as follows. For* $\theta\not\equiv 0, \pi\,(\mathrm{mod}\,2\pi)$,

$$\Theta_m^*(u_\theta) = \frac{e^{im\theta} - e^{-im\theta}}{e^{i\theta} - e^{-i\theta}}$$

while for $t\neq 0$

$$\Theta_m^*(a_t) = \frac{2e^{-m|t|}}{|e^t - e^{-t}|}$$

$$\Theta_m^*(\gamma a_t) = (-1)^{m-1}\frac{2e^{-m|t|}}{|e^t - e^{-t}|}$$

where $\gamma = -\mathrm{id}$.

Proof. Let $m\geqslant 0$ and let ε be of the *opposite* parity to m; let char(F_m) be the character of F_m, interpreted as 0 if $m=0$. Then, if $T_{\varepsilon,m}$ is the character of $\pi_{\varepsilon,m}$, we have

$$T_{\varepsilon,m} = \Theta_m^* + \mathrm{char}\,(F_m)$$

This shows that Θ_m^* is a locally summable function. But (cf. Theorem 19, Chapter 4),

$$T_{\varepsilon,m}(u_\theta) = 0,\ T_{\varepsilon,m}(\mathrm{diag}\,(a, a^{-1})) = \mathrm{sgn}\,(a)^\varepsilon\frac{|a|^m + |a|^{-m}}{|a - a^{-1}|}$$

where $\theta\not\equiv 0, \pi(\mathrm{mod}\,2\pi)$ and $a\neq\pm 1$ while (for $m\geqslant 1$)

$$\mathrm{char}\,(F_m)(u_\theta) = \frac{e^{im\theta} - e^{-im\theta}}{e^{i\theta} - e^{-i\theta}},\ \mathrm{char}\,(F_m)(\mathrm{diag}\,(a, a^{-1})) = \frac{a^m - a^{-m}}{a - a^{-1}}$$

The required formulae follow at once if we remember that $\operatorname{sgn}(\gamma e') = (-1)^{m-1} = \text{parity if } \operatorname{char}(F_m)$.

It would be important to determine the characters Θ_m and Θ_{-m} separately. This is however a much deeper question and we shall take it up in the next chapter.

We shall now turn to the equivalence between $\pi_{\varepsilon,\lambda}$ and $\pi_{\varepsilon,-\lambda}$ suggested by the equality $T_{\varepsilon,\lambda} = T_{\varepsilon,-\lambda}$ of characters. For $\lambda \in i\mathbb{R}$ this is certainly a consequence of the general theory but once again we can do it explicitly.

Theorem 35. *Let $\lambda \notin \mathbb{Z}$. Then there are constants $t_n \neq 0 \, (n \in \mathbb{Z})$ such that $t_n/t_{n+2} = (n+1-\lambda)/(n+1+\lambda)$. For any such (t_n), $T(e_n \mapsto t_n e_n)$ intertwines $\pi_{\varepsilon,\lambda}$ and $\pi_{\varepsilon,-\lambda}$ (on $\mathscr{F}_\varepsilon$): and all intertwining operators $\mathscr{F}_\varepsilon \to \mathscr{F}_\varepsilon$ are of this form. Moreover there is a unitary operator $U(\lambda)(\lambda \in i\mathbb{R}\backslash(0))$ leaving $\mathscr{F}_\varepsilon$ stable that intertwines $\pi_{\varepsilon,\lambda}$ and $\pi_{\varepsilon,-\lambda}$.*

Proof. Since $\pi_{\varepsilon,\lambda}(\Omega) = \lambda^2$, we need only look at T such that $T^{-1}\pi_{\varepsilon,\lambda}T = \pi_{\varepsilon,-\lambda}$. Since H acts as $e_n \mapsto n e_n$, T must be diagonal in the basis (e_n), $Te_n = t_n e_n$. For intertwining the condition is

$$T^{-1}\pi_{\varepsilon,\lambda}(Z)T = \pi_{\varepsilon,-\lambda}(Z), \quad Z = X', Y'$$

which reduces $t_n/t_{n+2} = (n+1-\lambda)/(n+1+\lambda)$. If we start with $t_{n_0} = 1$, the other t_n are uniquely determined; if $\lambda \in i\mathbb{R}\backslash(0)$, $|t_n| = |t_{n+2}|$ and so $|t_n| = 1$ for all n. The operator T thus extends to a *unitary* isomorphism $\mathscr{H}_\varepsilon \cong \mathscr{H}_\varepsilon$ intertwining the two representations.

In contrast with $SL(2, \mathbb{C})$, the unitary principal series of $SL(2, \mathbb{R})$ is not always irreducible but only generically. The point is that one should not think of the principal series of a real group as the analogue of the principal series of the complex group. For a real group there are in general other Cartan subgroups; and in particular there is a conjugacy class of Cartan subgroups, called *fundamental*, characterized by the property that their compact parts have maximum dimension. Harish-Chandra proved [H7, III] that *the unitary representations induced from the associated parabolic subgroups are always irreducible.* This is the 'correct' generalization of the irreducibility theorem for complex groups. The literature on the irreducibility questions is very vast. The irreducibility of the unitary principal series of general complex semisimple groups was first dealt with by Wallach [Wa2]; irreducibility for real groups when the inducing character is in general position was established by Bruhat [Br], while the case of spherical representations was studied by Parthasarathy, Ranga Rao and Varadarajan [PRV] and Kostant [Ko], just to mention a few.

All the principal series representations have infinitesimal characters. However, we can generalize the construction, replacing the quasicharacter $\eta_\lambda(a_t \mapsto e^{\lambda t})$ by a representation $\eta_E(a_t \mapsto \exp tE)$ of A in a finite-dimensional Hilbert space W, L being an endomorphism of W. The induced representation $\pi_{\varepsilon,E}$ can now be defined on the Hilbert space $\mathscr{H}_{\varepsilon,W} = L^2(K:W)$(Hilbert space of W-valued functions on K, of parity ε), the action being given by

$$(\pi_{\varepsilon,E}(x)f(k) = e^{-t(x^{-1}k)(E+1)}f(x^{-1}[k])$$

It can then be shown that

$$\pi_{\varepsilon,E}(\Omega)F = E^2F \quad ((E^2F)(k) = E^2F(k))$$

Since E is completely arbitrary, it is clear that $\pi_{\varepsilon,E}$ need not have an infinitesimal character. It is $\mathfrak{Z}$-finite however. This method is completely general.

5.6 Remarks on proofs

Giving complete proofs of the theorems of §4 is out of the question. Let me restrict myself to a discussion of the theorem of density of analytic vectors and the finiteness theorem for $(\mathfrak{g}, \mathfrak{k})$-modules. I follow Harish-Chandra [H2]

Analytic vectors. Their denseness in Banach-space representations of groups $S = A \cdot N$ must be established. Let π be a representation in a Banach space V, let $w(x) = \max(\|\pi(x)\|, \|\pi(x^{-1})\|)$. The idea is to imitate Gårding's construction and consider vectors of the form

$$\pi(f)v = \int_S f(x)\pi(x)v\,dx \quad (v \in V \text{ arbitrary})$$

for suitable *analytic functions* f on S. We consider $L^1_\pi = L^1(S, w\,dx)(dx = d_lx)$ and $f \in L^1_\pi$; then $\pi(f)v$ makes sense for all $f \in L^1_\pi$. If λ is the left regular representation of S in L^1_π, $\pi(y)\pi(f)v = \pi(\lambda(y)f)v$, so that if $y \mapsto \lambda(y)f$ is an analytic map of S into L^1_π, $y \mapsto \pi(y)\pi(f)v$ will be an analytic map of S into V. So we are reduced to finding analytic vectors $f \in L^1_\pi$ for λ. To prove the density we must find a sequence (f_n) with the following properties:

(i) $f_n \in L^1_\pi$ and analytic for λ;
(ii) $\int_S f_n\,dx = 1$;
(iii) for some constant $c > 0$ and each compact neighbourhood E of 1,
 (a) $\int_E |f_n|\,dx \leqslant c < \infty$ for all n,
 (b) $\int_{S\setminus E} |f_n|(1 + w)\,dx \to 0$ as $n \to \infty$.

where in the first relation, c_E is a constant. For then

$$\pi(f_n)v - v = \int_S f_n(x)(\pi(x)v - v)\,dx$$

and hence

$$\|\pi(f_n)v - v\| \leqslant \|v\| \int_{S\setminus E} |f_n|(1+w)\,dx + \sup_{x\in E} \|\pi(x)v - v\| \cdot \int_E |f_n|\,dx$$

From this we see that $\pi(f_n)v \to v$ as $n \to \infty$. In practice the f_n are $\geqslant 0$ so that (a) of (iii) may be dropped. To produce $f \in L^1_\pi$ that are analytic for λ, the following lemma gives a sufficient condition.

Lemma 36. *Assume that* S is a real form of a complex group $S_{\mathbb{C}}$. *Suppose there is a neighbourhood* $W = W^{-1}$ *of* 1 *in* $S_{\mathbb{C}}$ *and a function* f *holomorphic on* $W \cdot S \subset S_{\mathbb{C}}$ *such that, for some* $g \in L^1_\pi$, $|f(y^{-1}x)| \leqslant g(x)$ *for all* $x \in S$, $y \in W$. *Then the restriction of* f *to* S *is an analytic vector in* L^1_π.

Proof. Since $\lambda(y_0 y)f = \lambda(y_0)\lambda(y)f$, it is enough to check analyticity at $y_0 = 1$. Since the dual of L^1_π is $wL^\infty(S{:}dx)$ we must check analyticity at 1 of

$$b(y) = \int_S f(y^{-1}x)u(x)w(x)\,dx$$

if u is bounded on S. This is a standard result in the elementary theory of analytic functions of several complex variables.

As an example take $S = \mathbb{R}$, $S_{\mathbb{C}} = \mathbb{C}$. Then $w(x) \leqslant c e^{a|x|}$ for some constants $c > 0$, $a > 0$. We take $g_b(z) = e^{-bz^2}$, $z \in \mathbb{C}$, $b > 0$. If $\beta > 0$, we have, for all $x \in \mathbb{R}$, $|z| \leqslant \beta$,

$$|e^{-b(x-z)^2}| \leqslant C e^{-bx^2 + c|x|}$$

where $C > 0$, $c > 0$ are constants. Hence g_b is an analytic vector for λ by the lemma. To get an approximating sequence we take as usual $f_n = (n/\pi)^{1/2} g_n$. If $S = B$, the upper-triangular group in $SL(2, \mathbb{R})$, we take (u, v) for coordinates on S by writing

$$s(u,v) = \begin{pmatrix} 1 & u \\ 0 & 1 \end{pmatrix} \begin{pmatrix} e^v & 0 \\ 0 & e^{-v} \end{pmatrix}$$

Then $d_l s = e^{-2v}\,du\,dv$ while

$$\|\pi(s(u,v))\| \leqslant \text{const.}\, e^{c(|u|+|v|)}$$

and we take

$$f_n(s) = c_n e^{-n(u^2+v^2)} \quad (c_n = (\pi e^{1/n})^{-1} n)$$

The only nontrivial things to verify are the conditions of the lemma, for the function $f(s) = e^{-b(u^2+v^2)} (b > 0)$. But if $s_0 = s(u_0, v_0)$ where u_0, v_0 are *complex* and $|u_0| \leqslant \beta$, $|v_0| \leqslant \beta$

$$s_0 s(u,v) = s(u_0 + e^{2v_0}u, v + v_0)$$

and it is a question of verifying that, if β is small enough,

$$\mathrm{Re}((u_0 + e^{2v_0}u)^2 + (v_0 + v)^2) \geqslant \tfrac{1}{2}(u^2 + v^2)$$

for *all* $u, v \in \mathbb{R}$, u_0, $v_0 \in \mathbb{C}$ with $|u_0| \leqslant \beta$, $|v_0| \leqslant \beta$. This is an elementary exercise.

In the general case we may assume that A and N are simply connected and use the linear coordinates on $\mathfrak{a} = \mathrm{Lie}(A)$ and $\mathfrak{n} = \mathrm{Lie}(N)$. We may introduce the complex groups $A_{\mathbb{C}} = \exp(\mathfrak{a}_{\mathbb{C}})$ and $N_{\mathbb{C}} = \exp(\mathfrak{n}_{\mathbb{C}})$ as well as $S_{\mathbb{C}} = A_{\mathbb{C}} \cdot N_{\mathbb{C}} = N_{\mathbb{C}} \cdot A_{\mathbb{C}}$. For $s \in S_{\mathbb{C}}$, $n \in N_{\mathbb{C}}$ let $s[n] \in N_{\mathbb{C}}$ be defined by $sn = s[n]a$, $a \in A_{\mathbb{C}}$; then $s, n \mapsto s[n]$ is an action of $S_{\mathbb{C}}$ on $N_{\mathbb{C}}$, and if, $s = n_0 a_0$, $s[n] = n_0 a_0 n a_0^{-1}$. Now $A_{\mathbb{C}}$ acts on $\mathfrak{n}_{\mathbb{C}}$ linearly while the group law is (on $N_{\mathbb{C}}$) a polynomial map of $\mathfrak{n}_{\mathbb{C}} \times \mathfrak{n}_{\mathbb{C}}$, into $\mathfrak{n}_{\mathbb{C}}$, mapping $\mathfrak{n} \times \mathfrak{n}$ to $\mathfrak{n}$. So there is an integer $d \geqslant 1$ such that, for any linear function g on $\mathfrak{n}$, all the functions $u \mapsto g(s[\exp u])(u \in \mathfrak{n})$ are polynomials of degree $\leqslant d$. Hence there is a finite-dimensional space F of *real* polynomials on $\mathfrak{n}$ (of degree $\leqslant d$) containing the constants and all real linear functions, which is stable under S. Let 1, $Q_1, \ldots, Q_m$ be a basis for F where Q_j vanish at 0 such that a basis for $\mathfrak{n}^*$ occurs among the Qs, say $Q_1, \ldots, Q_r$. Let $|\cdot|$ be a norm on $\mathfrak{a}_{\mathbb{C}} \oplus \mathfrak{n}_{\mathbb{C}}$. We may clearly assume that, for $u \in \mathfrak{n}$, $|u|^2 = Q_1(u)^2 + \cdots + Q_r(u)^2$. For $u \in \mathfrak{n}$, $v \in \mathfrak{a}$, let $s(u, v) = \exp(u) \cdot \exp(v)$. Then

$$d_l s = ds = e^{\mu(v)}\, du\, dv \quad (\mu \in \mathfrak{a}^*)$$

$$\| \pi(s) \| \leqslant C e^{c(|u| + |v|)} \quad (C, c > 0 \text{ constants})$$

We define

$$F(s) = \sum_j Q_j(u)^2 + |v|^2$$

$$f_n(s) = c_n e^{-nF(s)}$$

where $c_n > 0$ is such that $\int_S f_n\, ds = 1$. Now we have the estimate

$$\sum Q_j(u)^2 \leqslant C(|u|^2 + |u|^4 + \cdots + |u|^{2d})$$

for all $u \in \mathfrak{n}$, for some constants $C > 0$. Hence if

$$F^*(u) = C(|u|^2 + |u|^4 + \cdots + |u|^{2d})$$

then

$$n \sum_j Q_j(u)^2 \leqslant nF^*(u) \leqslant F^*(\sqrt{n} \cdot u)$$

Hence, as $|\mu(v)| \leqslant |\mu|\,|v| \leqslant |\mu|\,|\sqrt{n} \cdot v|$

$$\int f_n\, ds \geqslant c_n \int e^{-F^*(\sqrt{n} \cdot u) - |\sqrt{n} \cdot v|^2 - |\mu||\sqrt{n} \cdot v|}\, du\, dv$$

$$\geqslant n^{-(p+q)/2} c_n \int e^{-F^*(u') - |v'|^2 - |\mu|\,|v'|}\, du'\, dv'$$

where $p = \dim(\mathfrak{a})$, $q = \dim(\mathfrak{n})$. Hence

$$c_n = O(n^{(p+q)/2})$$

On the other hand, as

$$e^{-(nx^2/4)+bx} \quad (x \geqslant 0)$$

reaches its maximum at $x = 2b/n$, the maximum is $e^{b^2/n} \leqslant e^{b^2}$ for $n \geqslant 1$. Hence for any $\delta > 0$, $b > 0$, as $F(s) \geqslant |u|^2 + |v|^2$, we have

$$\begin{aligned}\int_{|u|^2+|v|^2 \geqslant \delta^2} & f_n(s) e^{b(|u|+|v|)} \, du \, dv \\ & \leqslant e^{-n\delta^2/4} c_n \int e^{-(n/2)(|u|^2+|v|^2)} e^{-(n/4)(|u|^2+|v|^2)+b(|u|+|v|)} \, du \, dv \\ & \leqslant \text{const.}\, c_n n^{-(p+q)/2} e^{-n\delta^2/4} \to 0\end{aligned}$$

when $n \to \infty$.

Finally, to prove that the f_n are analytic vectors in L^1_π it is enough to prove the following: if $\beta > 0$ is sufficiently small, then for all $u_0 \in \mathfrak{n}_\mathbb{C}$, $v_0 \in \mathfrak{n}_\mathbb{C}$, with $|u_0| \leqslant \beta$, $|v_0| \leqslant \beta$, we have

$$\operatorname{Re} F(s_0 s) \geqslant \tfrac{1}{2}(|u|^2 + |v|^2) - 1$$

for all $u \in \mathfrak{n}$, $v \in \mathfrak{n}$; here $s_0 = s_0(u_0, v_0)$, $s = s(u, v)$. If we write $Q'_j(s) = Q_j(s_0 s)$ this comes down to showing, for β small enough,

$$\operatorname{Re} \sum Q'^2_j \geqslant \tfrac{1}{2}|u|^2 \tag{*}$$

Since the Q_j(and 1) span the space F stable under S,

$$Q'_j = \sum_i a_{ij}(s_0) Q_i + b_j(s_0)$$

where $b_j(s_0) \to 0$ and $a_{ij}(s_0) \to \delta_{ij}$ as $s_0 \to 1$. So (*) follows by continuity, in the range $0 \leqslant |u|^2 \leqslant 1$, if β is small enough. On the other hand,

$$\left|\sum Q'^2_j - \sum Q^2_j\right| \leqslant \delta_1 \sum Q^2_i + \delta_2 \sum |Q_i| + \delta_3$$

where δ_1, δ_2, δ_3 depend only on s_0 and $\to 0$ when $s_0 \to 1$. Hence, for $\beta > 0$ small enough,

$$\left|\sum Q'^2_j - \sum Q^2_j\right| < \tfrac{1}{2} \sum Q^2_j$$

if $\sum Q^2_j \geqslant 1$, in particular if $|u|^2 \geqslant 1$, as $\sum Q^2_j \geqslant |u|^2$. Hence, for any $b > 0$, and $\beta > 0$ sufficiently small,

$$|e^{-bF(s_0 s)}| \leqslant e^{-b(|u|^2+|v|^2)/2+1}$$

for all $s_0 = s(u_0, v_0)$, $u_0 \in \mathfrak{n}_\mathbb{C}$, $v_0 \in \mathfrak{a}_\mathbb{C}$, $|u_0| \leqslant \beta$, $|v_0| \leqslant \beta$, $s \in S$. This completes the proof of the theorem.

Finiteness theorem. We sketch the main idea of the proof which already appears in the proof for the special case $G = SL(2, \mathbb{R})$. We follow the notation of Lemma 11 and Proposition 12 and prove that, for $V = U/(U \cdot M)$, the $V_\xi(\xi \in \hat{\mathfrak{k}})$ are finite $\mathfrak{Z}$-modules; here $U = U(\mathfrak{g})$. Let $\mathfrak{g} = \mathfrak{k} \oplus \mathfrak{p}$ where $\mathfrak{p}$ is $\mathfrak{k}$-stable. Since $U(\mathfrak{k})/M \cong W \subsetneq V$ we may identify $U(\mathfrak{k})/M$ with its image in V. Let $a \mapsto \bar{a}$ be the natural map of U onto V. If $\mathfrak{P}$ is the symmetric algebra over $\mathfrak{p}_{\mathbb{C}}$ and λ is the symmetrizer map $\mathfrak{P} \to U$, the map

$$a \otimes b \mapsto \lambda(a)b$$

extends to a *linear isomorphism* of $\mathfrak{P} \otimes U(\mathfrak{k})$ with U that carries $\mathfrak{P} \otimes M$ to $U \cdot M$. Now $\mathfrak{k}$ acts on $\mathfrak{p}$, hence on $\mathfrak{P}$, by the adjoint representation, and on $U(\mathfrak{k})$ by left multiplication (giving rise to the action on $U(\mathfrak{k})/M$). This action goes over to the action of $\mathfrak{k}$ on U by left multiplication; indeed, for $X \in \mathfrak{k}$, $X \cdot (a \otimes b)$ goes to $(X\lambda(a) - \lambda(a)X)b + \lambda(a)Xb$. We thus have a linear isomorphism

$$\gamma : \mathfrak{P} \otimes (U(\mathfrak{k})/M) \to V = U/(U \cdot M)$$

commuting with $\mathfrak{k}$; $a \otimes \bar{b}$ goes over to $\lambda(a)\bar{b} = \overline{\lambda(a)b}$. In particular $\gamma((\mathfrak{P} \otimes (U(\mathfrak{k})/M))_\xi) = V_\xi$. Note also that $\mathfrak{P} \otimes (U(\mathfrak{k})/M)$ is a $\mathfrak{P}$-module by the action a', $a \otimes \bar{b} \mapsto a'a \otimes \bar{b}$.

Let Ω be the algebra of $\mathfrak{k}$-invariants of $\mathfrak{P}$ (i.e., elements killed by $\mathfrak{k}$) and J the algebra of $\mathfrak{g}$-invariants of $S(\mathfrak{g})$. Let $a \mapsto a_\mathfrak{p}$ be the 'restriction' map of $S(\mathfrak{g})$ onto $\mathfrak{P}$; $a_\mathfrak{p}$ is the element of $\mathfrak{P}$ such that $a - a_\mathfrak{p}$ is in the ideal $S(\mathfrak{g}) \cdot \mathfrak{k}$. We write $J_\mathfrak{p}$ for the image of J under this; obviously $J_\mathfrak{p} \subset \Omega$.

We regard $\mathfrak{P}$ as a graded algebra in the usual way with homogeneous spaces $\mathfrak{P}_d$; then $\mathfrak{P} \otimes (U(\mathfrak{k})/M)$ may be graded with $\mathfrak{P}_d \otimes (U(\mathfrak{k})/M)$ as the homogeneous spaces. The action of $\mathfrak{k}$ respects the gradings and hence the decomposition with respect to ξ also respects the gradings. The key step of the argument is then the following: if $j \in S(\mathfrak{g})$ is homogeneous of degree d, and $A \in \mathfrak{P}_{d'} \otimes (U(\mathfrak{k})/M)$, then

$$\gamma(j_\mathfrak{p} \cdot A) = \lambda(j) \cdot \gamma(A) \mod \gamma\left(\sum_{0 \leqslant e < d + d'} \mathfrak{P}_e \otimes (U(\mathfrak{k})/M) \right)$$

This is quite simple to prove. If $j \in J$, $j_\mathfrak{p} \in J_\mathfrak{p}$, and so the above observation leads to the following conclusion: if $(\mathfrak{P} \otimes (U(\mathfrak{k})/M))_\xi$ is a finite $J_\mathfrak{p}$-module, V_ξ is a finite $\mathfrak{Z}$-module. On the other hand, *by classical invariant theory*, $(\mathfrak{P} \otimes (U(\mathfrak{k})/M))_\xi$ is a finite Ω-module. So the proof would be complete *if we prove that Ω is a finite $J_\mathfrak{p}$-module.*

This last fact is proved as follows. Extend $\mathfrak{a}_{\mathbb{C}} (\subset \mathfrak{p}_{\mathbb{C}})$ to a Cartan subalgebra $\mathfrak{h}_{\mathbb{C}}$ of $\mathfrak{g}_{\mathbb{C}}$. Then J restricts to $\mathfrak{h}_{\mathbb{C}}$ isomorphically onto the Weyl group invariants. Similarly Ω restricts isomorphically to $\mathfrak{a}_{\mathbb{C}}$ (onto the 'little' Weyl group invariants). This leads to the inclusions

$$J_{\mathfrak{a}_{\mathbb{C}}} \subset \Omega_{\mathfrak{a}_{\mathbb{C}}} \subset S(\mathfrak{a}_{\mathbb{C}}), \quad J_{\mathfrak{h}_{\mathbb{C}}} = S(\mathfrak{h}_{\mathbb{C}})^{\text{inv}}$$

$J_{\mathfrak{h}_{\mathbb{C}}}$, being finitely generated, is noetherian, and $S(\mathfrak{h}_{\mathbb{C}})$ is integral over $J_{\mathfrak{h}_{\mathbb{C}}}$ Hence $J_{\mathfrak{a}_{\mathbb{C}}}$ is noetherian and $S(\mathfrak{a}_{\mathbb{C}})$ is integral over it. So $S(\mathfrak{a}_{\mathbb{C}})$ is a finite $J_{\mathfrak{a}_{\mathbb{C}}}$-module, showing that $\Omega_{\mathfrak{a}_{\mathbb{C}}}$ is a finite $J_{\mathfrak{a}_{\mathbb{C}}}$-module.

$SL(2, \mathbb{C})$

From the Lie algebra point of view, it is a question of studying ($\tilde{\mathfrak{g}}$, $\mathfrak{g}$)-modules where $\mathfrak{g}$ is a complex semisimple Lie algebra, $\tilde{\mathfrak{g}} = \mathfrak{g} \times \mathfrak{g}$ with $\mathfrak{g}$ imbedded diagonally in $\tilde{\mathfrak{g}}$ (Here $\mathfrak{g} = \mathfrak{sl}(2, \mathbb{C})$). The theory can therefore be developed in great depth. I shall not spend time on it here, but just confine myself to a brief discussion of the supplementary series of representations.

The idea of Gel'fand and Naimark as well as Bargmann, who first discovered the existence of the supplementary series, is very simple and beautiful. The principal series acts on a space of functions of z and the idea is to choose a scalar product that will be invariant for this action. The scalar product will depend on some parameters and *when these are appropriately restricted one obtains a positive-definite scalar product.*

Let us choose (for $G = SL(2, \mathbb{C})$) the scalar product in the form

$$(f_1 | f_2) = \iint k(z_1, z_2) f_1(z_1) \overline{f_2(z_2)} \, dx_1 \, dy_1 \, dx_2 \, dy_2$$

Here k is a kernel which should be thought of as a generalized function, i.e., a distribution on $\mathbb{C} \times \mathbb{C}$; the usual scalar product is obtained when $k(z_1, z_2) = \delta(z_1 - z_2)$. The action of G is by

$$f(z) \mapsto f\left(\frac{az + c}{bz + d}\right)(bz + d)^{-m} |bz + d|^{m - \rho - 2}$$

where $m \in \mathbb{Z}$ and $\rho \in \mathbb{C}$. In particular, the element

$$\begin{pmatrix} 1 & c \\ 0 & 1 \end{pmatrix}$$

acts by translation: $f(z) \mapsto f(z + c)$, and so k has to be a function of $z_1 - z_2$ only. Let us now write the condition of invariance under

$$\begin{pmatrix} 0 & 1 \\ -1 & 0 \end{pmatrix}$$

which acts by

$$f(z) \mapsto f\left(-\frac{1}{z}\right) z^{-m} |z|^{m - \rho - 2}$$

It is easily calculated to be

$$k(z_1 - z_2) = k\left(\frac{z_1 - z_2}{z_1 z_2}\right) z_1^m z_2^m |z_1|^{-m-\rho-2} |z_2|^{-m+\bar{\rho}-2}$$

The invariance under

$$\begin{pmatrix} a & 0 \\ 0 & a^{-1} \end{pmatrix}$$

which acts by

$$f(z) \mapsto f(a^2 z) a^m |a|^{-m+\rho+2}$$

leads to

$$k(z_1 - z_2) = k(a^{-2}(z_1 - z_2)) |a|^{2(\mathrm{Re}\rho - 2)}$$

This gives

$$k(t) = \text{const.} \, |t|^{\mathrm{Re}\rho - 2}$$

Substituting in the previous relation we get

$$z_1^{m-\mathrm{Re}\rho+2} |z_1|^{-m+\rho-2} z_2^{m-\mathrm{Re}\rho+2} |z|^{-m+\bar{\rho}-2} = 1$$

This is possible only if ρ is real and $m = 0$. Thus

$$k(z_1, z_2) = \text{const.} \, |z_1 - z_2|^{\rho-2} \quad (\rho \in \mathbb{R}).$$

We shall now examine more carefully the scalar product

$$(f_1 | f_2) = \int\!\!\int |z_1 - z_2|^{\rho-2} f_1(z_1) \overline{f_2(z_2)} \, \mathrm{d}x_1 \, \mathrm{d}y_1 \, \mathrm{d}x_2 \, \mathrm{d}y_2$$

where $0 < \rho < 2$, and $f_1, f_2 \in C_c^\infty(\mathbb{C})$. For $f \in C_c^\infty(\mathbb{C})$ we introduce the Fourier transform

$$\hat{f}(w) = \frac{1}{2\pi} \int\!\!\int_{\mathbb{C}} f(z) \mathrm{e}^{-\mathrm{i}\mathrm{Re}(\bar{z}w)} \, \mathrm{d}x \, \mathrm{d}y$$

We rewrite $(f | f)(f \in C_c^\infty(\mathbb{C}))$ as

$$(f | f) = \int |z|^{\rho-2} F(z) \, \mathrm{d}x \, \mathrm{d}y$$

where

$$F(z) = \int f(z_1) \overline{f(z + z_1)} \, \mathrm{d}x_1 \, \mathrm{d}y_1$$

Now $F \in C_c^\infty(\mathbb{C})$ while $|z|^{\rho-2} \, \mathrm{d}x \, \mathrm{d}y$ is a tempered distribution whose Fourier transform is classically known to be

$$2^\rho \pi \frac{\Gamma(\rho/2)}{\Gamma(1 - (\rho/2))} |w|^{-\rho}$$

More precisely, this function is locally integrable and the Fourier transform in question is the distribution defined by it. Hence

$$(f\,|\,f) = 2^{\rho}\pi \frac{\Gamma(\rho/2)}{\Gamma(1-\rho/2)} \int |w|^{-\rho} \hat{F}(-w)\, du\, dv$$

But $F^0 = f * f^0$ $(h^0(z) = h(-z))$ so that $\hat{F^0} = |\hat{f}|^2$. Thus

$$(f\,|\,f) = 2\rho\pi \frac{\Gamma(\rho/2)}{\Gamma(1-\rho/2)} \int |w|^{-\rho} |\hat{f}(w)|^2\, du\, dv$$

for $f \in C_c^{\infty}(\mathbb{C})$, $0 < \rho < 2$. The positivity of the scalar product is beautifully manifest in this formula. The complete Hilbert space is then the Hilbert space of equivalence classes of measurable functions f on $\mathbb{C}$ such that

$$\|f\|^2 = \iint |z_1 - z_2|^{\rho-2} f(z_1)\overline{f(z_2)}\, dx_1\, dy_1\, dx_2\, dy_2 < \infty$$

and G acts on it as follows: $g = \begin{pmatrix} a & b \\ c & d \end{pmatrix} \in SL(2, \mathbb{C})$,

$$L_{\chi_{0,\rho}}(g){:}\, f(z) \mapsto f\left(\frac{az+c}{bz+d}\right)|bz+d|^{-\rho-2} \quad (0 < \rho < 2)$$

I leave it to the reader to finish off the remaining details of this construction. The unitary representations thus obtained are the so-called *supplementary series of* $SL(2, \mathbb{C})$.

The same method works for SL $(2, \mathbb{R})$; and, interestingly, it was exactly the one used by Bargmann [Ba] for constructing the supplementary series. The functions are now defined on $\mathbb{R}$ and the scalar product is

$$(f_1\,|\,f_2) = \iint_{\mathbb{R}\times\mathbb{R}} |x_1 - x_2|^{s-1} f_1(x_1) f_2(x_2)\, dx_1\, dx_2 \quad (0 < s < 1)$$

The representation is

$$L_{\chi_{0,s}}(g){:}\, f(x) \mapsto f\left(\frac{ax+c}{bx+d}\right)|bx+d|^{-s-1} \qquad g = \begin{pmatrix} a & b \\ c & d \end{pmatrix} \in SL(2, \mathbb{R})$$

Gel'fand and Naimark extended the construction of the supplementary series to $SL(n, \mathbb{C})$. For $G = SL(2, \mathbb{C})$, they proved that the principal series, supplementary series and trivial representations exhaust $\hat{G}$. For $G = SL(2, \mathbb{R})$ Bargmann proved that the principal, discrete and supplementary series and trivial representations exhaust $\hat{G}$.

It is a fundamental question to understand the structure of the category of Harish-Chandra modules. For results in the case of interest to us, namely $G = SL(2, \mathbb{C})$, see [GP].

Irreducibility of the principal series. I shall now give the classical proof of Gel'fand and Naimark [GN] of the irreducibility of the unitary principal series. I have of course modified it so as to use Mackey's theory of unitary (induced) representations of semidirect products. It is actually simpler to work with $G_n = GL(n, F)$; here $F = \mathbb{R}$, or $\mathbb{C}$, and the reader will see that the proof works over any local field. B_n is the Borel subgroup of upper-triangular matrices of G_n. As usual characters (quasi characters) of the diagonal subgroup of G_n may be viewed as one-dimensional characters of B_n. The family (L_{χ_n}),

$$L_{\chi_n} = \mathrm{Ind}_{B_n}^{G_n} \chi_n \quad (G_n = GL(n, F))$$

where χ_n runs through the characters of the diagonal group, is the *unitary principal series* of G_n. Everything done in Chapter 4 for $\mathrm{SL}(n, F)$ goes through essentially without change for G_n. We write Q_n for the parabolic subgroup of matrices of the form

$$\begin{pmatrix} & & 0 \\ & * & \vdots \\ & & 0 \\ * & * & * \end{pmatrix} \quad (n = (n-1) + 1)$$

Theorem 37. All representations of the unitary principal series of G_n are already irreducible when restricted to Q_n.

We shall prove this by induction on n. For $n = 1$ there is nothing to prove. So we assume $n \geqslant 2$ and the validity of the theorem for G_{n-1}. The argument is built on the three following lemmas from the general theory of induced representations [Ma3].

Lemma 38. Let H be a locally compact separable group; H_1, H_2, two closed subgroups such that $H \backslash (H_2 \cdot H_1)$ has measure 0. Then, for any unitary representation σ of H_1.

$$\mathrm{Res}_{H_2}^{H}(\mathrm{Ind}_{H_1}^{H} \sigma) = \mathrm{Ind}_{H_1 \cap H_2}^{H_2} \sigma_{12}$$

where $\sigma_{12} = \sigma|_{H_1 \cap H_2}$, and $\mathrm{Res}_{H_2}^{H}$ denotes restriction from H to H_2.

Proof. This is immediate if we write $\mathrm{Ind}_{H_1}^{H} \sigma$ in the cocycle form, using a quasi-invariant measure μ on $X = H/H_1$; $\pi = \mathrm{Ind}_{H_1}^{H} \sigma$ acts on the Hilbert space $L^2(X : \mathscr{H}(\sigma) : \mu)(\mathscr{H}(\sigma) = \text{space of } \sigma)$ by

$$(\pi(h)f)(x) = \rho(h : h^{-1}[x])^{1/2} \quad c(h : h^{-1}[x]) f(h^{-1}[x])$$

where $\rho(\cdot:\cdot)$ are appropriate Radon–Nikodym derivatives, c the cocycle defined by σ, so that $c(h_1:x_0)=\sigma(h_1)(h_1\in H_1$, x_0 is the coset $H_1)$. Since $H\backslash(H_2\cdot H_1)$ has measure zero, $\mu(X\backslash(H_2\cdot x_0))=0$, and so we may work on $H_2\cdot x_0$ instead of X. The above formula makes sense for $h\in H_2$ and defines the induced representation $\mathrm{Ind}_{H_1\cap H_2}^{H_2}\sigma_{12}$, since $H_1\cap H_2$ is the stabilizer of x_0 in H_2.

For the next lemma we assume that $H=U\cdot S$ where U, S are closed subgroups, $U\cap S=(1)$, U normal in H and abelian. Let du be a Haar measure on U, $d\hat{u}$ the dual Haar measure on $\hat{U}$, and $j(S\to\mathbb{R}_+^\times)$ the positive quasicharacter such that $\mathrm{d}(sus^{-1})=j(s)\,du$ for $s\in S$. We denote by $s,\hat{u}\mapsto s[\hat{u}]$ the dual action of S on $\hat{U}$; $\langle u, s[\hat{u}]\rangle=\langle s^{-1}us, \hat{u}\rangle$. For $h\in H$, $s(h)\in S$ and $u(h)\in U$ are defined by $h=u(h)s(h)$.

Lemma 39. *Suppose σ is a unitary representation of S in a (separable) Hilbert space $\mathscr{K}$. If $\pi=\mathrm{Ind}_S^H\sigma$, then π may be realized as follows: π acts on $\mathscr{H}$, the Hilbert space $L^2(\hat{U}:\mathscr{K})$ of $\mathscr{K}$-valued square-integrable (equivalence classes of) functions on $\hat{U}$; for $h\in H$, $f\in\mathscr{H}$,*

$$(\pi(h)f)(\hat{u})=j(s(h))^{1/2}\langle u(h),\hat{u}\rangle\sigma(s(h))f(s(h)^{-1}[\hat{u}])$$

Proof. First we describe the standard model π' for π. We identify H/S with U so that H acts on U by $h[u]=u(h)s(h)us(h)^{-1}$, $u\in U$. The map $(h,u)\mapsto\sigma(s(h))$ is a cocycle with values in the unitary group of $\mathscr{K}$, giving rise to the representation σ of S when $h\in S$, $u=1$. So π' acts on $L^2(U:\mathscr{K})$ by

$$(\pi'(h)f)(u)=j(s(h))^{-1/2}\sigma(s(h))f(h^{-1}[u])$$

We now introduce the Fourier transform $f\mapsto\hat{f}$ which is the unique unitary isomorphism from $L^2(U:\mathscr{K})$ to $L^2(\hat{U}:\mathscr{K})$ such that, for all $f\in L^1(U:\mathscr{K})\cap L^2(U:\mathscr{K})$,

$$\hat{f}(\hat{u})=\int_U\langle u,\hat{u}\rangle f(u)\,du$$

The required form of π is now obtained if we transfer π' to $L^2(\hat{U}:\mathscr{K})$ by the Fourier transform.

Lemma 40. *Let assumptions be as in the preceding lemma. Suppose there is a $\hat{u}_0\in\hat{U}$ such that*

(i) *$\hat{U}\backslash S[\hat{u}_0]$ has measure 0,*
(ii) *the restriction of σ to the stabilizer $S_{\hat{u}_0}$ (of $\hat{u}_0$ in S) is irreducible.*

Then $\pi=\mathrm{Ind}_S^H\sigma$ is irreducible.

Proof. In view of (i), the representation π as described in the preceding lemma is actually acting on $L^2(S[\hat{u}_0]:\mathscr{H})$, and is the unitary representation of the semidirect product $U\cdot S$ that the Mackey theory associates with the orbit $S[\hat{u}_0]$ and the cocyle s, $\hat{u}\mapsto c(s:\hat{u})=\sigma(s)(s\in S,\ \hat{u}\in S[\hat{u}_0])$. Since $c(s:\hat{u}_0)=\sigma(s)$, $s\in S_{\hat{u}_0}$, assumption (ii) guarantees that this is irreducible.

Proof of theorem. Fix a character χ_n of the diagonal subgroup of G_n. We imbed G_{n-1} in G_n by

$$g\mapsto\begin{pmatrix} & & & 0\\ & g & & \vdots\\ & & & 0\\ 0 & \cdots & 0 & 1\end{pmatrix}$$

and write χ_{n-1} for the characters of B_{n-1} obtained by restriction of χ_n. By the induction hypothesis we know that $L_{\chi_{n-1}}|_{Q_{n-1}}$ is irreducible. Now Q_n has the Levy–Langlands decomposition $Q_n=U\cdot S$ where

$$S=\begin{pmatrix} & & & 0\\ & G_{n-1} & & \vdots\\ & & & 0\\ 0 & \cdots & 0 & G_1\end{pmatrix}\qquad U=\begin{pmatrix} & & & 0\\ & & & \vdots\\ & & & 0\\ * & \cdots & * & 1\end{pmatrix}$$

U is abelian and normal in Q_n; and if we identify S with $G_{n-1}\times G_1$ and U with F^{n-1} (space of row vectors), the action of S on U is given by

$$(g,c):x\mapsto cxg^{-1}\quad(c\in\mathbb{C}^{\times},g\in G_{n-1})$$

We identify $\hat{U}$ with F^{n-1} in such a way that $\xi=\begin{pmatrix}\xi_1\\ \vdots\\ \xi_{n-1}\end{pmatrix}$ corresponds to the character

$$(x_1,\dots,x_n)\mapsto\exp\mathrm{i}(\xi_1x_1+\cdots+\xi_{n-1}x_{n-1})$$

when $F=\mathbb{R}$, and to the character

$$(x_1,\dots,x_{n-1})\mapsto\exp\mathrm{i}\,\mathrm{Re}\,(\xi_1x_1+\cdots+\xi_{n-1}x_{n-1})$$

when $F=\mathbb{C}$. Then S acts on $\hat{U}$ by

$$(g,c):\xi\mapsto g\xi c^{-1}$$

Now it is not difficult to show that, if $\bar{N}$ is the group of lower-triangular matrices with 1's on the diagonal, $G_n\backslash\bar{N}B$ has measure zero; in fact it is a proper algebraic subvariety of G_n. Hence $G_n\backslash Q_nB$ has measure zero. So, by the first of the above lemmas,

$$L_{\chi_n}|_{Q_n}=\mathrm{Ind}_{Q_n\cap B_n}^{Q_n}\chi_n',\quad \chi_n'=\chi_n|_{Q_n\cap B_n}$$

But $Q_n \cap B_n \subset S$ and so, by the theorem of 'inducing in stages'

$$L_{\chi_n}|_{Q_n} = \mathrm{Ind}_S^{Q_n}(\mathrm{Ind}_{Q_n \cap B_n}^S \chi_n')$$

Let us write σ for $\mathrm{Ind}_{Q_n \cap B_n}^S \chi_n'$. We shall now apply the third lemma above to deduce the irreducibility of $L_{\chi_n}|_{Q_n}$. We must therefore check the two conditions of that lemma. The orbital condition is obvious: there are only two orbits, (0) and $F^{n-1}\backslash(0)$. If $\xi_0 = \begin{pmatrix} 0 \\ \vdots \\ 0 \\ 1 \end{pmatrix} \in F^{n-1}$, the stabilizer of ξ_0 is the subgroup of $S = G_{n-1} \times G_1$ of all (g, c) such that $g\xi_0 = c\xi_0$, namely the subgroup Q^* of all matrices

$$\begin{pmatrix} & & & 0 & 0 \\ & * & & & \\ & & & 0 & 0 \\ * & \cdots & * & c & 0 \\ 0 & & 0 & 0 & c \end{pmatrix}$$

We must therefore show that the restriction of σ to Q^* is irreducible. But $Q_n \cap B_n = B_{n-1} \times G_1$, $S = G_{n-1} \times G_1$, and $1 \times G_1$ is in the centre of S. Hence $\sigma = L_{\chi_{n-1}} \boxtimes \chi_{(n)}$ where $\chi_{(n)}$ is the restriction of χ_n to $1 \times G_1$. So $\sigma|_{Q_{n-1} \times G_1} = (L_{\chi_{n-1}}|_{Q_{n-1}}) \boxtimes \chi_{(n)}$, and consequently, by the induction hypothesis, $\sigma|_{Q_{n-1} \times G_1}$ is irreducible. However, it is obvious that $Q_{n-1} \times G_1$ is generated by Q^* and $1 \times G_1$, and hence $\sigma|_{Q^*}$ is irreducible. This completes the proof of the theorem.

We can now deduce

Theorem 41. *All representations of the unitary principal series of $SL(n, \mathbb{C})$ are irreducible when restricted to the subgroup $SQ_n = Q_n \cap SL(n, \mathbb{C})$. All representations of the unitary principal series of the group $G_n^{\pm}$ of $n \times n$ real matrices of determinant ± 1 are irreducible when restricted to $Q_n^{\pm} = G_n^{\pm} \cap Q_n$. For the unitary principal series of $SL(n, \mathbb{R})$, the representations are irreducible when n is odd (even if restricted to $Q_n \cap SL(n, \mathbb{R})$), and are either irreducible or split into two irreducible pieces, when n is even.*

Proof. Let Z_n be the group of scalar matrices of G_n. It is then trivial to check that $Q_n = (Q_n \cap SL(n, \mathbb{C})) \cdot Z_n$. Hence the representations L_{χ_n} above are still irreducible when restricted to $Q_n \cap SL(n, \mathbb{C})$. But $SL(n, \mathbb{C}) \cdot B_n = GL(n, \mathbb{C})$ and so, by the first of the above lemmas, $L_{\chi_n}|_{SL(n,\mathbb{C})}$ is the principal series representation of $SL(n, \mathbb{C})$ corresponding to the character $\chi_n|_{B_n \cap SL(n,\mathbb{C})}$. Since any character of the diagonal subgroup of $SL(n, \mathbb{C})$ is the restriction of a character χ_n, the first assertion is clear. When the ground field is $\mathbb{R}$, the above argument goes through for $G_n^{\pm}$ if n is arbitrary, and for $SL(n, \mathbb{R})$ if n is odd. The problem for even n is that the scalars all have determinants > 0 and hence $(Q_n \cap SL(n, \mathbb{R}))\cdot Z_n \neq Q_n$, although it is still true that $SL(n, \mathbb{R})\cdot B_n = GL(n, \mathbb{R})$ so that $L_{\chi_n}|_{SL(n, \mathbb{R})}$ is still the principal series representation defined by $\chi_n|_{B_n \cap SL(n, \mathbb{R})}$. We also know that when $n = 2$ the unitary-principal-series representation $\pi_{1,0}$ actually splits into two inequivalent irreducible pieces.

Suppose finally that n is even. It is easy to see that the principal-series representations define Harish-Chandra modules and so, by Theorem 19, they have irreducible subrepresentations. Let L_1 be a unitary-principal-series representation of $SL(n, \mathbb{R})$ and let us write it as the restriction to $SL(n, \mathbb{R})$ of a unitary-principal series representation L' of $G_n^{\pm}$. Now $[G_n^{\pm} : SL(n, \mathbb{R})] = 2$ and it is clear that we can choose $x \in G_n^{\pm}$ such that $x^2 = 1$ and $G_n = SL(n, \mathbb{R}) \coprod x SL(n, \mathbb{R})$. Let $\mathscr{H}'$ be the space on which L' acts, $\mathscr{H}_1$ an irreducible invariant subspace for L_1. Then $L'(x)\cdot\mathscr{H}_1$ is also invariant and irreducible under L_1, and $\mathscr{H}'' = \mathscr{H}_1 + L'(x)\cdot\mathscr{H}_1$ is stable under L'. Hence $\mathscr{H}''$ is dense in $\mathscr{H}'$. It is not difficult to deduce from this that $\mathscr{H}'$ splits into at most two irreducible pieces for L_1.

We have already remarked that the principal series for $SL(n, \mathbb{R})$ is generically irreducible. Indeed, let us write $\chi = (\xi, \eta)$ where $\xi \in \hat{M}$ and $\eta \in \hat{A}$ is the character

$$\operatorname{diag}(e^{t_1}, \ldots, e^{t_n}) \mapsto e^{(i\lambda_1 t_1 + \cdots + i\lambda_n t_n)}$$

where $\lambda_1, \ldots, \lambda_n \in \mathbb{R}$; then Bruhat [Br] proved that, *if the* λ_j *are distinct*, L_χ is irreducible. Of course for $SL(n, \mathbb{R})$ or $GL(n, \mathbb{R})$ the principal series is just *one* of the series contributing to the Plancherel formula; and the real question is the proof of the (generic) irreducibility of *all* the series.

Notes and comments

For a detailed treatment of $SL(2, F)(F = \mathbb{R}$ or $\mathbb{C})$ from the point of view of distributions see [GGV].

Problems

1. Explicitly determine a set of (two) generators for the centre $\mathfrak{z}$ of $U(\mathfrak{g})$ where $\mathfrak{g} = \mathfrak{sl}(3, \mathbb{C})$.
2. Prove that if G is a connected simply connected Lie group and $G_{\mathbb{R}}$ its underlying real group, the finite-dimensional irreducible representations L of $G_{\mathbb{R}}$ are precisely those of the form $L_1 \otimes \bar{L}_2$ where L_1 and L_2 are *holomorphic* irreducibles of G, and that the correspondence $L \mapsto (L_1, L_2)$ is bijective up to isomorphism.

3. Prove the equivalence of the weak and strong analyticity for Banach-space-valued functions on any analytic manifold.
4. Prove the equivalence of weak and strong differentiability for Fréchet-space-valued functions on any smooth manifold.
5. Try to carry out for $SL(2, \mathbb{C})$ both the classification of irreducible Harish-Chandra modules as well as the splitting of the principal series (use the models in [GGV]).
6. Prove, using Theorem 14 of this chapter, that, for a *complex* semisimple group, an irreducible *spherical* Harish-Chandra module is determined up to equivalence by its infinitesimal character.
7. Let L be an irreducible unitary representation of $G = SL(2, \mathbb{R})$ with infinitesimal character χ. Prove that $\chi(\Omega)$ is real. Hence deduce that $\pi_{\varepsilon,\lambda}$ is not unitarizable if λ is not in $\mathbb{R} \cup i\mathbb{R}$.
8. Let $\mathfrak{g} = \mathfrak{sl}(2, \mathbb{C})$ and P the algebra of polynomials on $\mathfrak{g}$ regarded as a module $G = SL(2, \mathbb{C})$. Let P^m be the subspace of elements of P transforming according to the irreducible holomorphic representations of G of dimension m. Prove that $P^m = 0$ for even m and P^m is a free I-module of rank m for odd m, I being the algebra of invariants in P.
9. For $G = SL(n, F)$ prove that the representations L_χ and $L_{\chi^{-1}}$ are in duality for any quasicharacter χ, the G-invariant bilinear form being

$$(f, g) \mapsto \int_K fg \, dk$$

10. Prove the statements on representations of finite type made on p. 117, just before Theorem 13.

6

The Plancherel formula: character form

6.1 Plancherel formula for complex groups

From a very general point of view the Plancherel formula may be viewed as a completeness theorem for L^2 harmonic analysis. Let G be a locally compact unimodular separable group and $\hat{G}$ its unitary dual. Then the Plancherel formula is a relation of the form

$$\int_G |f(x)|^2 \, dx = \int_{\hat{G}} \operatorname{tr}_\omega(\tilde{f} * f) \, d\mu(\omega)$$

for a dense (in $L^2(G)$) convolution algebra of functions f in $C_c(G)$; here $\tilde{f}(x) = f(x^{-1})^{\mathrm{conj}}$, tr_ω denotes the trace calculated for $\omega \in \hat{G}$, i.e.,

$$\operatorname{tr}_\omega(\tilde{f} * f) = \operatorname{tr}(\pi(f)^\dagger \pi(f)) \quad (\pi \in \omega)$$

and μ is a measure on $\hat{G}$. Since

$$\int_G |f|^2 \, dx = (\tilde{f} * f)(1)$$

this can also be written as

$$g(1) = \int_{\hat{G}} \operatorname{tr}_\omega(g) \, d\mu(\omega) \quad (g = \tilde{f} * f)$$

In this form, it may be viewed as the relation

$$\delta = \int_{\hat{G}} \operatorname{tr}_\omega \, d\mu(\omega)$$

which expresses the Dirac delta measure at the origin as a linear combination of the traces of the irreducible unitary representations.

If G is a connected semisimple Lie group with finite centre, its irreducible unitary representations have characters, and the representations themselves are very explicitly parametrized. Let $\mathscr{D}(G)$ be the space of distributions on G and $\mathscr{D}_0(G)$ the subspace of those which are invariant eigendistributions. If π is an irreducible unitary representation, its character lies in $\mathscr{D}_0(G)$; we shall write $\mathrm{Ch}(G)$ for the set of all elements of $\mathscr{D}_0(G)$ obtained as π varies, and think of it as $\hat{G}$. Let

$$A_1, \ldots, A_r$$

be a complete set of mutually nonconjugate Cartan subgroups of G. Then we have explained in Chapter 4 how Harish-Chandra was able to associate, for any $i = 1, 2, \ldots, r$, a family of unitary representations parametrized by the characters of A_i (the 'series' attached to A_i), generically irreducible, in terms of which the regular representation may be explicitly decomposed. This process may therefore be viewed as the construction of a map

$$\coprod_{1 \leqslant i \leqslant r} \hat{A}_i \to \mathscr{D}_0(G)$$

with finite fibres such that, for each i, for points in a dense open subset of $\hat{A}_i$ whose complement in $\hat{A}_i$ has measure zero, their images are in $\mathrm{Ch}(G)$. For $\chi \in \hat{A}_i$ let Θ_χ denote its image in $\mathscr{D}_0(G)$; then the Plancherel formula may be regarded as the explicit construction of positive measures μ_i on the $\hat{A}_i$ such that

$$\delta = \sum_{1 \leqslant i \leqslant r} \int_{\hat{A}_i} \Theta_\chi \, \mathrm{d}\mu_i(\chi)$$

It will turn out that the μ_i are absolutely continuous with respect to the Haar measure on A_i,

$$\mathrm{d}\mu_i(\chi) = \mu_i(\chi) \, d_i\chi$$

so that this relation becomes

$$f(1) = \sum_{1 \leqslant i \leqslant r} \int_{\hat{A}_i} \Theta_\chi(f)\mu_i(\chi)\mathrm{d}_i\chi \quad (f \in C_c^\infty(G))$$

Given $f = C_c^\infty(G)$ let us write

$$\hat{f}_i(\chi) = \Theta_\chi(f) \quad (\chi \in \hat{A}_i)$$

By analogy with what happens in the classical case one may view the vector

$$(\hat{f}_1, \ldots, \hat{f}_r) = \hat{f}$$

as a 'Fourier transform' of f, actually an 'invariant Fourier transform'. Then the Plancherel formula may be viewed as the 'Fourier inversion formula'

$$f(1) = \sum_{1 \leqslant i \leqslant r} \int_{\hat{A}_i} \hat{f}_i(\chi)\mu_i(\chi)\mathrm{d}_i\chi \quad (f \in C_c^\infty(G))$$

The simplest case of all this is when $r = 1$. This is the case for example if G is a complex group. Let us write $H = A_1$. Then Θ_χ is the character of the principal series representation which is always irreducible, so that we actually have a map

$$\hat{H} \to \mathrm{Ch}(G)$$

and the fibres are precisely the orbits for the Weyl group. Then

$$f(1) = \int_{\hat{H}} \hat{f}(\chi)\mu(\chi) \, \mathrm{d}\chi$$

where μ is a Weyl-group-invariant nonnegative continuous function. An explicit formula for μ, together with the explicit formula for the character Θ_χ, would then be a complete and far-reaching generalization of the compact theory; *actually the formula is essentially the same as in the compact case*! Indeed, if $G = SL(n, \mathbb{C})$, $H =$ diagonal subgroup, and

$$\chi_{m_1,\ldots,m_n:i\rho_1,\ldots,i\rho_n}(\operatorname{diag}(a_1,\ldots,a_n)) = \prod_{1\leqslant j\leqslant n} (a_j/|a_j|)^{m_j}|a_j|^{i\rho_j}$$

then, for a suitable choice of the Haar measure on G,

$$\mu(\chi_{m_1,\ldots,m_n:i\rho_1,\ldots,i\rho_n}) = \prod_{r<s} [(m_r - m_s)^2 + (\rho_r - \rho_s)^2]$$

The corresponding formula for $U = SU(n)$, $T =$ diagonal subgroup, is

$$\mu(\chi_{m_1,\ldots,m_n}) = \prod_{r<s} |m_r - m_s|$$

$\chi_{m_1,\ldots,m_n}$ being the character

$$\operatorname{diag}(a_1,\ldots,a_n) \mapsto a_1^{m_1}\cdots a_n^{m_n}$$

The 'squaring' that appears when we go from $SU(n)$ to $SL(n, \mathbb{C})$ is due to the fact that we are working with the real group underlying $SL(n, \mathbb{C})$.

Gel'fand and Naimark proved such a Plancherel formula when $G = SL(n, \mathbb{C})$ [GN]. Their method was the remarkable and original one of applying Fourier transform theory on H to the limit formula for orbital integrals on G. If we write, for any $f \in C_c^\infty(G)$, F_f for the suitably normalized integral of f on the regular conjugacy classes of G, then the essence of this method consists in establishing the following results:

(i) F_f defines an element of $C_c^\infty(H)$;
(ii) the Fourier transform of F_f, viewed as a function on $\hat{H}$, is just the character of the principal series, i.e, $\hat{F}_f(\chi) = \Theta_\chi(f) = \hat{f}(\chi)$;
(iii) there is a differential operator ϖ on H, invariant under translations, such that, for all $f, f(1) = (\partial(\varpi)F_f)(1)$.

The operator $\partial(\varpi)$ is explicitly computable, and is given by the product of positive roots, not of the original complex Lie algebra, but of the complexification of its underlying real Lie algebra. This is the *limit formula*; and by applying Fourier transform to (iii) one obtains, in view of (ii), the Plancherel formula; the function μ is the polynomial (symbol) which is the *Fourier transform of* $\partial(\varpi)$. If $\partial(\varpi_0)$ is the *holomorphic* differential operator on H corresponding to the product of positive roots of the *complex Lie algebra*, $\partial(\varpi) = \partial(\varpi_0)\partial(\varpi_0^{\text{conj}})$, so that

$$\mu = |\mu_0|^2, \quad \mu_0 = \text{complex symbol of } \partial(\varpi_0)$$

The proof of the limit formula which Gel'fand and Naimark gave turned out to be extremely complicated. Harish-Chandra introduced in [H9]

[H10] a very simple and beautiful way to do this. Let me explain briefly his method. The exponential map from the Lie algebra to the Lie group is very well behaved near the identity of G and carries over the conjugacy classes of $\mathfrak{g}$ to those in G. So the orbital integral F_f may be transferred to the Lie algebra and viewed as the orbital integral ψ_f on the Lie algebra. The limit formula for F_f then becomes the limit formula for ψ_f:

$$f(0) = (\partial(\varpi)\psi_f)(0)$$

But now the analogue for ψ_f of the formula $\hat{F}_f = \Theta(f)$ is the following remarkable one discovered by Harish-Chandra:

$$\psi_{\hat{f}} = \text{const.}\, \hat{\psi}_f$$

where $\hat{f}$ is the Fourier transform of f on $\mathfrak{g}$! One can then follow Harish-Chandra and apply the Fourier transform to both sides and reduce the limit formula to the relation

$$\int_{\mathfrak{g}} \hat{f}\,\mathrm{d}X = \text{const.} \int_{\mathfrak{h}} |\mu_0|^2 \psi_{\hat{f}}\,\mathrm{d}\mathfrak{h}$$

which is just the Weyl integration formula!

It is impossible to overemphasize the importance of Harish-Chandra's method. Although its extension to real groups turned out to be extremely difficult, it firmly established the astounding fact that *the character theory on the group corresponds under the exponential map to the Fourier transform theory on the Lie algebra.* This was to be the guiding principle for all subsequent work of Harish-Chandra, especially his construction of the discrete series.

Let us now proceed to give the details of this method for $G = SL(n, \mathbb{C})$, $\mathfrak{g} = \mathfrak{sl}(n, \mathbb{C})$.

Orbital integral on $\mathfrak{g}$ and its relation to orbital integral on *G*. Let H be the group of diagonal matrices in G and $\mathfrak{h}$ the abelian Lie algebra of diagonal matrices in $\mathfrak{g}$. Let $\mathfrak{h}'$ be the set of elements $X \in \mathfrak{h}$ with distinct diagonal entries. For $X \in \mathfrak{h}'$, H is the centralizer of X in G, and the map

$$x \mapsto xXx^{-1} = X^x \quad (x \in G)$$

is a bijection of G/H with the conjugacy class X^G of X in $\mathfrak{g}$. Obviously X^G is the set of all matrices Y whose characteristic polynomial is the same as that of X. Hence X^G is closed in $\mathfrak{g}$ and the above map is a homeomorphism. So, for any $f \in C_c(\mathfrak{g})$, the function

$$xH \mapsto f(X^x)$$

lies in $C_c(G/H)$ and hence is integrable on G/H. The invariant integral ψ_f is then the integral of this function on G/H. However, since we are interested in *explicit* formulae, we need to be precise about the choice of Haar and Lebesgue measures.

Let V be any real Euclidean space of dimension r with scalar product $(\cdot,\cdot)$; ω_V will denote the form $\mathrm{d}t_1 \wedge \cdots \wedge \mathrm{d}t_r$ where $t_i (1 \leqslant i \leqslant r)$ are the linear coordinates on V relative to an orthonormal basis of V; this is canonical and we write $\mathrm{d}V$ for the corresponding Lebesgue measure. If S is a Lie group with Lie algebra $\mathfrak{s}$ on which a scalar product (not necessarily invariant) is defined converting $\mathfrak{s}$ into a Euclidean space, we write ω_S for the left-invariant form on S which corresponds to $\omega_{\mathfrak{s}}$ under the exponential map at the identity, and $\mathrm{d}S$ for the corresponding Haar measure. If $S_1 \subset S$ is a closed subgroup, and $\mathfrak{s}_1 = \mathrm{Lie}(S_1)$, $\mathfrak{s}_1$ is a Euclidean space, being a subspace of $\mathfrak{s}$; hence $\omega_{\mathfrak{s}_1}$, $\mathrm{d}\mathfrak{s}_1$, and consequently ω_{S_1}, $\mathrm{d}S_1$ also, are canonically defined. If S is *compact*, we put

$$\mathrm{vol}(S) = \int_S \mathrm{d}S$$

The reader should note that $\mathrm{vol}(S)$ need not be 1; take $S = \mathbb{T}^n$, $\mathfrak{s} = \mathbb{R}^n$ with the usual Euclidean structure, with exp as the map $(a_1, \ldots, a_n) \mapsto (\mathrm{e}^{ia_1}, \ldots, \mathrm{e}^{ia_n})$; then $\mathrm{vol}(S) = (2\pi)^n$. For a compact Lie group S we shall write $d_0 S$ for the normalized Haar measure so that $\int_S \mathrm{d}_0 S = 1$, and $\mathrm{d}S = \mathrm{vol}(S)\,\mathrm{d}_0 S$.

In our case we take

$$(u, v) = \mathrm{Re}(\mathrm{tr}(uv^{\dagger})) \quad (u, v \in \mathfrak{g} = \mathfrak{sl}(n, \mathbb{C}))$$

as the scalar product. Put

$$\langle u, v \rangle = \mathrm{Re}(\mathrm{tr}(uv)) \quad (u, v \in \mathfrak{g})$$

and let

$$\lambda(v) = v^{\dagger}$$

Then $\langle \cdot,\cdot \rangle$ is G-invariant and

$$(u, v) = \langle u, \lambda v \rangle$$

The preceding remarks then allow us to speak without any ambiguity of the Haar measure $\mathrm{d}S$ (resp. Lebesgue measure $\mathrm{d}\mathfrak{s}$) for any closed subgroup S (resp. linear subspace $\mathfrak{s}$) of G (resp. $\mathfrak{g}$). For instance, on $\mathfrak{n}$ (resp. N), if $z_{rs} (r < s)$ are matrix entries,

$$\mathrm{d}N = \mathrm{d}\mathfrak{n} = \prod_{r<s} \frac{i}{2} (\mathrm{d}z_{rs} \wedge \mathrm{d}\bar{z}_{rs})$$

If $S \subset G$ is a unimodular closed subgroup, $\omega_{G/S}$ and $\mathrm{d}(G/S)$ are then defined by requiring that $\mathrm{d}G = \mathrm{d}S\,\mathrm{d}(G/S)$ and $\omega_G = \omega_S \omega_{G/S}$ at 1. Also $\mathrm{vol}(K)$ and $\mathrm{vol}(M)$ are well defined, K, M being as in Ch. 4, §4.4 (p. 81). Let

$$\pi(X) = \prod_{r<s} (x_r - x_s) \quad (X = \mathrm{diag}(x_1, \ldots, x_n))$$

Then, the *orbital or invariant integral of f on* $\mathfrak{g}$ is defined by

$$\begin{aligned}\psi_f(X) &= |\pi(X)|^2 \int_{G/H} f(xXx^{-1})\,\mathrm{d}(G/H) \quad (X \in \mathfrak{h}') \\ &= \mathrm{vol}(M)^{-1} |\pi(X)|^2 \int_{G/A} f(xXx^{-1})\,\mathrm{d}(G/A)\end{aligned}$$

Actually we shall define ψ_f by the above formula for all continuous functions f for which the integral converges absolutely for all $X \in \mathfrak{h}'$.

The first step is to rewrite this integral as we did for F_f in Proposition 12 of Chapter 4. For any Euclidean space V let $\mathscr{S}_0(V)$ be the space of continuous functions f on V such that

$$\sup_{v \in V} (1 + \|v\|^2)^q |f(v)| < \infty$$

for each $q \geqslant 0$, and let $\mathscr{S}(V)$ be the space of C^∞ functions f on V such that all derivatives of f are in $\mathscr{S}_0(V)$; $\mathscr{S}(V)$ is of course the usual Schwartz space of V.

The analogue of Lemma 11 of Chapter 4 is then as follows.

Lemma 1. *Fix $X \in \mathfrak{h}'$. Then the map $n \mapsto X^n - X$ is a polynomial isomorphism (in the complex coordinates z_{rs}) of N with $\mathfrak{n}$; and*

$$\mathrm{d}(X^n - X) = |\pi(X)|^2\,\mathrm{d}n$$

More precisely, for any $g \in \mathscr{S}_0(\mathfrak{n})$,

$$\int g(Z)\,\mathrm{d}\mathfrak{n} = |\pi(X)|^2 \int g(X^n - X)\,\mathrm{d}N$$

Proof. let $Z = (z_{ij}) \in \mathfrak{n}$ be given and let us search for $n = (a_{ij}) \in N$ so that $nXn^{-1} = X + Z$ or $nX = Xn + Zn$. We are led to the equations

$$(x_j - x_i)a_{ij} = z_{ij} + \sum_{i<l<j} z_{il}a_{lj}$$

We now proceed exactly as in Lemma 11 of Chapter 4 to conclude that

$$a_{ij} = (x_j - x_i)^{-1} z_{ij} + b_{ij}$$

where b_{ij} is a polynomial in all the z_{rs} with $i \leqslant r < s \leqslant j$ except z_{ij}. So, as in that lemma,

$$\prod_{i<j} \mathrm{d}a_{ij}\,\mathrm{d}\bar{a}_{ij} = |\pi(X)|^{-2} \prod_{i<j} \mathrm{d}z_{ij}\,\mathrm{d}\bar{z}_{ij}$$

finishing the proof.

Proposition 2. *For any $f \in \mathscr{S}_0(\mathfrak{g})$, ψ_f is well defined on $\mathfrak{h}'$, and*

$$\psi_f(X) = \mathrm{vol}(K)\,\mathrm{vol}(M)^{-1} b(G) \int \bar{f}(X + Z)\,\mathrm{d}\mathfrak{n}(Z)$$

where $\bar{f} = \int_K \mathrm{Ad}(u) f \, \mathrm{d}_0 K$ *and* $\mathrm{d}G = b(G)\,\mathrm{vol}(K)\,\mathrm{d}_0 K\,\mathrm{d}N\,\mathrm{d}A$ $(x = una)$. *In particular,* ψ_f *is the restriction to* $\mathfrak{h}'$ *of an element (also denoted by* ψ_f*) of* $\mathscr{S}_0(\mathfrak{h})$. *Moreover* ψ_f *is symmetric, i.e., invariant with respect to* W. *If* $f \in C_c^\infty(\mathfrak{g})(\text{resp.}\,\mathscr{S}(\mathfrak{g}))$, $\psi_f \in C_c^\infty(\mathfrak{h})(\text{resp.}\,\mathscr{S}(\mathfrak{h}))$.

Proof. We have, for $X \in \mathfrak{h}', f \in C_c(\mathfrak{g})$,

$$\begin{aligned}
\psi_f(X) &= \mathrm{vol}(M)^{-1} |\pi(X)|^2 \int_{G/A} f(xXx^{-1})\,\mathrm{d}(G/A) \\
&= \mathrm{vol}(M)^{-1} b(G)\,\mathrm{vol}(K) |\pi(X)|^2 \int\!\!\int f(unXn^{-1}u^{-1})\,\mathrm{d}_0 K\,\mathrm{d}N \\
&= \mathrm{vol}(M)^{-1} b(G)\,\mathrm{vol}(K) |\pi(X)|^2 \int \bar{f}(X^n)\,\mathrm{d}N \\
&= \mathrm{vol}(M)^{-1} b(G)\,\mathrm{vol}(K) \int \bar{f}(X + Z)\,\mathrm{d}\mathfrak{n} \quad \text{(Lemma 1)}
\end{aligned}$$

Since $\bar{f} \in C_c(\mathfrak{g})$ also, it is trivial to see that this last integral is defined for *all* $X \in \mathfrak{h}$ and defines an element of $C_c(\mathfrak{h})$. The proof that ψ_f is W-invariant is the same as for F_f. If we observe that the integral over $\mathfrak{n}$ is well defined for all $f \in \mathscr{S}_0(\mathfrak{g})$ and defines an element of $\mathscr{S}_0(\mathfrak{h})$, the first statement is clear. If $f \in \mathscr{S}(\mathfrak{g})$, the integral over $\mathfrak{n}$ is clearly in $\mathscr{S}(\mathfrak{h})$.

We shall consider next the Weyl integration formula. The treatment is very close to Lemmas 14–16 of Chapter 4. We shall therefore omit most of the details. We set up the conjugacy map

$$\varphi:(G/H) \times \mathfrak{h}' \to \mathfrak{g}', \quad xH, X \mapsto xXx^{-1}$$

which is holomorphic; if we write $L \mapsto \bar{L}$ for the natural map $\mathfrak{g} \to \mathfrak{g}/\mathfrak{h}$ and identify the tangent space to G/H at $\bar{1}$ ($=$ the coset H) with $\mathfrak{g}/\mathfrak{h}$, the key calculation is the following:

$$(\mathrm{d}\varphi)_{\bar{1},X}(\bar{L}, L') = \overline{[L, X]} + L' \quad (L \in \mathfrak{g}, L' \in \mathfrak{h})$$

The map $\bar{L} \mapsto \overline{[L, X]}$ is an endomorphism of $\mathfrak{g}/\mathfrak{h}$ with determinant $\pm \pi(X)^2$. Hence we see that φ has bijective differential everywhere; in particular, $\mathfrak{g}'$ is open, and as in Chapter 4 we show that φ is proper and all its fibres have the cardinality $n!$. For the corresponding volume forms we have

$$\varphi^* \omega_{\mathfrak{g}} = \pm |\pi(X)|^4 \,\mathrm{d}(G/H)\,\mathrm{d}\mathfrak{h}$$

Hence we obtain the following.

Lemma 3. *For any* $f \in \mathscr{S}_0(\mathfrak{g})$,

$$\int_{\mathfrak{g}} f \, d\mathfrak{g} = \frac{1}{n!} \int_{\mathfrak{h}'} |\pi(X)|^2 \psi_f(X) \, d\mathfrak{h}$$

We now come to the key step in the proof. If we forget the constants for the moment, $\psi_f(X)$ is just const. $\int_{\mathfrak{n}} \bar{f}(X+Z)\, d\mathfrak{n}$, which can be viewed as const. $\langle \bar{f}, t_X \rangle$ where t_X *is the Lebesgue measure on the affine space* $X + \mathfrak{n}$. If we now take Fourier transforms (with respect to the G-invariant bilinear forms), $\hat{t}_X$ will just be $e^{i\langle X,\cdot\rangle} d\mathfrak{n}^{\perp} = e^{i\langle X,\cdot\rangle} d\mathfrak{h}\, d\mathfrak{n}$ since $\mathfrak{n}^{\perp} = \mathfrak{h} \oplus \mathfrak{n}$. Thus the map $f \mapsto \psi_f$ *will commute with Fourier transforms up to a constant factor.* The theorem below is just a more precise version of this argument.

For any $g \in \mathscr{S}(\mathfrak{g})$ we define its Fourier transform $\hat{g}$ by

$$\hat{g}(Y) = \int_{\mathfrak{g}} g(Z) e^{i\langle Y,Z\rangle} \, d\mathfrak{g}(Z) \quad (Y \in \mathfrak{g})$$

It is obvious that the map $g \mapsto \hat{g}$ commutes with the adjoint of G on $\mathfrak{g}$. In particular, averaging over $\mathrm{Ad}(K)$, we have

$$\hat{\bar{g}} = \bar{\hat{g}}$$

The forms $\langle\cdot,\cdot\rangle$, $(\cdot,\cdot)$ are nonsingular on $\mathfrak{h}$ also and hence we can also introduce the Fourier transform on $\mathfrak{h}$: for any $h \in \mathscr{S}(\mathfrak{h})$

$$\hat{h}(Y) = \int_{\mathfrak{h}} e^{i\langle Y,Z\rangle} h(Z) \, d\mathfrak{h}(Z) \quad (Y \in \mathfrak{h})$$

Finally, if V is a Euclidean space and $V_1 \subset V$ a linear subspace, we write

$$V_1(g) = \int_{V_1} g \, dV_1 \quad (g \in \mathscr{S}(V)).$$

Theorem 4. *We have the following:*

(i) $\mathfrak{n}(\hat{g}) = (2\pi)^{\dim_{\mathbb{R}}(\mathfrak{n})} (\mathfrak{h} \oplus \mathfrak{n})(g)$ *for all* $g \in \mathscr{S}(\mathfrak{g})$;
(ii) *for any* $X \in \mathfrak{h}$, *and* $f \in \mathscr{S}(\mathfrak{g})$, $\psi_{\hat{f}}(X) = (2\pi)^{\dim_{\mathbb{R}}(\mathfrak{n})} \hat{\psi}_f(X)$.

Proof. (i) If V is a Euclidean space with scalar product $(\cdot,\cdot)$ and V_1 a linear subspace of V, it is an immediate consequence of the Fourier inversion formula that, for any $v \in \mathscr{S}(V)$,

$$V_1(\tilde{v}) = (2\pi)^{\dim_{\mathbb{R}}(V_1)} V_1^{\perp}(v)$$

where $\tilde{v}$ is the Fourier transform of v. Let us apply this to the case $V = \mathfrak{g}$, $V_1 = \mathfrak{n}$, $(\cdot,\cdot)$ being as defined earlier. Then $V_1^{\perp} = \mathfrak{h} \oplus \mathfrak{n}^{\mathrm{t}}$, $\mathfrak{n}^{\mathrm{t}}$ being the space of lower-triangular nilpotent matrices. On the other hand, $\langle Y, Z \rangle =$

$(Y, \lambda Z)$ where $\lambda Z = Z^{\dagger}$; as λ is an involutive isometry of the Euclidean structure of $\mathfrak{g}$ we see that

$$\hat{g} = (g \circ \lambda)^{\sim}$$

Hence

$$\begin{aligned} \mathfrak{n}(\hat{g}) &= \mathfrak{n}((g \circ \lambda)^{\sim}) \\ &= (2\pi)^{\dim_{\mathbb{R}}(\mathfrak{n})} \int_{\mathfrak{h} \oplus \mathfrak{n}^{\dagger}} g \circ \lambda \, d(\mathfrak{h} \oplus \mathfrak{n}^{\dagger}) \\ &= (2\pi)^{\dim_{\mathbb{R}}(\mathfrak{n})} \int_{\mathfrak{h} \oplus \mathfrak{n}} g \, d(\mathfrak{h} \oplus \mathfrak{n}) \end{aligned}$$

proving (i).

(ii) Let $X \in \mathfrak{h}$. Then for $f \in \mathscr{S}(\mathfrak{g})$, as $\hat{\bar{f}} = \bar{\hat{f}}$, we have

$$\psi_{\hat{f}}(X) = \mathrm{vol}(M)^{-1} b(G) \, \mathrm{vol}(K) \int_{\mathfrak{u}} \hat{\bar{f}}(X + Z) \, d\mathfrak{n}(Z)$$

But $\hat{\bar{f}}(X + Y) = \hat{g}(Y)(Y \in \mathfrak{g})$ where $g(Y') = e^{i\langle X, Y' \rangle} \bar{f}(Y')$. Hence, writing $c = \mathrm{vol}(M)^{-1} b(G) \, \mathrm{vol}(K)$, as $\langle X, Z' \rangle = 0 \forall Z' \in \mathfrak{n}$,

$$\psi_{\hat{f}}(X) = c\mathfrak{n}(\hat{g}) = c(2\pi)^{\dim_{\mathbb{R}}(\mathfrak{n})} \int_{\mathfrak{h} \times \mathfrak{n}} e^{i\langle X, Y' + Z' \rangle} \bar{f}(Y' + Z') \, d\mathfrak{h}(Y') \, d\mathfrak{n}(Z')$$

$$\begin{aligned} (D(e)u)\,(Y) &= (2\pi)^{\dim_{\mathbb{R}}(\mathfrak{n})} \int_{\mathfrak{h}} \psi_f(Y') e^{i\langle X, Y' \rangle} \, d\mathfrak{h}(Y') \\ &= (2\pi)^{\dim_{\mathbb{R}}(\mathfrak{n})} \hat{\psi}_f(X) \end{aligned}$$

This proves the thorem.

The Fourier transform operator may now be applied to the integration formula to get the limit formula. To this end we must calculate the differential operator whose Fourier transform is the polynomial $\pi\bar{\pi}$. For any $e \in \mathfrak{h}$ define $\partial(e)$ as the operator which differentiates in the direction e and write

$$D(e) = \partial(e) - i\partial(ie), \quad \overline{D(e)} = \partial(e) + i\partial(ie)$$

If $u \in \mathscr{S}(\mathfrak{h})$, it is then immediate that

$$\begin{aligned} (D(e)u)^{\wedge}(Y) &= -i(\langle Y, e \rangle - i\langle Y, ie \rangle)\hat{u}(Y) \\ &= -i\{\mathrm{Re}(\mathrm{tr}(Ye)) - i\,\mathrm{Re}(i\,\mathrm{tr}(Ye))\}\hat{u}(Y) \\ &= -i\,\mathrm{tr}(Ye)\hat{u}(Y) \end{aligned}$$

So,

$$(\overline{D(e)}\, D(e)u)^{\wedge}(Y) = -|\mathrm{tr}(Ye)|^2 \hat{u}(Y)$$

If $e_{r,s} = \mathrm{diag}(\ldots,\overset{\overset{r}{\downarrow}}{1},\ldots,-\overset{\overset{s}{\downarrow}}{1},\ldots)$ $(r<s)$, and

$$\square = \prod_{r<s} D(e_{r,s})$$

we have

$$(-1)^{n(n-1)/2}(\square\bar{\square}u)\hat{}\,(Y) = |\pi(Y)|^2\hat{u}(Y)$$

We now have the following.

Theorem 5 (limit formual for $\mathfrak{g}$***).*** *Let*

$$\psi_f(X) = |\pi(X)|^2 \int_{G/H} f(xXx^{-1})\,\mathrm{d}(G/H) \quad (X\in\mathfrak{h}')$$

where $\mathrm{d}G = \mathrm{d}\mathfrak{g}$, $\mathrm{d}H \equiv \mathrm{d}\mathfrak{h}$, $\mathrm{d}(G/H)\,\mathrm{d}H = \mathrm{d}G$. *Then, for all* $f\in\mathscr{S}(\mathfrak{g})$,

$$f(0) = \frac{(-1)^{n(n-1)/2}}{(2\pi)^{n(n-1)}\cdot n!}(\square\bar{\square}\psi_f)(0)$$

Proof. We have, for $f\in\mathscr{S}(\mathfrak{g})$,

$$\begin{aligned} f(0) &= \frac{1}{(2\pi)^{2(n^2-1)}}\int \hat{f}\,\mathrm{d}\mathfrak{g} \\ &= \frac{1}{(2\pi)^{2(n^2-1)}\cdot n!}\int |\pi|^2\psi_{\hat{f}}\,\mathrm{d}\mathfrak{h} \\ &= \frac{1}{(2\pi)^{2(n-1)+n^2-n}\cdot n!}\int |\pi|^2\hat{\psi}_f\,\mathrm{d}\mathfrak{h} \\ &= \frac{(-1)^{n(n-1)/2}}{(2\pi)^{2(n-1)+n^2-n}\cdot n!}\int (\square\bar{\square}\psi_f)\hat{}\,\mathrm{d}\mathfrak{h} \\ &= \frac{(-1)^{n(n-1)/2}}{(2\pi)^{n^2-n}\cdot n!}(\square\bar{\square}\psi_f)(0) \end{aligned}$$

which proves what we want [H10].

We shall now transfer this to the group via the exponential map. This requires some preparation. I would like to be a little more general than needed and examine also the structure of the invariant open neighbourhoods of the origin in $\mathfrak{g}$ and the identity in G. Fix any norm $\|\cdot\|$ on $\mathfrak{g}$. For any $\varepsilon > 0$ let

$$\mathfrak{g}(\varepsilon) = \{u\in\mathfrak{g}\,|\,|\lambda| < \varepsilon \text{ for all eigenvalues } \lambda \text{ of } u\}$$
$$G(\varepsilon) = \{z\in G\,|\,|\lambda| < \varepsilon \text{ for all eigenvalues } \lambda \text{ of } z-1\}$$

Then $\mathfrak{g}(\varepsilon)$ (resp. $G(\varepsilon)$) is an invariant open neighbourhood of 0 in $\mathfrak{g}$ (resp. 1 in G).

Lemma 6. *The* $\mathfrak{g}(\varepsilon)$ (resp. $G(\varepsilon)$) *form a basis for the family of invariant open neighbourhoods of* 0 *in* $\mathfrak{g}$ (resp. 1 *in* G). *If* $\varepsilon > 0$ *is sufficiently small,* exp *is a diffeomorphism of* $\mathfrak{g}(\varepsilon)$ *onto an invariant open neighbourhood of* 1 *in* G. *Then, for all sufficiently small* ε *we can find* ε', $0 < \varepsilon' < \varepsilon$, $G(\varepsilon') \subset \exp(\mathfrak{g}(\varepsilon))$, *and an invariant* $j \in C^\infty(G)$(resp. $k \in C^\infty(\mathfrak{g})$) *such that* $j = 1$ *on* $G(\varepsilon')$, $\operatorname{supp}(j) \subset G(\varepsilon/2)$(resp. $k = 1$ *on* $\mathfrak{g}(\varepsilon)$, $\operatorname{supp}(k) \subset \mathfrak{g}(\varepsilon/2)$).

Proof. Assume the first assertion to have been established. If ε is small enough, exp is one-one on $\mathfrak{g}(\varepsilon)$ while $d(\exp)$ is bijective on an invariant open neighbourhood of 0 in $\mathfrak{g}$. The second statement follows from this. Let $p_1, \ldots, p_n$ be the smooth invariant functions on G such that $T^n + p_1(x)T^{n-1} + \cdots + p_n(x)$ is the characteristic polynomial of $x - 1$. For any $\delta > 0$ let $g_\delta \in C_c^\infty(C^n)$ be such that $\operatorname{supp}(g_\delta) \subset \{(t_1, \ldots, t_n) \mid |t_j| < \delta, 1 \leqslant j \leqslant n\}$ and $g_\delta = 1$ in a neighbourhood of the origin. Then $j = g_\delta(p_1, \ldots, p_n)$ satisfies our requirements for sufficiently small δ. The argument on $\mathfrak{g}$ is similar.

To prove the first statement we first work over $\mathfrak{g}$. First we note that if N is *nilpotent* and $t \in \mathbb{C}^\times$, then tN is conjugate to N. In particular, $0 \in \mathrm{Cl}(N^G)$. Indeed, it is enough to do this for

$$N = \begin{pmatrix} 0 & 1 & & & 0 \\ & 0 & 1 & & \\ & & 0 & & \\ & 0 & & & 1 \\ & & & & 0 \end{pmatrix}$$

Let $h = \operatorname{diag}(h_1, \ldots, h_n)$; we wish to choose the $h_i \in \mathbb{C}$ so that $h_1 \cdots h_n = 1$ and $hNh^{-1} = tN$. If we choose $h = \operatorname{diag}(ct^{n-1}, ct^{n-2}, \ldots, ct, c)$, then $hNh^{-1} = tN$; and c should be such that $c^n t^{n(n-1)/2} = 1$. It is enough to take $c = t^{-(n-1)/2}$. This said, let ω be an invariant open neighbourhood of 0 in $\mathfrak{g}$. Choose $\delta > 0$ such that the ball $\{z \mid \|z\| < \delta\} = B \subset \omega$. If $\mathfrak{h}$ is the subspace of diagonal matrices in $\mathfrak{g}$, we can find $\varepsilon > 0$ such that $\mathfrak{h} \cap \mathfrak{g}(\varepsilon) \subset B \subset \omega$. We claim that $\mathfrak{g}(\varepsilon) \subset \omega$. For, suppose $Z \in \mathfrak{g}(\varepsilon)$. Then we can find $x \in G$ such that $xZx^{-1} = D + N$ where D is diagonal and N is a nilpotent commuting with D. Thus $D \subset B$. But now N stabilizes each eigensubspace of D, and so, by the remark made earlier, we can find $y \in G$ such that $yDy^{-1} = D$ and yNy^{-1} is arbitrarily close to 0. So $yxZ(yx)^{-1}$ is arbitrarily close to D, and so must lie in B, hence in ω. For the result in G, we remark that for any *unipotent* $u \in G$, $1 \in \mathrm{Cl}(u^G)$ since $0 \in \mathrm{Cl}((u-1)^G)$; the argument then remains the same, except that we use the multiplicative Jordan decomposition. I leave the details of the proof to the reader.

Let ε and j be as in Lemma 6. Then, for any $f \in C_c^\infty(G)$, $jf \in C_c^\infty(G)$, $\operatorname{supp}(jf) \subset \exp(\mathfrak{g}(\varepsilon))$, $jf = f$ in an invariant neighbourhood of 1. So, first of all, $F_{jf} = jF_f = F_f$ in a neighbourhood of 1 in H; moreover, $(jf)\circ\exp$ is an element of $C_c^\infty(\mathfrak{g})$. We now have the following.

***Theorem* 7.** *Define F_f by*

$$F_f(h) = |\Delta(h)|^2 \int_{G/H} f(xhx^{-1})\,\mathrm{d}(G/H) \quad (h \in H')$$

where $\mathrm{d}G = \mathrm{d}\mathfrak{g}$, $\mathrm{d}H = \mathrm{d}\mathfrak{h}$, $\mathrm{d}(G/H) = \mathrm{d}G$. Then

$$f(1) = \frac{(-1)^{n(n-1)/2}}{(2\pi)^{n^2-n}\cdot n!}(\square\bar{\square}F_f)(1) \quad (f \in C_c^\infty(G))$$

Proof. In view of the remark made above, we may prove this for jf in place of f. In other words, we may assume that $\operatorname{supp}(f) \subset \exp(\mathfrak{g}(\varepsilon))$. Let $g(Z) = f(\exp Z)$, $Z \in \Omega$. Then $g \in C_c^\infty(\mathfrak{g}(\varepsilon))$ and so lies in $C_c^\infty(\mathfrak{g})$. Moreover, for $X \in \mathfrak{g}(\varepsilon) \cap \mathfrak{h}'$, $h = \exp X \in H' \cap \exp(\mathfrak{g}(\varepsilon))$. Hence, for such X,

$$\begin{aligned} F_f(\exp X) &= |\Delta(\exp X)|^2 \int_{G/H} f(x \exp X x^{-1})\,\mathrm{d}(G/H) \\ &= |w(X)|^2 \psi_g(X) \end{aligned}$$

where

$$w(X) = \Delta(\exp X)/\pi(X) \quad (X \in \mathfrak{h}')$$

We now claim that there is an invariant entire function w_1 on $\mathfrak{g}$ which restricts to $w(X)^2$ for $X \in \mathfrak{h}'$. Indeed,

$$\begin{aligned} w(X)^2 &= \prod_{r \neq s} \left(\frac{e^{x_r} - e^{x_s}}{x_r - x_s} \right) \\ &= \prod_{r \neq s} e^{(x_r + x_s)/2} \frac{e^{(x_r - x_s)/2} - e^{-(x_r - x_s)/2}}{x_r - x_s} \\ &= \prod_{r \neq s} \varphi((x_r - x_s)) \end{aligned}$$

where φ is the entire function $\sinh(z/2)/(z/2)$. Now ad X has $x_r - x_s (r \neq s)$ and 0 as its eigenvalues, the former occurring with multiplicity 1. Hence

$$w(X)^2 = \det \varphi(\operatorname{ad} X)$$

and hence the desired entire function is

$$w_1(Z) = \det \varphi(\operatorname{ad} Z) \quad (Z \in \mathfrak{g})$$

Since $\varphi(0)=1$, $\varphi(z)\neq 0$ for $|z|<\delta$ if δ is small enough; moreover, if $Z\in\mathfrak{g}(\varepsilon)$ the eigenvalues of ad Z are less than 2ε. Hence, for sufficiently small ε, w_1 is never zero on $\mathfrak{g}(\varepsilon)$, so that $|w_1|$ is a real-analytic function. Hence $|w_1|g\in C_c^\infty(\mathfrak{g}(\varepsilon))$ for all $g\in C_c^\infty(\mathfrak{g}(\varepsilon))$ and we can write

$$F_f(\exp X)=\psi_{|w_1|g}(X)\quad (X\in\mathfrak{h}'\cap\mathfrak{g}(\varepsilon))$$

Applying $\Box\bar{\Box}$ to both sides and letting $X\to 0$ we get Theorem 7 from Theorem 5 as $|w_1|(0)=1$.

We shall apply the Fourier transform, defined with respect to dH, to Theorem 7. We need one more lemma.

Lemma 8. *Let μ be the function on $\hat{H}$ given by*

$$\mu(\chi_{m_1,\ldots,m_n:i\rho_1,\ldots,i\rho_n})=\prod_{r<s}[(m_r-m_s)^2+(\rho_r-\rho_s)^2]$$

where $\chi_{m_1,\ldots,m_n:i\rho_1,\ldots,i\rho_n}$ is the character

$$\mathrm{diag}(z_1,\ldots,z_n)\mapsto\prod_{j=1}^{n}(z_j/|z_j|)^{m_j}|z_j|^{i\rho_j}$$

the m_j being integers and the ρ_j being real numbers, with $\sum m_j=0, \sum\rho_j=0$. Then, for any $v\in C_c^\infty(H)$,

$$(-1)^{n(n-1)/2}(\Box\bar{\Box}v)\hat{}(x)=\mu(\chi)\hat{v}(\chi)$$

Proof. The desired property of μ does not depend on the choice of Haar measure in computing the Fourier transform. Now, for any $e\in\mathfrak{h}$, $\partial(e)\chi=c(e)\chi$ and hence $(\partial(e)v)\hat{}(\chi)=-c(e)\hat{v}(\chi)$. But if $\xi=\mathrm{diag}(\rho_1-im_1,\ldots,\rho_n-im_n)$, and $\chi_\xi=\chi_{m_1,\ldots,m_n:i\rho_1,\ldots,i\rho_n}$, we see that $\xi\in\mathfrak{h}$ and $\chi_\xi\circ\exp$ is the character

$$X=\mathrm{diag}(x_1,\ldots,x_n)\mapsto e^{i\langle X,\xi\rangle}$$

A simple calculation then gives

$$\mathrm{D}(e)(\chi_\xi)=\mathrm{i}\,\mathrm{tr}(\xi e)\chi_\xi$$

So

$$(-1)^{n(n-1)/2}\Box\bar{\Box}(\chi_\xi)=\mu(\chi_\xi)\chi_\xi$$

from which we get the lemma easily.

Theorem 9 (Plancherel formula). *For any character χ of H let T_χ be the distribution character of the principal series representation L_χ, computed with respect to the Haar measure d$G\cong$ d$\mathfrak{g}$, and let dχ be the measure on $\hat{H}$ dual to d$H\cong$ d$\mathfrak{h}$. Then, for any $f\in C_c^\infty(H)$, we have*

$$f(1)=\frac{1}{(2\pi)^{n^2-n}\cdot n!}\int_{\hat{H}}T_\chi(f)\mu(\chi)\,\mathrm{d}\chi$$

where

$$\mu(\chi_{m_1,\ldots,m_n;i\rho_1,\ldots,i\rho_n}) = \prod_{r<s} [(m_r - m_s)^2 + (\rho_r - \rho_s)^2]$$

Proof. We have, by Theorem 7,

$$\begin{aligned} f(1) &= \frac{(-1)^{n(n-1)/2}}{(2\pi)^{n^2-n}\cdot n!}(\square\bar{\square}F_f)(1) \\ &= \frac{(-1)^{n(n-1)/2}}{(2\pi)^{n^2-n}\cdot n!}\int_{\hat{H}}(\square\bar{\square}F_f)\hat{}(\chi)\,\mathrm{d}\chi \\ &= \frac{1}{(2\pi)^{n^2-n}\cdot n!}\int_{\hat{H}}\hat{F}_f(\chi)\mu(\chi)\,\mathrm{d}\chi \quad \text{(Lemma 8)} \\ &= \frac{1}{(2\pi)^{n^2-n}\cdot n!}\int_{\hat{H}}T_\chi(f)\mu(\chi)\,\mathrm{d}\chi \end{aligned}$$

by Theorem 13 of Chapter 5, since $\mathrm{d}(G/H)\,\mathrm{d}H = \mathrm{d}G$.

The above treatment can be extended to *all* complex groups; Harish-Chandra did this in [H10]. It turns out in fact that the key ingredient for the success of the method is the fact that *there is a single conjugacy class of Cartan subgroups.* Now there are some real groups with this property, such as $SO(1, 2k+1)$, and the same method works for all of them. But in the general case, when there are several conjugacy classes of Cartan subgroups, the problem of the Plancherel formula becomes very difficult. In order to understand the nature and source of these difficulties it is useful to study the case of $SL(2,\mathbb{R})$ where they are already encountered.

6.2 Orbital integral for $G = SL(2,\mathbb{R})$ and $\mathfrak{g} = \mathfrak{sl}(2,\mathbb{R})$

It is clear from the definition of F_f that we can associate an orbital integral to any Cartan subgroup; when the Cartan subgroup is moved by conjugacy, the orbital integral transforms in the obvious way. However the orbital integrals associated to nonconjugate Cartan subgroups behave in completely different ways. In general, given a Cartan subgroup L in a real semisimple group G, the orbital integral associated to L is a map

$$f \mapsto F_{f,L}$$

taking functions $f \in C_c^\infty(G)$ to functions F_f defined on L', given by

$$F_{f,L}(h) = \nu_L(h)\int_{G/L} f(xhx^{-1})\,\mathrm{d}\dot{x}$$

where ν_L is a normalizing factor and $\mathrm{d}\dot{x}$ is an invariant measure on G/L.

The basic questions concerning the orbital integrals are the following.

(i) What is the behaviour of $F_{f,L}$ near the singular points of L, namely, the points where the centralizer in the group is bigger than L?
(ii) What is the structure of the Fourier transform of $F_{f,L}$ and its relationship to irreducible characters of G?
(iii) What is the relationship $F_{,L_1}$ and $F_{,L_2}$ for two Cartan subgroups L_1 and L_2 (in general nonconjugate), at points of $L_1 \cap L_2$?
(iv) Is there a limit formula for $F_{,L}$ and, if so, what is its relationship to the Plancherel formula for G?

Obviously one is led to these questions from the theory for $SL(n, \mathbb{C})$; and one should add to these the following.

(v) Develop the theory of orbital integrals $\psi_{f,\mathfrak{l}}$ on the Lie algebra ($\mathfrak{l}$ a Cartan subalgebra), with special attention to the questions (i)–(iv).

In a series of beautiful papers in the late 1950s Harish-Chandra constructed a decisive theory of orbital integrals which answered all these questions and formed the foundation for his theory of characters and the Plancherel formula. Before looking into his theory in more detail it is essential to understand the case of $SL(2, \mathbb{R})$ which is, in a certain fundamental sense, typical. Let us turn to it now.

From now on, $G = SL(2, \mathbb{R})$, $\mathfrak{g} = \mathfrak{sl}(2, \mathbb{R})$. We have the standard basis for $\mathfrak{g}$, and the subgroups $K, A, N, M = \{1, \gamma\}$. We take the same notation as in Chapter 5 in our discussion of $SL(2, \mathbb{R})$. We write B for K often, especially when we want to think of it in the role of a Cartan subgroup. Let $\mathfrak{a}$, $\mathfrak{b}$ be the Lie algebras of A and B so that

$$\mathfrak{a} = \mathbb{R}\cdot H, \quad \mathfrak{b} = \mathbb{R}\cdot(X - Y) = \mathfrak{k}$$

The Cartan subgroups are always defined as centralizers in the group of Cartan subalgebras. Thus they are B, which is the centralizer of $\mathfrak{b}$, and $L = A \coprod \gamma A = MA$ which is the centralizer of $\mathfrak{a}$.

Haar measures and Lebesgue measures. We use the *Riemannian measures.* On $\mathfrak{g}$ we have the scalar product

$$(u|v) = \tfrac{1}{2}\,\mathrm{tr}\,(uv^t) \quad (t = \text{transpose})$$

Note that $\{H, \sqrt{2}\cdot X, \sqrt{2}\cdot Y\}$ and $\{H, X - Y, X + Y\}$ are orthonormal bases. We thus have $d\mathfrak{g}$, the usual Lebesgue measure in the linear coordinates with respect to any orthonormal basis. The measure dG is then induced by the differential form that corresponds at the identity to the form on $\mathfrak{g}$ inducing $d\mathfrak{g}$, via the exponential map. For any subspace $\mathfrak{s} \subset \mathfrak{g}$ we have the measure $d\mathfrak{s}$; for any closed subgroup $S \subset G$ *connected or not*, we transfer ds to dS at the identity by exp, and then by left translation. Thus, as $\|H\| = \|X - Y\| = 1$, we have

$$dA = dt \quad (a = a_t), \quad dB = d\theta \quad (b = u_\theta)$$

while $\|X\| = 1/\sqrt{2}$ gives

$$dN = \frac{1}{\sqrt{2}} ds \quad (n = n_s)$$

The Iwasawa and polar decompositions. Let ψ be the map $u_\theta, a_t, n_s \mapsto u_\theta a_t n_s$. To calculate the differential $d\psi$ we notice that $x \exp Zy = xy \exp(Z^{y^{-1}})$ and obtain the following:

$$d\psi: \begin{cases} \partial/\partial s \mapsto X, \quad \partial/\partial t \mapsto H^{n_{(-s)}} = 2sX + H \\ \partial/\partial\theta \mapsto (X - Y)^{n_{(-s)}a_{(-t)}} = (e^{-2t} + s^2 e^{2t})X + se^{2t}H - e^{2t}Y \end{cases}$$

Hence $d\psi$ is invertible and

$$(d\psi)^{-1}: \begin{cases} X \mapsto \partial/\partial s, \quad H \mapsto -2s\partial/\partial s + \partial/\partial t \\ Y \mapsto e^{-2t}(-\partial/\partial\theta + se^{2t}\partial/\partial t + (e^{-2t} - s^2 e^{2t})\partial/\partial s) \end{cases}$$

Obviously

$$dG = \tfrac{1}{2} e^{2t} \, d\theta \, dt \, ds$$

For the polar decomposition the map is $\varphi((u_{\theta_1}, a_t, u_{\theta_2}) \mapsto u_{\theta_1} a_t u_{\theta_2})$. Once again we have the following easily calculated formulae:

$$d\varphi: \begin{cases} \partial/\partial\theta_2 \mapsto X - Y, \quad \partial/\partial t \mapsto H^{u_{(-\theta_2)}} = \cos 2\theta_2 \cdot H + \sin 2\theta_2 \cdot (X + Y) \\ \partial/\partial\theta_1 \mapsto (X - Y)^{u_{(-\theta_2)}a_{(-t)}} = \sin 2\theta_2 \cdot \sinh 2t \cdot H \\ \qquad\qquad - \cos 2\theta_2 \cdot \sinh 2t \cdot (X + Y) + \cosh 2t \cdot (X - Y) \end{cases}$$

Then $d\varphi$ is invertible precisely when $t \neq 0$; and then

$$(d\varphi)^{-1}: \begin{cases} X - Y \mapsto \partial/\partial\theta_2 \\ H \mapsto \cos 2\theta_2 \cdot \partial/\partial t + \dfrac{\sin 2\theta_2}{\sinh 2t}(\partial/\partial\theta_1 - \cosh 2t \cdot \partial/\partial\theta_2) \\ X + Y \mapsto \sin 2\theta_2 \cdot \partial/\partial t - \dfrac{\cos 2\theta_2}{\sinh 2t}(\partial/\partial\theta_1 - \cosh 2t \cdot \partial/\partial\theta_2) \end{cases}$$

We have

$$G = K \coprod KA^+K = K\mathrm{Cl}(A^+)K, A^+ = \{a_t | t > 0\}$$

and

$$dG = \tfrac{1}{2} \sinh 2t \cdot (d\theta_1)(d\theta_2) \, dt$$

The factor $\frac{1}{2}$ comes in because *φ is a map of degree 2.*

Orbital integrals for L, $\mathfrak{a}$. We have already seen these for $SL(n, \mathbb{R})$; but the details here are very elementary and we go over them briefly. In view of our definition of $\mathrm{d}L$,

$$\mathrm{d}L = \mathrm{d}M\,\mathrm{d}t \quad (h = ma)$$

Here $\mathrm{d}M$ has masses 1 at 1 and γ. So

$$\int_L f\,\mathrm{d}L = \int_{\mathbb{R}} f(a_t)\,\mathrm{d}t + \int_{\mathbb{R}} f(\gamma a_t)\,\mathrm{d}t$$

Recall that centre $(G) = \{1, \gamma\}$. For any function g defined on G or on a subgroup containing γ, let

$$g_\gamma(x) = g(\gamma x)$$

Then

$$\int_L f\,\mathrm{d}L = \int_{\mathbb{R}} (f + f_\gamma)\,\mathrm{d}t$$

We define $\mathrm{d}(G/L)$ and $\mathrm{d}(G/A)$ respectively by $\mathrm{d}G = \mathrm{d}(G/L)\,\mathrm{d}L = \mathrm{d}(G/A)\,\mathrm{d}A$. Then for $t \neq 0$, $m \in \{1, \gamma\}$,

$$\begin{aligned} F_{f,L}(ma_t) &:= |\mathrm{e}^t - \mathrm{e}^{-t}| \int_{G/L} f(xma_tx^{-1})\,\mathrm{d}(G/L) \\ &= \tfrac{1}{2}|\mathrm{e}^t - \mathrm{e}^{-t}| \int_{G/A} f(xma_tx^{-1})\,\mathrm{d}(G/A) \end{aligned}$$

So

$$F_{f,L}(\gamma a_t) = F_{f_\gamma,L}(a_t)$$

Moreover, as $\mathrm{d}G = \frac{1}{2}\mathrm{d}\theta\,\mathrm{d}s\,\mathrm{d}t\ (x = u_\theta n_s a_t)$,

$$\begin{aligned} F_{f,L}(a_t) &= \tfrac{1}{2}|\mathrm{e}^t - \mathrm{e}^{-t}| \iint f(u_\theta n_s a_t n_{-s} u_{-\theta})\,\mathrm{d}\theta\,\mathrm{d}s \\ &= \tfrac{1}{4}|\mathrm{e}^t - \mathrm{e}^{-t}| \int \bar{f}(n_s a_t n_{-s})\,\mathrm{d}s \quad \left(\bar{f}(x) = \int f(u_\theta x u_{-\theta})\,\mathrm{d}\theta\right) \\ &= \tfrac{1}{4}|\mathrm{e}^t - \mathrm{e}^{-t}| \int \bar{f}(a_t n_{s'})\,\mathrm{d}s \quad (s' = s(\mathrm{e}^{-2t} - 1)) \\ &= \tfrac{1}{4}\mathrm{e}^t \int \bar{f}(a_t n_{s'})\,\mathrm{d}s' \end{aligned}$$

Hence

$$F_{f,L}(a_t) = \tfrac{1}{4}\mathrm{e}^t \int f(a_t n_s)\,\mathrm{d}s$$

This formula shows that $F_{f,L}$ extends to an element of $C_c^\infty(L)$ and that $f \mapsto F_{f,L}$ is a continuous map of $C_c^\infty(G)$ to $C_c^\infty(L)$; the continuity means that, if E is any compact subset of G, there is a compact subset $E' \subset L$ such that, for $f \in C_c^\infty(E)$, $F_{f,L} \in C_c^\infty(E')$, and if $f_n \to f$ in $C_c(E)$, i.e., the convergence is uniform for each derivative, $F_{f_n,L} \to F_{f,L}$ in $C_c^\infty(E')$. The original definition shows that $F_{f,L}$ is even in t, i.e.,

$$F_{f,L} \in C_c^\infty(L)^W \quad (W = \{1, \sigma\}, \quad \sigma : ma_t \mapsto ma_{-t})$$

Let $\hat{M} = \{0, 1\}$ where $\varepsilon = 0$ or 1 represents the character of M which sends γ to $(-1)^\varepsilon$. If $\hat{}$ denotes the Fourier transform on L with respect to $\mathrm{d}M\,\mathrm{d}t = \mathrm{d}L$, then, as $\mathrm{d}G = \mathrm{d}(G/L)\,\mathrm{d}L$, we have, for $T_{\varepsilon,\lambda} = \mathrm{ch}(\pi_{\varepsilon,\lambda})$, the character of $\pi_{\varepsilon,\lambda}$, the formula

$$T_{\varepsilon,\lambda}(f) = \hat{F}_{f,L}(\varepsilon : \lambda) \quad (f \in C_c^\infty(G))$$

Finally $T_{\varepsilon,\lambda}$ is given by a locally integrable class function $\theta_{\varepsilon,\lambda}$, vanishing outside the set of hyperbolic elements (i.e., elements conjugate to ma_t, $t \neq 0$) and given for these by

$$\theta_{\varepsilon,\lambda}(ma_t) = \varepsilon(m) \frac{\mathrm{e}^{\lambda t} + \mathrm{e}^{-\lambda t}}{|\mathrm{e}^t - \mathrm{e}^{-t}|} \quad (\varepsilon(\gamma) = (-1)^\varepsilon)$$

By letting $t \to 0 \pm$ and remembering that $F_{f,L}$ is even, we get the following

Theorem 10. *We have*

(a) $F_{f,L}(1) = \frac{1}{4} \int_{-\infty}^{\infty} \bar{f}(n_s)\,\mathrm{d}s$,
(b) $F'_{f,L}(1) = 0 \quad (F'_{f,L}(a_t) = F_{f,L}(a_t; H) = (\mathrm{d}/\mathrm{d}t) F_{f,L}(a_t))$.

I wish to draw attention to (b). By analogy with $SU(2)$ or $SL(2, \mathbb{C})$ it is clear that it is the limit formula for L. *So we cannot recover* $f(1)$ *from* $F_{f,L}$ *by applying* $\mathrm{d}/\mathrm{d}t$ at $t = 0$. One may ask however whether $f(1)$ may be recovered by applying a more complicated operator to $F_{f,L}$. Even this is not possible. We shall see why this is so later.

The corresponding orbital integral on the Lie algebra is denoted by $\psi_{f,\mathfrak{a}}$. It is defined on $\mathfrak{a}$ by

$$\psi_{f,\mathfrak{a}}(tH) = 2|t| \int_{G/L} f(tH^x)\,\mathrm{d}(G/L) \quad (t \neq 0)$$
$$= |t| \int_{G/A} f(tH^x)\,\mathrm{d}(G/A)$$

As in the case of $F_{f,L}$ we have

$$\psi_{f,\mathfrak{a}}(tH) = \frac{|t|}{2} \iint f(u_\theta n_s(tH) n_{-s} u_{-\theta})\,\mathrm{d}\theta\,\mathrm{d}s$$
$$= \frac{|t|}{2} \int \bar{f}(t(H - 2sX))\,\mathrm{d}s \quad \left(\bar{f}(Z) = \int f(u_\theta Z u_{-\theta})\,\mathrm{d}\theta\right)$$

Thus, making the substitution $s' = -2st$, we get

$$\psi_{f,\mathfrak{a}}(tH) = \tfrac{1}{4}\int \bar{f}(tH + sX)\,\mathrm{d}s$$

This formula shows that $\psi_{f,\mathfrak{a}}$ makes sense for all f in $\mathscr{S}(\mathfrak{g})$, lies in $\mathscr{S}(\mathfrak{a})$, and $f \mapsto \psi_{f,\mathfrak{a}}$ is continuous.

Let us now introduce Fourier transforms. For $g \in \mathscr{S}(\mathfrak{g})$(resp. $h \in \mathscr{S}(\mathfrak{h})$) let

$$\hat{g}(u) = \int_{\mathfrak{g}} g(v)\mathrm{e}^{\mathrm{i}\langle u,v\rangle}\,\mathrm{d}\mathfrak{g}(v) \quad (u \in \mathfrak{g})$$

$$\hat{h}(u) = \int_{\mathfrak{a}} h(v)\mathrm{e}^{\mathrm{i}\langle u,v\rangle}\,\mathrm{d}\mathfrak{a}(v) \quad (u \in \mathfrak{a})$$

where

$$\langle u, v\rangle = \tfrac{1}{2}\operatorname{tr}(uv)$$

Theorem 11. *For any $f \in \mathscr{S}(\mathfrak{g})$, $\psi_{f,\mathfrak{a}}$ lies in $\mathscr{S}(\mathfrak{a})$, and is in $C_c^\infty(\mathfrak{a})$ for $f \in C_c^\infty(\mathfrak{a})$. The map $f \mapsto \psi_{f,\mathfrak{a}}$ from $\mathscr{S}(\mathfrak{g})$ to $\mathscr{S}(\mathfrak{a})$ is continuous. Moreover,*

(i) *$\psi_{f,\mathfrak{a}}$ is even and $\psi'_{f,\mathfrak{a}} := (\mathrm{d}/\mathrm{d}t)\psi_{f,\mathfrak{a}}$ is odd (with respect to $tH \mapsto -tH$),*
(ii) *$\psi_{f,\mathfrak{a}}(0) = \frac{1}{4}\int_{-\infty}^{\infty} \bar{f}(sX)\,\mathrm{d}s$, $\psi'_{f,\mathfrak{a}}(0) = 0$*
(iii) *$\psi_{\hat{f},\mathfrak{a}}(tH) = 2\pi\hat{\psi}_{f,\mathfrak{a}}(tH)$.*

Proof. Only the third formula needs a proof. Observe first that H, $\sqrt{2}\cdot X$, $\sqrt{2}\cdot Y$ is an orthonormal basis for $\mathfrak{g}$ and hence

$$\mathrm{d}\mathfrak{g}(Z) = \mathrm{d}t\,\mathrm{d}s\,\mathrm{d}s' \quad Z = tH + s\sqrt{2}\cdot X + s'\sqrt{2}\cdot Y$$

As in Theorem 4 we find, for any $g \in \mathscr{S}(\mathfrak{g})$,

$$\int \hat{g}(s\sqrt{2}\cdot X)\,\mathrm{d}s = 2\pi \int\int g(t'H + s\sqrt{2}\cdot X)\,\mathrm{d}t'\,\mathrm{d}s$$

Take $g(Z) = \overline{\bar{f}(Z)\mathrm{e}^{\mathrm{i}\langle tH,Z\rangle}}$; then $\hat{g}(Z') = \bar{\hat{f}}(tH + Z')$ and so,

$$\begin{aligned}\psi_{\hat{f},\mathfrak{a}}(tH) &= (1/2\sqrt{2})\int \bar{\hat{f}}(tH + s\sqrt{2}\cdot X)\,\mathrm{d}s \\ &= (\pi/\sqrt{2})\int\int \bar{f}(t'H + s\sqrt{2}\cdot X)\mathrm{e}^{\mathrm{i}\langle tH,t'H\rangle}\,\mathrm{d}t'\,\mathrm{d}s \\ &= 2\pi\int \mathrm{e}^{\mathrm{i}\langle tH,t'H\rangle}\psi_{f,\mathfrak{a}}(t'H)\,\mathrm{d}t' \\ &= 2\pi\hat{\psi}_{f,\mathfrak{a}}(tH)\end{aligned}$$

Finally, $F_{f,L}$ and $\psi_{f,\mathfrak{a}}$ are connected naturally via the exponential map. As in §1 let us put

$$\mathfrak{g}(\varepsilon)=\{u\in\mathfrak{g}\,|\,|\lambda|<\varepsilon \text{ for each eigenvalue } \lambda \text{ of } u\}$$
$$G(\varepsilon)=\{z\in G\,|\,|\lambda|<\varepsilon \text{ for each eigenvalue } \lambda \text{ of } z-1\}$$

By a slight modification of the arguments of Lemma 6 we can prove that $(\mathfrak{g}(\varepsilon))$ (resp. $(G(\varepsilon))$) is a basis for the invariant neighbourhoods of 0 (resp. 1) in $\mathfrak{g}$ (resp. G). If $\varepsilon>0$ is sufficiently small, exp is a diffeomorphism of $\mathfrak{g}(\varepsilon)$ with an invariant open neighbourhood of 1 in G. For all sufficiently small ε we can find as in Lemma 6, $\varepsilon', 0<\varepsilon'<\varepsilon$, $G(\varepsilon')\subset \exp\mathfrak{g}(\varepsilon)$, and a $j\in C^\infty(G)$ such that j is invariant, $j=1$ on $G(\varepsilon')$ and $\operatorname{supp}(j)\subset G(\varepsilon/2)$. If $f\in C_c^\infty(G(\varepsilon'))$, we can then associate to f the function $g\in C_c^\infty(\mathfrak{g}(\varepsilon))$ given by

$$g(Z)=(jf)(\exp Z)\quad (Z\in\mathfrak{g}(\varepsilon))$$

We have, for all sufficiently small t,

$$F_{f,L}(a_t)=|(e^t-e^{-t})/2t|\psi_{g,\mathfrak{a}}(tH)$$

Orbital integrals for B, $\mathfrak{b}$. We begin with

Proposition 12. (a) *For any compact set $E\subset B'=B\backslash\{1,\gamma\}$, the map $(x,u_\theta)\mapsto xu_\theta x^{-1}$ from $G\times E$ to G is proper. In particular, E^G is closed in G.*

(b) *For any $u_\theta\in B'$, i.e. $\theta\not\equiv 0,\pi \pmod{2\pi}$, u^G is closed in G and meets B precisely at u_θ, and $xB\mapsto xu_\theta x^{-1}$ is a homeomorphism of G/B with u_θ^G.*

(c) *For any compact set $E\subset\mathfrak{b}'=\mathfrak{b}\backslash(0)$, the map $(x,\theta(X-Y))\mapsto\theta(X-Y)^x$ from $G\times E$ to $\mathfrak{g}$ is proper. In particular, E^G is closed in $\mathfrak{g}$. For any $\theta\neq 0$, $\theta(X-Y)^G$ is closed in $\mathfrak{g}$ and $xB\mapsto\theta(X-Y)^x$ is a homeomorphism of G/B with this class. Finally $z\in Z$ (resp. $Z\in\mathfrak{g}$) is conjugate to some u_θ, $\theta\not\equiv 0,\pi$ (resp. $\theta(X-Y)$, $\theta\neq 0$) if and only if $|\operatorname{tr}(z)|<2$ (resp. $\det(Z)<0$).*

Proof. It is enough to prove (a); (b) is an easy consequence of (a) and (c) is similar to (a) and (b). If

$$x=\begin{pmatrix} a & b\\ c & d\end{pmatrix}$$

we have

$$xu_\theta x^{-1}=\begin{pmatrix}\cos\theta-(ac+bd)\sin\theta & (a^2+b^2)\sin\theta\\ -(c^2+d^2)\sin\theta & \cos\theta+(ac+bd)\sin\theta\end{pmatrix}$$

Since $|\sin\theta|\geqslant \text{const.}>0$ for $u_\theta\in E$, it is clear that a^2+b^2 and c^2+d^2 are bounded if $xu_\theta x^{-1}$ varies in a compact subset of G. The argument in $\mathfrak{g}$ is similar, θ replacing $\sin\theta$.

Remark. We should note that u_θ and $u_{-\theta}$ are *not* conjugate in $SL(2,\mathbb{R})$ unlike what happens in $SL(2,\mathbb{C})$.

Let us write

$$\Delta(u_\theta) = \Delta_B(u_\theta) = e^{i\theta} - e^{-i\theta}$$

$$F_{f,B}(u_\theta) = \Delta(u_\theta) \int_{G/B} f(xu_\theta x^{-1})\, d(G/B)$$

$$= (2\pi)^{-1}\Delta(u_\theta) \int_G f(xu_\theta x^{-1})\, dG$$

The function $x, u_\theta \mapsto f(xu_\theta x^{-1})$ is C^∞ on $G \times B'$, and Proposition 12 shows that the integral for $F_{f,B}$ may be differentiated under the integral sign. So, $F_{f,B} \in C^\infty(B')$; and if $E \subset B'$ is a compact set, the map

$$f \mapsto F_{f,B}|_E$$

is continuous from $C_c^\infty(G)$ to $C_c^\infty(E)$. Note that

$$F_{f,B}(u_{\theta+\pi}) = -F_{f_y,B}(u_\theta)$$

We take up the integration formula. It is a little more complicated than the case of the complex groups, for there are now two Cartan subgroups. Let G_{ell} (resp. G_{hyp}) be the open invariant set of element z of G such that $|\text{tr}(z)| < 2$ (resp. $|\text{tr}(z)| > 2$). Clearly

$$G' = G_{\text{ell}} \coprod G_{\text{hyp}}$$

where, as usual, G' is the set of elements with distinct eigenvalues. Then $G \backslash G'$ is the set of matrices $\pm u$, u unipotent, which has measure zero. We set up the maps

$$\varphi_{\text{ell}}: G/B \times B' \to G_{\text{ell}}, \quad \varphi_{\text{hyp}}: G/L \times L' \to G_{\text{hyp}}$$

in the usual way by

$$\varphi_{\text{ell}}(xB, u_\theta) = xu_\theta x^{-1}, \quad \varphi_{\text{hyp}}(xL, h) = xhx^{-1}$$

It is obvious that these are analytic, surjective, with φ_{ell} bijective and φ_{hyp} of degree 2; moreover,

$$\varphi_{\text{ell}}(yxB, u_\theta) = y\varphi_{\text{ell}}(xB, u_\theta)y^{-1}, \quad \varphi_{\text{hyp}}(yxL, h) = y\varphi_{\text{hyp}}(xL, h)y^{-1}.$$

Now,

$$(d/ds)_{s=0}(\exp sZx \cdot \exp(-sZ)) = Z^{x^{-1}} - Z$$

from which we get the following formula for the differentials of these maps. (We write $\tilde{1}$ for the coset B in $G = G/B$ and $\dot{1}$ for the coset L in $G = G/L$; moreover we identify the respective tangent spaces with $\mathbb{R}\cdot H \oplus \mathbb{R}\cdot(X+Y)$ and $\mathbb{R}\cdot(X+Y) \oplus \mathbb{R}\cdot(X-Y)$):

$$(d\varphi_{\text{ell}})_{\tilde{1},u_\theta}: \begin{cases} H \mapsto (\cos 2\theta - 1)H + \sin 2\theta \cdot (X+Y) \\ X+Y \mapsto -\sin 2\theta \cdot H + (\cos 2\theta - 1)(X+Y) \\ \partial/\partial\theta \mapsto X - Y \end{cases}$$

$$(\mathrm{d}\varphi_{\mathrm{hyp}})_{1,ma_t} \quad \begin{array}{l} X+Y \mapsto (\cosh 2t-1)(X+Y) - \sinh 2t\cdot(X-Y) \\ X-Y \mapsto -\sinh 2t\cdot(X+Y) + (\cosh 2t-1)(X-Y) \\ \partial/\partial t \mapsto H \end{array}$$

The differentials are thus seen to be bijective everywhere. A determinant calculation then gives

$$\varphi^*_{\mathrm{ell}}\,\mathrm{d}G = |\mathrm{e}^{\mathrm{i}\theta} - \mathrm{e}^{-\mathrm{i}\theta}|^2\,\mathrm{d}(G/B)\,\mathrm{d}\theta$$
$$\varphi^*_{\mathrm{hyp}}\,\mathrm{d}G = |\mathrm{e}^{t} - \mathrm{e}^{-t}|^2\,\mathrm{d}(G/L)\,\mathrm{d}L$$

Proposition 13 (integration formula). *If f is a Borel function on G, then $f \in L^1(G)$ if and only if $F_{|f|,B}$ and $F_{|f|,L}$ exist almost everywhere on B and L respectively, and*

$$\int_B |\Delta(u_\theta)|\,F_{|f|,B}(u_\theta)\,\mathrm{d}\theta < \infty, \qquad \int_L |\mathrm{e}^t - \mathrm{e}^{-t}|\,F_{|f|,L}(h)\,\mathrm{d}L < \infty$$

In this case

$$\int_G f\,\mathrm{d}G = \int_B \overline{\Delta(u_\theta)}\,F_{f,B}(u_\theta)\,\mathrm{d}\theta + \frac{1}{2}\int_L |\mathrm{e}^t - \mathrm{e}^{-t}|\,F_{f,L}(h)\,\mathrm{d}L$$
$$= -\int_B \Delta(u_\theta)F_{f,B}(u_\theta)\,\mathrm{d}\theta + \int_{\mathbb{R}} |\mathrm{e}^t - \mathrm{e}^{-t}|\,F_{h,L}(a_t)\,\mathrm{d}t$$

where $h = (f + f_\gamma)/2$.

Proof. This follows from the differential analysis sketched above; the factor $\frac{1}{2}$ comes because the map φ_{hyp} has degree 2.

The companion integral on the Lie algebra is defined by

$$\psi_{g,\mathfrak{b}}(\theta(X-Y)) = 2\mathrm{i}\theta \int_{G/B} g(\theta(X-Y)^x)\,\mathrm{d}(G/B) \quad (\theta \neq 0)$$
$$= (2\pi)^{-1}2\mathrm{i}\theta \int_G g(\theta(X-Y)^x)\,\mathrm{d}G$$

For any Euclidean space V and any open set $U \subset V$ let $\mathscr{S}(U)$ denote the space of all $f \in C^\infty(U)$ such that each derivative of f is $O((1+|\cdot|)^{-n})$ for every $n \geqslant 1$. Exactly as one does for $\mathscr{S}(V)$, we shall view $\mathscr{S}(U)$ as a Fréchet space.

Proposition 14. *$\psi_{g,\mathfrak{b}}$ is well defined for all $g \in \mathscr{S}(\mathfrak{g})$. If $\mathfrak{b}_\delta = \{\theta(X-Y)\,|\,|\theta| > \delta\}$, $\psi_{g,\mathfrak{b}} \in \mathscr{S}(\mathfrak{b}_\delta)$ for any fixed $\delta > 0$, and the map $g \mapsto \psi_{g,\mathfrak{b}}$ from $\mathscr{S}(\mathfrak{g})$ to $\mathscr{S}(\mathfrak{b}_\delta)$ is continuous.*

Proof. The integral for $\psi_{g,\mathrm{b}}$ certainly makes sense for $g \in C_c^\infty(\mathfrak{g})$. We now rewrite the integral using the polar decomposition. We have

$$\psi_{g,\mathrm{b}}(\theta(X-Y)) = (2\pi)^{-1}\mathrm{i}\theta \iiint_{\substack{0 \leqslant \theta_i < 2\pi \\ t>0}} g(\theta(X-Y)^{u_{\theta_1}a_t}) \sinh 2t \, \mathrm{d}\theta_1 \, \mathrm{d}\theta_2 \, \mathrm{d}t$$

$$= \frac{\mathrm{i}\theta}{2}\int_0^\infty \bar{g}(\theta(\mathrm{e}^{2t}X - \mathrm{e}^{-2t}Y))(\mathrm{e}^{2t} - e^{-2t})\,\mathrm{d}t$$

where $\bar{g} = \int g^{u_\theta}\mathrm{d}\theta$. So

$$\psi_{g,\mathrm{b}}(\theta(X-Y)) = \frac{\mathrm{i}\theta}{2} I(\bar{g}:\theta)$$

where

$$I(\bar{g}:\theta) = \int_0^\infty \bar{g}(\theta(\mathrm{e}^{2t}X - \mathrm{e}^{-2t}Y))(\mathrm{e}^{2t} - \mathrm{e}^{-2t})\,\mathrm{d}t$$

We now consider $g \in \mathscr{S}_0(\mathfrak{g})$ and show that this integral makes sense. Since $g \mapsto \bar{g}$ is continuous map of $\mathscr{S}_0(\mathfrak{g})$ to itself,

$$\mu_n(g) = \sup|\bar{g}(Z)|(1 + \|Z\|)^n < \infty$$

for $g \in \mathscr{S}_0(\mathfrak{g})$ and μ_n is a continuous seminorm. Now we use the estimate

$$\|\theta(\mathrm{e}^{2t}X - \mathrm{e}^{-2t}Y)\| \geqslant \tfrac{1}{2}|\theta|\mathrm{e}^{2t}$$

to conclude that, for any $n \geqslant 1$,

$$\int_0^\infty |\bar{g}(\theta(\mathrm{e}^{2t}X - \mathrm{e}^{-2t}Y))|(\mathrm{e}^{2t} - \mathrm{e}^{-2t})\,\mathrm{d}t$$

is majorized by

$$\mu_{n+1}(g)2^{n+1}|\theta|^{-(n+1)}\int_0^\infty \mathrm{e}^{-2(n+1)t}(\mathrm{e}^{2t} - \mathrm{e}^{-2t})\,\mathrm{d}t$$

$$= c_n\mu_{n+1}(g)|\theta|^{-(n+1)} \quad (c_n > 0 \text{ a constant})$$

So I is well defined for $g \in \mathscr{S}_0(\mathfrak{g})$. We next assume g to be in the Schwartz space and estimate the derivatives of I with respect to θ by the same method. Since

$$(\mathrm{d}/\mathrm{d}\theta)\bar{g}(\theta(\mathrm{e}^{2t}X - \mathrm{e}^{-2t}Y)) = (\mathrm{e}^{2t}\partial(X) - \mathrm{e}^{-2t}\partial(Y))\bar{g}(\theta(e^{2t}X - \mathrm{e}^{-2t}Y))$$

it follows that $(\mathrm{d}/\mathrm{d}\theta)^r\bar{g}(\theta(\mathrm{e}^{2t}X - \mathrm{e}^{-2t}Y))$ equals

$$\sum_{\substack{p,q \geqslant 0 \\ p+q=r}} (-1)^q\binom{r}{p}(\partial(X)^p\partial(Y)^q\bar{g})(\theta(\mathrm{e}^{2t}X - \mathrm{e}^{-2t}Y))\mathrm{e}^{2(p-q)t}$$

The worst exponential is e^{2rt}; in addition there is the overall density factor $(\mathrm{e}^{2t} - \mathrm{e}^{-2t})$. So if we use the same method as before and write, for $n \geqslant 1$,

$$\nu_{n,r}(g) = \sum \binom{r}{p} \mu_{n+r+1}(\partial(X)^p \partial(Y)^q \bar{g})$$

we find that

$$\int_0^\infty |(d/d\theta)^r \bar{g}(\theta(e^{2t}X - e^{-2t}Y))|(e^{2t} - e^{-2t})\, dt \leqslant c_{n,r}\nu_{n,r}(g)|\theta|^{-(n+r+1)}$$

where $c_{n,r}$ is a constant >0 independent of g, θ. This estimate shows that $\psi_{g,\mathfrak{b}}$ is well defined on $\mathfrak{b}' = \mathfrak{b}\backslash(0)$ for all $g \in \mathscr{S}(\mathfrak{g})$, lies in $\mathscr{S}(\mathfrak{b}_\delta)$ for any $\delta > 0$, and that the maps $g \mapsto \psi_{g,\mathfrak{b}}$ from $\mathscr{S}(\mathfrak{g})$ to $\mathscr{S}(\mathfrak{b}_\delta)$ are continuous.

The link between $F_{f,B}$ and $\psi_{g,\mathfrak{b}}$ is exactly the same as in the hyperbolic case. If j is, as before, an invariant element of $C^\infty(G), = 1$ on $G(\varepsilon'/4)$ and with support contained in $G(\varepsilon/2)$, we have, for all small $\theta \neq 0$

$$F_{f,B}(u_\theta) = ((e^{i\theta} - e^{-i\theta})/2i\theta)\psi_{g,\mathfrak{b}}(\theta(X - Y))$$

where $f \in C_c^\infty(G(\varepsilon'))$ and $g \in C_c^\infty(\mathfrak{g}(\varepsilon))$ is defined by

$$g(Z) = (jf)(\exp Z) \quad (Z \in \mathfrak{g}(\varepsilon))$$

The problem now is to investigate the behaviour of $F_{f,B}$ and its derivatives near 1, and relate this to $F_{f,L}$. This of course comes down to the corresponding questions for $\psi_{g,\mathfrak{b}}$ and $\psi_{g,\mathfrak{a}}$. The analysis given in Proposition 14 is not adequate to handle this question; the reason is that in the formula for $\psi_{g,\mathfrak{b}}$ as an integral over $(0, \infty)$, θ and e^{2t} are coupled, so that the estimates that are good for large t are not so good for small θ. So a deeper study has to be made. The most natural way to do this is to exploit certain differential equations satisfied by the $F_{f,B}$ and $\psi_{g,\mathfrak{b}}$. For this we need to introduce and study the 'radial components' of the invariant differential operators on G and $\mathfrak{g}$.

6.3 Radial components of invariant differential operators on G and $\mathfrak{g}$ and their relation to orbital integrals

If one takes the Laplace operator Δ in $\mathbb{R}^3$, expresses it in spherical coordinates and retains the terms that involve only the radial coordinate r, one obtains a second-order differential operator $\tilde{\Delta}$ in r which we call the *radial component of* Δ: it is characterized uniquely by the requirement that the action of Δ on the rotation-invariant functions is the same as that of $\tilde{\Delta}$. We shall now generalize this to the two contexts we are working in: the action of G on itself and on $\mathfrak{g}$.

Let D be any smooth differential operator defined on $G'_L = (L')^G$

(resp. $G'_B = (B')^G$) and invariant with respect to the action of G. A smooth differential operator D_L on L' (resp. D_B on B') is said to be a *radial component of D on L'* (resp. on B') if for all open sets $E \subset L'$ (resp. $E \subset B'$) and all invariant C^∞ functions f on E^G we have, writing f_E for the restriction of f to E,

$$(Df)_E = D_L f_E \quad (\text{resp.}\, (Df)_E = D_B f_E)$$

It follows from rather general principles that such an operator can always be constructed, is unique, and the map which takes D to D_L (or D_B) is an algebra homomorphism.

Let me briefly explain the idea behind this. Given a point x of L' (resp. B') there is an open set E containing x in L' (resp. B') such that the conjugacy map $G/L \times E \to (E)^G$ (resp. $G/B \times E \to E^G$) is a diffeomorphism. For B this is clear since $G/B \times B' \to G_{\text{ell}}$ is already a diffeomorphism; for L, since the fibres of $G/L \times L' \to (L')^G$ are the orbits for the Weyl group, this will be true as soon as E and its image under the Weyl reflection are disjoint. Then the invariant functions may be identified with functions on $G/L \times E$ (resp. $G/B \times E$) that depend only on the second coordinate; the radial component D_E is then the part of D, when viewed on $G/L \times E$ (resp. $G/B \times E$), which involves only differentiations along E. The uniqueness of D_E is obvious and hence one has the homomorphism property also. The uniqueness implies compatibility of the different D_Es and hence leads to a globally defined operator.

Theorem 15 (radial components on G). *Let*

$$\Omega = H^2 + 2H + 1 + 4YX$$

Then

(a) $\Omega_B = -\Delta_B^{-1}(\mathrm{d}^2/\mathrm{d}\theta^2)\Delta_B \quad (\Delta_B(u_\theta) = \mathrm{e}^{\mathrm{i}\theta} - \mathrm{e}^{-\mathrm{i}\theta})$,
(b) $\Omega_L = |\Delta_L|^{-1}(\mathrm{d}^2/\mathrm{d}t^2)|\Delta_L| \quad (\Delta_L(ma_t) = \mathrm{e}^t - \mathrm{e}^{-t})$.

Proof. We have already observed the existence and uniqueness of the radial components. So it is only a question of checking that the two sides of these equations agree on an algebra $\mathscr{A}$ of functions on L (or B) which contains a coordinate function at each point of L' (or B'); this is because a differential operator which kills all polynomials in the coordinate variables is necessarily 0. For $\mathscr{A}$ we take the *linear span of the irreducible finite-dimensional characters* of G; $\mathscr{A}$ is then an *algebra* and so we shall show that the two sides of (a) and (b) agree on each irreducible character. If ϕ_m (m integer > 0) is the character of the irreducible representation of dimension m,

$$\phi_2(u_\theta) = e^{i\theta} + e^{-i\theta}, \quad \phi_2(ma_t) = \varepsilon(m)(e^t + e^{-t})$$

Since $(d/d\theta)\phi_2(u_\theta) = i(e^{i\theta} - e^{-i\theta})$, and $(d/dt)\phi_2(ma_t) = \varepsilon(m)(e^t - e^{-t})$, these derivatives never vanish on B' and L', and so ϕ_2 may be used as a local coordinate variable on B' as well as L'.

Let us now verify that, for any integer $m > 0$, the two sides of (a) and (b) give the same result when applied to ϕ_m. Now ϕ_m is an eigenfunction for Ω with eigenvalue m^2; in fact, if π_m is the representation in question with infinitesimal character χ_m, $\Omega\phi_m = \chi_m(\Omega)\phi_m$; and, if v is the highest-weight vector for π_m, we have $\pi_m(X)v = 0$, $\pi_m(H)v = (m-1)v$, so that $\pi_m(\Omega)v = m^2 v$, showing $\chi_m(\Omega) = m^2$. Hence

$$\Omega_B\phi_{m,B} = m^2\phi_{m,B}, \quad \Omega_L\phi_{m,L} = m^2\phi_{m,L}$$

On the other hand, by Weyl's formula,

$$\begin{aligned} -\Delta_B^{-1}\frac{d^2}{d\theta^2}\Delta_B(\phi_{m,B}) &= -\Delta_B^{-1}\frac{d^2}{d\theta^2}(e^{im\theta} - e^{-im\theta}) \\ &= m^2\Delta_B^{-1}(e^{im\theta} - e^{-im\theta}) \\ &= m^2\phi_{m,B} \end{aligned}$$

Similarly we find that

$$\Omega_L = \Delta_L^{-1}\frac{d^2}{dt^2}\Delta_L$$

However L' is precisely the set where Δ_L (which is real) is $\neq 0$, so that $|\Delta_L| = \operatorname{sgn}(\Delta_L)\cdot\Delta_L$, and hence

$$\Omega_L = |\Delta_L|^{-1}\frac{d^2}{dt^2}|\Delta_L|$$

For the action of G on $\mathfrak{g}'$, the radial components are defined in an entirely analogous manner. The invariant differential operator considered now is the Casimir operator on $\mathfrak{g}$. Let

$$\square = H^2 + 4XY = -(X-Y)^2 + (X+Y)^2 + H^2$$

We view $\square$ as an element of degree 2 in the symmetric algebra over $\mathfrak{g}$ and hence as a differential operator on $\mathfrak{g}$. It is obviously invariant on $\mathfrak{g}$.

Theorem 16 (radial components on $\mathfrak{g}$). *Let $\square$ be as above. Then the radial components $\square_{\mathfrak{a}}$ and $\square_{\mathfrak{b}}$ of $\square$ on $\mathfrak{a}'$ and $\mathfrak{b}'$ are given by*

(a) $\square_{\mathfrak{b}} = -\pi_{\mathfrak{b}}^{-1}d^2/d\theta^2\pi_{\mathfrak{b}} \quad \pi_{\mathfrak{b}}(\theta(X-Y) = 2i\theta)$

(b) $\square_{\mathfrak{a}} = -|\pi_{\mathfrak{a}}|^{-1}d^2/dt^2|\pi_{\mathfrak{a}}| \quad (\pi_{\mathfrak{a}}(tH) = 2t).$

Proof. Let ω be the polynomial

$$\omega(Z) = \tfrac{1}{2}\operatorname{tr}(Z^2)$$

Then ω is invariant; $\omega(tH) = t^2$, $\omega(\theta(X-Y)) = -\theta^2$ and so $\omega_{\mathfrak{a}}$ and $\omega_{\mathfrak{b}}$ may be used as local coordinates on $\mathfrak{a}'$ and $\mathfrak{b}'$ respectively. So, as in the proof of the previous theorem it is enough to prove that the two sides of (a) and (b) agree on ω^m for all integers $m \geqslant 1$. Denote the right sides of (a) and (b) by $\Box'_{\mathfrak{b}}$ and $\Box'_{\mathfrak{a}}$. Then

$$\Box'_{\mathfrak{b}}\omega^m_{\mathfrak{b}} = (2m+1)(2m)\omega^{m-1}_{\mathfrak{b}}$$
$$\Box'_{\mathfrak{a}}\psi^m_{\mathfrak{a}} = (2m+1)(2m)\omega^{m-1}_{\mathfrak{a}}$$

(in deriving the second we must note, as in Theorem 15, that $|\pi_{\mathfrak{a}}|^{-1} \times (\mathrm{d}^2/\mathrm{d}t^2)|\pi_{\mathfrak{a}}| = \pi_{\mathfrak{a}}^{-1}(\mathrm{d}^2/\mathrm{d}t^2)\pi_{\mathfrak{a}}$ on $\mathfrak{a}'$). On the other hand, in the linear coordinates a, b, c relative to the basis H, $X-Y$, $X+Y$, we have

$$\Box = \partial^2/\partial a^2 - \partial^2/\partial b^2 + \partial^2/\partial c^2$$

$$\omega = a^2 - b^2 + c^2$$

Hence, after a simple calculation we get

$$\Box\omega^m = (2m+1)(2m)\omega^{m-1}$$

This shows that $\Box_{\mathfrak{h}}\omega^m_{\mathfrak{h}} = (2m+1)(2m)\omega^{m-1}_{\mathfrak{h}}$ for $\mathfrak{h} = \mathfrak{a}, \mathfrak{b}$.

These formulae lead at once to the following remarkable consequence for the orbital integrals.

Theorem 17. *For any $f \in C_c^\infty(G)$, we have*

$$F_{\Omega f, L} = \Omega_L f_{f,L} = \frac{\mathrm{d}^2}{\mathrm{d}t^2} F_{f,L}$$

$$F_{\Omega f, B} = \Omega_B F_{f,B} = -\frac{\mathrm{d}^2}{\mathrm{d}\theta^2} F_{f,B}$$

For any $f \in \mathcal{S}(\mathfrak{g})$, we have likewise

$$\psi_{\Box g, \mathfrak{a}} = \Box_{\mathfrak{a}}\psi_{g,\mathfrak{a}} = \frac{\mathrm{d}^2}{\mathrm{d}t^2}\psi_{g,\mathfrak{a}}$$

$$\psi_{\Box g, \mathfrak{b}} = \Box_{\mathfrak{b}}\psi_{g,\mathfrak{b}} = -\frac{\mathrm{d}^2}{\mathrm{d}\theta^2}\psi_{g,\mathfrak{b}}$$

Proof. These are statements on the regular sets of the appropriate Cartan subgroups and the subalgebras. By the continuity of ψ it is enough to prove the relations for $g \in C_c^\infty(\mathfrak{g})$. Moreover, it is easy to see that if $x_0 \in L'$ (resp. B') and E is an open neighbourhood of x_0 in L' (or B'), there

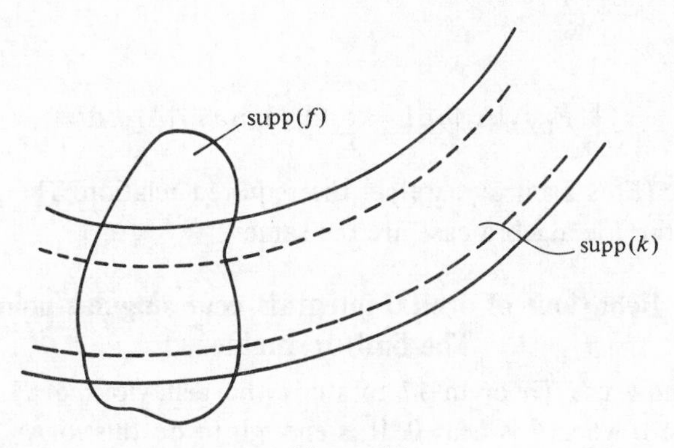

Figure 6.1

is an invariant C^∞ function k on G which is 1 in an invariant neighbourhood V of x_0 and whose support lies within E^G; for instance we may write $u(x) = \mathrm{tr}\,(x) - \mathrm{tr}\,(x_0)$ and take $k(x) = g(u(x))$ where $g \in C_c^\infty(\mathbb{R})$ is 1 around 0 and vanishes outside an arbitrarily small neighbourhood of 0. Then, for any $f \in C_c^\infty(G)$, $f = kf$ on V and $F_{f,\cdot} = F_{kf,\cdot}$ around x_0 in L (resp. B) (see Figure 6.1). As kf is in $C_c(E^G)$, it is clear that we need only prove the first set of relations for f in $C_c(E^G)$.

The argument is now essentially formal. Consider the case of L for instance. By taking E sufficiently small we may assume that any $h \in C_c^\infty(E)$ is the restriction to E of a unique invariant $h^* \in C^\infty(E^G)$. Then, for any $f \in C_c^\infty(E^G)$,

$$\begin{aligned}\int \Omega f \cdot h^* \,\mathrm{d}G &= \int f \cdot \Omega h^* \,\mathrm{d}G \quad (\text{as } \Omega \text{ has order 2})\\ &= \frac{1}{2}\int_E |\Delta_L| F_{f,L} (\Omega h^*)_L \,\mathrm{d}L\\ &= \frac{1}{2}\int_E F_{f,L} (\mathrm{d}^2/\mathrm{d}t^2)(|\Delta_L| h)\,\mathrm{d}L\end{aligned}$$

since $|\Delta_L|(\Omega h^*)_L = |\Delta_L|\Omega_L h = (\mathrm{d}^2/\mathrm{d}t^2)(|\Delta_L| h)$. Since $F_{f,L}$ vanishes outside a compact set inside E, this becomes

$$= \frac{1}{2}\int_E (\mathrm{d}^2/\mathrm{d}t^2) F_{f,L} |\Delta_L| h \,\mathrm{d}L$$

On the other hand, we find by the same reasoning that

$$\int \Omega f \cdot h^* \,\mathrm{d}G = \frac{1}{2}\int_E F_{\Omega f,L} |\Delta_L| h \,\mathrm{d}L$$

Hence

$$\int_E F_{\Omega f,L}|\Delta_L|h\,\mathrm{d}L = \int_E (\mathrm{d}^2/\mathrm{d}t^2)F_{f,L}|\Delta_L|h\,\mathrm{d}L$$

Since $h \in C_c^\infty(E)$ is arbitrary, we get the required relation. The proofs for B and for the Lie algebra case are the same.

6.4 Behaviour of orbital integrals near singular points. The limit formulae

We shall now use Theorem 17 to study the behaviour of $F_{f,B}(u_\theta)$ and $\psi_{g,\mathrm{b}}(\theta(X-Y))$ when θ is near 0. It is enough to do this for $\psi_{g,\mathrm{b}}$ since we have already seen how to express $F_{f,B}$ in terms of $\psi_{g,\mathrm{b}}$.

Theorem 18. *For any $f \in C_c^\infty(G)$, $F_{f,B}(u_\theta)$ and all its derivatives with respect to θ are bounded over B'. Moreover, for any integer $m \geqslant 0$ there is a continuous seminorm ν_m on $C_c^\infty(G)$ such that for all $f \in C_c^\infty(G)$*

$$\sup_{u_\theta \in B'} |(\mathrm{d}/\mathrm{d}\theta)^m F_{f,B}(u_\theta)| \leqslant \nu_m(f)$$

Theorem 19. *For any $g \in \mathscr{S}(\mathfrak{g})$, $\psi_{g,\mathrm{b}} \in \mathscr{S}(\mathrm{b}\backslash(0))$; and the map $g \mapsto \psi_{g,\mathrm{b}}$ from $\mathscr{S}(\mathfrak{g})$ to $\mathscr{S}(\mathrm{b}\backslash(0))$ is continuous.*

In view of the preceding remarks it is enough to prove Theorem 19. Since we already know that $g \mapsto \psi_{g,\mathrm{b}}$ is continuous from $\mathscr{S}(\mathfrak{g})$ to $\mathscr{S}(\mathrm{b}_\delta)$ for any $\delta > 0$ where $\mathrm{b}_\delta = \{\theta(X-Y) \mid |\theta| > \delta\}$, it is enough to prove that, for any integer $m \geqslant 0$ and fixed $\delta > 0$,

$$\sup_{0<|\theta|\leqslant\delta} |(\mathrm{d}/\mathrm{d}\theta)^m \psi_{g,\mathrm{b}}(\theta(X-Y))| \leqslant \nu_m(g)$$

for all $g \in \mathscr{S}(\mathfrak{g})$, ν_m being a continuous seminorm on $\mathscr{S}(\mathfrak{g})$.

From Proposition 14 we know that, for any integers $r \geqslant 0$ and $n \geqslant 1$, we have

$$|(\mathrm{d}/\mathrm{d}\theta)^r \psi_g(\theta)| \leqslant \nu_{n,r}(g)|\theta|^{-(n+r+1)}$$

for all $\theta \neq 0$, $g \in \mathscr{S}(\mathfrak{g})$; here $\psi_g(\theta) = \psi_{g,\mathrm{b}}(\theta(X-Y))$, and $\nu_{n,r}$ is a continuous seminorm on $\mathscr{S}(\mathfrak{g})$. We fix n, say $n=1$, write $\nu_r = \nu_{1,r}$, so that the estimate becomes

$$|(\mathrm{d}/\mathrm{d}\theta)^r \psi_g(\theta)| \leqslant \nu_r(g)|\theta|^{-(r+2)}$$

Since the power of $|\theta|^{-1}$ appearing on the right grows with r, this estimate is very poor. *We now use Theorem 17 to get an estimate where there is a fixed power of θ on the right*; indeed, by Theorem 17,

$$|(\mathrm{d}/\mathrm{d}t)^{2r}\psi_g(\theta)| = |\psi_{\square^r g}(\theta)| \leqslant \nu_0(\square^r g)|\theta|^{-2}$$

We can now repeatedly integrate these estimates with respect to θ. This will reduce the term involving $|\theta|$ from $|\theta|^{-2}$ to $|\theta|^{-1}$ to $|\log|\theta||$, and finally to a constant, since

$$\int_{0<|\theta|<\delta} |\log|\theta|| \, d\theta < \infty$$

In other words, we have an estimate

$$|(d/d\theta)^r \psi_g(\theta)| \leqslant v_r'(g) \quad (0<|\theta|<\delta)$$

for all $r \geqslant 0$, $g \in \mathscr{S}(\mathfrak{g})$, v_r' being a continuous seminorm on $\mathscr{S}(\mathfrak{g})$. This proves Theorem 19.

Theorem 20. *If r is any integer $\geqslant 0$, the limits*

$$J_{r,\pm}(f) = \lim_{\theta \to 0\pm} (d/d\theta)^r F_{f,B}(u_\theta)$$

exist for all $f \in C_c^\infty(G)$; and $J_{r,\pm}$ are invariant distributions on G. Similarly the limits

$$j_{r,\pm}(g) = \lim_{0 \to 0\pm} \psi_{g,\mathfrak{b}}^{(r)}(\theta(X-Y))$$

exist for all $g \in \mathscr{S}(\mathfrak{g})$; and $j_{r,\pm}$ are tempered invariant distributions on $\mathfrak{g}$.

Proof. By definition $F_{f,B} = F_{f^x,B}$ and $\psi_{g,\mathfrak{b}} = \psi_{g^x,\mathfrak{b}}$ where $f^x(y) = f(x^{-1}yx)$ and $g^x(Z) = g(x^{-1}Zx)$. The existence of the limits follows from the boundedness of *all* the derivatives when $\theta \to 0\pm$, and their continuity in f (resp. g) from the continuity of the maps $f \mapsto F_{f,B}$ (resp. $g \mapsto \chi_{g,\mathfrak{b}}$).

For the actual calculation we shall follow Harish-Chandra and first establish the following lemma.

Lemma 21. *Fix $u \in C_c^\infty(\mathbb{R}^2)$ and let*

$$U(\theta) = \theta \int_0^\infty u(\theta e^{2t}, \theta e^{-2t})(e^{2t} - e^{-2t}) \, dt \quad (\theta \neq 0)$$

Then

(a) *$U(0\pm)$ and $((d/d\theta)U)(0\pm)$ exist,*
(b) *$(d/d\theta)U$ is continuous at $\theta = 0$, and $((d/d\theta)U)(0) = -u(0,0)$,*
(c) *$U(0\pm) = \pm\frac{1}{2}\int_0^\infty u(\pm s, 0) \, ds$,*
(d) *$|((d/d\theta)U)(\theta) + u(0,0)| \leqslant \text{const.}(1 + |\log|\theta||)(0 < |\theta| \leqslant 2)$.*

Proof. The key point is the formal identity

$$\theta(d/d\theta)u(\theta e^{2t}, \theta e^{-2t}) = \tfrac{1}{2}(d/dt)u(\theta e^{2t}, \theta e^{-2t}) + 2\theta u_y(\theta e^{2t}, \theta e^{-2t})$$

($u_y = (\partial/\partial y)u$) which is trivial to check. First, for $\theta > 0$, if $\tau = \theta e^{2t}$, we have

$$U(\theta) = \frac{1}{2}\int_\theta^\infty u(\tau, \theta^2/\tau)\,d\tau - \theta\int_0^\infty u(\theta e^{2t}, \theta e^{-2t})e^{-2t}\,dt$$

As $\theta \to 0+$, the second term $\to 0$ as the integral has a finite limit. The first term $\to \frac{1}{2}\int_0^\infty u(\tau, 0)\,d\tau$ since $|u| \leqslant$ constant and the integration is over a fixed range $0 \leqslant \tau \leqslant T$. The argument for $\theta \to 0-$ is similar, giving the first half of (a) and (c). For the derivative we need a more delicate argument and in particular use the formal identity given above. For $\theta \neq 0$, writing U' for $(d/d\theta)U$ and using the above identity,

$$\begin{aligned} U'(\theta) = &\int_0^\infty u(\theta e^{2t}, \theta e^{-2t})(e^{2t} - e^{-2t})\,dt \\ &+ \frac{1}{2}\int_0^\infty (d/dt)u(\theta e^{2t}, \theta e^{-2t})\cdot(e^{2t} - e^{-2t})\,dt \\ &+ 2\theta\int_0^\infty u_y(\theta e^{2t}, \theta e^{-2t})(e^{2t} - e^{-2t})\,dt \end{aligned}$$

The second term on the right may now be integrated by parts. The boundary terms vanish; and the part involving e^{2t} cancels with the part involving e^{2t} in the first term. Hence

$$\begin{aligned} U'(\theta) = &-2\int_0^\infty u(\theta e^{2t}, \theta e^{-2t})e^{-2t}\,dt \\ &+ 2\theta\int_0^\infty u_y(\theta e^{2t}, \theta e^{-2t})(e^{2t} - e^{-2t})\,dt \end{aligned}$$

The second term $\to 0$ since the integral (without θ) $\to$ finite limits (when $\theta \to 0\pm$) by the first part of (a). The first term tends to

$$-2\int_0^\infty u(0,0)e^{-2t}\,dt = -u(0,0)$$

Finally,

$$\begin{aligned} |U'(\theta) + u(0,0)| \leqslant &\, 2\int_0^\infty |u(\theta e^{2t}, \theta e^{-2t}) - u(0,0)|e^{-2t}\,dt \\ &+ 2|\theta|\int_0^\infty |u_y(\theta e^{2t}, \theta e^{-2t})|(e^{2t} - e^{-2t})\,dt \end{aligned}$$

Since u is bounded, the first term is at most $2\sup|u|$. To estimate the second term let $C > 0$ be such that $u(x, y) = 0$ if $|x| > C$. Then the integration need be only in the range $|\theta|e^{2t} \leqslant C$ or $t \leqslant (\log C - \log|\theta|)/2 = T(\theta)$, and so it is majorized by

$$2\sup|u|\cdot\int_0^{T(\theta)} |\theta|(e^{2t} + 1)\,dt \leqslant 2\sup|u|(C + |\theta|\,T(\theta)) \leqslant \text{const.}\,(1 + |\log|\theta||)$$

Theorem 22 (Limit formula). (a) *For all* $f \in C_c^\infty(G)$, $F'_{f,B}(u_\theta) = (d/d\theta) f_{f,B}(u_\theta)$ *is continuous at* $\theta = 0$; *and*

$$\frac{1}{i} F'_{f,B}(1) = -\pi f(1)$$

(b) *For all* $g \in \mathscr{S}(\mathfrak{g})$, $((d/d\theta)\psi_{g,\mathfrak{b}})(\theta(X - Y))$ *is continuous at* $\theta = 0$; *and*

$$\frac{1}{i}\left(\frac{d}{d\theta}\psi_{g,\mathfrak{b}}\right)(0) = -\pi g(0)$$

Proof. It is clearly enough to prove (b). We have

$$\psi_{g,\mathfrak{b}}(\theta(X - Y)) = \frac{i\theta}{2}\int_0^\infty \bar{g}(\theta e^{2t} X - \theta e^{-2t} Y)(e^{2t} - e^{-2t})\, dt$$

Hence, by the above lemma,

$$\lim_{\theta \to 0\pm}\left(\frac{d}{d\theta}\psi_{g,\mathfrak{b}}\right)(\theta(X - Y)) = -\frac{i}{2}\bar{g}(0) = -\frac{i}{2}g(0)\int d\theta = -i\pi g(0).$$

The orbital integral $F_{f,B}(u_\theta)$ itself will not be continuous at $\theta = 0$. Its jump at $\theta = 0$ is in fact the orbital integral $F_{f,L}(1)$, up to a multiplicative constant. This is the celebrated *jump relation of Harish-Chandra.*

Theorem 23 (Harish-Chandra jump relation). *For all* $f \in C_c^\infty(G)$,

$$\left[\frac{1}{i} F_{f,B}(u_\theta)\right]_{\theta = 0-}^{\theta = 0+} = F_{f,L}(1)$$

For all $g \in \mathscr{S}(\mathfrak{g})$,

$$\left[\frac{1}{i}\psi_{g,\mathfrak{b}}(\theta(X - Y))\right]_{\theta = 0-}^{\theta = 0+} = \psi_{g,\mathfrak{a}}(0)$$

(*Here* $[v(\theta)]_{\theta=0-}^{\theta=0+}$ *denotes* $v(0+) - v(0-)$.)

Proof. We have seen that one can choose an invariant C^∞ function j on G which is 1 on some $G(\varepsilon')$ and with support contained in some $G(\varepsilon'')$ where $0 < \varepsilon' < \varepsilon''$, $G(2\varepsilon'') \subset \exp \mathfrak{g}(\varepsilon)$. Then, for any $f \in C_c^\infty(G)$, $Z \mapsto (jf)(\exp Z)$, $Z \in \mathfrak{g}(\varepsilon)$ defines an element $g \in C_c^\infty(\mathfrak{g}(\varepsilon))$, and

$$F_{f,B}(u_\theta) = \frac{e^{i\theta} - e^{-i\theta}}{2i\theta}\psi_{g,\mathfrak{b}}(\theta(X - Y))$$

$$F_{f,L}(a_t) = \frac{|e^t - e^{-t}|}{2|t|}\psi_{g,\mathfrak{a}}(tH)$$

for $0 < |\theta| < \delta$, $0 < |t| < \delta$, δ being sufficiently small. It is therefore enough to establish the jump relations for $\psi_{g,\mathfrak{b}}$. In view of Lemma 21, and the integral representation for $\psi_{g,\mathfrak{b}}$, we have

$$\psi_{g,\mathfrak{b}}(0\pm) = \pm\frac{\mathrm{i}}{4}\int_0^\infty \bar{g}(\pm sX)\,\mathrm{d}s$$

On the other hand, we know from Theorem 11 that

$$\psi_{g,\mathfrak{a}}(0) = \frac{1}{4}\int_{-\infty}^\infty \bar{g}(sX)\,\mathrm{d}s$$

Hence

$$\frac{1}{\mathrm{i}}\psi_{g,\mathfrak{b}}(0+) - \frac{1}{\mathrm{i}}\psi_{g,\mathfrak{b}}(0-) = \frac{1}{4}\int_{-\infty}^\infty \bar{g}(sX)\,\mathrm{d}s = \psi_{g,\mathfrak{a}}(0)$$

We note the following direct expressions for the limits in the group for later use:

$$F_{f,B}(1\pm) = \pm\frac{\mathrm{i}}{4}\int_0^\infty \bar{f}(n_{\pm s})\,\mathrm{d}s$$

In harmonic analysis of general semisimple groups the jump relations play a crucial role. It may therefore be worthwhile to understand them geometrically. If we identify $\mathfrak{g}$ with $\mathbb{R}^3$ using the basis H, $X+Y$, $X-Y$ we have the picture shown in Figure 6.2.

The orbits $E_{\pm\theta}$ of $\pm\theta(X-Y)$ are the elliptic hyperboloids which lie

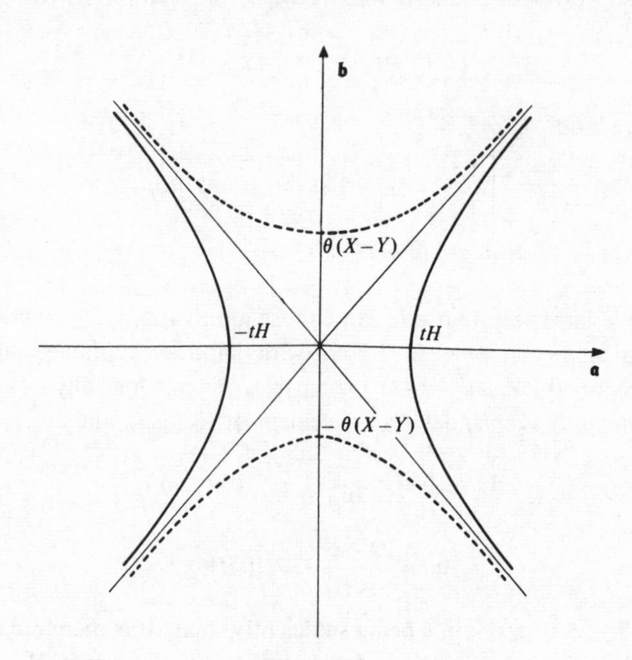

Figure 6.2

within the 'light' cone C of nilpotents, while the orbit H_t of tH (same as the orbit of $-tH$) is the hyperbolic hyperboloid that *wraps around the light cone.* Then $\psi_{g,a}(tH)$ represents the mean value of g on H_t, and, in the limit when $t \to 0$, it will tend to the mean value of g over the *full* cone C; however, $\psi_{g,b}(\pm\theta(X-Y))$ are the mean values of g over $E_{\pm\theta}$ and will tend to the (signed) mean values over the two half-cones $C_{\pm}$. The relation $C = C_+ \coprod C_-$ then leads to the jump relation.

Theorem 23 leads at once to the following result.

Theorem 24. *The derivatives of $F_{f,B}(u_\theta)$ (resp. $\psi_{g,b}(\theta(X-Y))$ of odd order with respect to θ are continuous at $\theta = 0$ and*

$$(1/\mathrm{i})(\mathrm{d}/\mathrm{d}\theta)_{\theta=0}^{2r+1} F_{f,B}(u_\theta) = (-1)^{r+1}\pi(\Omega^r f)(1)$$

$$(1/\mathrm{i})(\mathrm{d}/\mathrm{d}\theta)_{\theta=0}^{2r+1} \psi_{g,b}(\theta(X-Y)) = (-1)^{r+1}\pi(\Box^r g)(0)$$

The jumps of the even order derivatives are given by

$$\left[\frac{1}{\mathrm{i}}(\mathrm{d}/\mathrm{d}\theta)^{2r} F_{f,B}\right]_{\theta=0-}^{\theta=0+} = (-1)^r F_{\Omega^r f,L}(1)$$

$$\left[\frac{1}{\mathrm{i}}(\mathrm{d}/\mathrm{d}\theta)^{2r} \psi_{g,b}\right]_{\theta=0-}^{\theta=0+} = (-1)^r \psi_{\Box^r g,a}(0)$$

Remarks. With this we have completed the basic theory of orbital integrals. The reader should observe the profound maner in which the theory differs from that of complex groups. Although $F_{f,L}$ is smooth and its Fourier transform is the principal series character, $f(1)$ can be recovered only from $F_{f,B}$. However, the discontinuities of $F_{f,B}$ make the study of its Fourier transform harder, and therein lies the source of the real difficulties of obtaining the Plancherel formula. The reader should keep these remarks in mind while going over the derivation of the Plancherel formula in §6.

It is not difficult to show by examples, at least for the Lie algebra, that $g(0)$ cannot be recovered in general form $\psi_{g,a}$. In fact, let $h \in C_c^\infty(\mathfrak{g})$ be such that $\int h\,\mathrm{d}\mathfrak{g} > 0$ but $\mathrm{supp}(h) \subset \mathfrak{g}_{\mathrm{ell}} = (\mathfrak{b}')^G$. Obviously $\psi_{h,a} = 0$. If now $g = \hat{h}$, then $\psi_{g,a} = 2\pi\hat{\psi}_{h,a} = 0$; but $g(0) = \int h\,\mathrm{d}\mathfrak{g} > 0$.

6.5 The discrete series. The modules D_m and their characters

To obtain the Plancherel formula it is necessary to construct the discrete series. We shall show that for $G = SL(2,\mathbb{R})$ the discrete series are the equivalence classes of irreducible unitary representations whose modules of K-finite vectors are the D_m introduced in Chapter 5, §5. We shall also calculate their characters; our method rests on the Harish-Chandra regularity theorem which we shall establish in the next chapter. The derivation of the Plancherel formula will of course make use of the explicit formula for the characters of the D_m.

Let S be a separable locally compact unimodular group. Let π be an irreducible unitary representation of S, and $[\pi]=\omega$ the equivalence class of π. We shall say π or ω *belongs to the discrete series of* S if it can be realized as a direct summand of the regular representation of S. We write $\hat{S}_{\mathrm{d}}$ for the set of all such ω. If S is compact, $\hat{S}_{\mathrm{d}}=\hat{S}$. In this case, the matrix elements of the irreducible representations satisfy the well-known orthogonality relations. It turns out that for general S the discrete-series representations constitute a striking generalization of the finite-dimensional representations of compact groups. I shall begin by describing without proofs the basic facts concerning the discrete series of an arbitrary separable locally compact unimodular group S (cf. [V3]).

The first fact is that, if π is an irreducible unitary representation of S in a Hilbert space $\mathscr{H}$, $[\pi]\in\hat{S}_\alpha$ if and only if, for *some nonzero* $u,v\in\mathscr{H}$, the function

$$x\mapsto(\pi(x)u\,|\,v)$$

lies in $L^2(S)$. In this case, *all* the matrix elements $x\to(\pi(x)u'\,|\,v')(u',v'\in\mathscr{H})$ are in $L^2(S)$; moreover, there is a constant $d(\omega)>0$ such that

$$\int_S(\pi(x)u\,|\,v)\overline{(\pi(x)u'\,|\,v')}\,\mathrm{d}x=(u\,|\,u')\overline{(v\,|\,v')}\cdot d(\omega)^{-1}$$

for all $u,v,u',v'\in\mathscr{H}$. By analogy with the compact case we call $d(\omega)$ the *formal degree* of ω. Of course when S is noncompact $d(\omega)$ depends on the normalization of $\mathrm{d}x$. The discrete series representations are often called *square integrable* because of the above property of square integrability of matrix elements. The closed linear span of the matrix elements of π in $L^2(S)$ is denote by $A(\omega)$. It is stable under both right and left translations. Under right (resp. left) translations $A(\omega)$ decomposes as a direct sum of $\dim(\omega)$ copies of ω (resp. ω^*), and $A(\omega)$ contains every closed subspace of $L^2(S)$ transforming according to ω (resp. ω^*) under right (resp. left) translations. If $\omega,\omega'\in\hat{S}_\alpha$ and $\omega\neq\omega'$, $A(\omega)\perp A(\omega')$. The orthogonal direct sum

$${}^\circ L^2(S)=\bigoplus_{\omega\in\hat{S}_\alpha}A(\omega)$$

is the closed linear span of all subspaces which are irreducible under right (left) translations. It is often called the *discrete part* of $L^2(S)$.

For any representation π of the discrete series and any $f\in C_c(S)$, the operator $\pi(f)$ is of Hilbert–Schmidt class; if $|||\cdot|||$ denotes the Hilbert–Schmidt norm, we have, with $E_{\check{\omega}}$ as the orthogonal projection $L^2(S) \to A(\check{\omega})$,

$$|||\pi(f)|||^2 = \|E_{\check{\omega}} f\|^2 \quad (\|\cdot\| = L^2 \text{ norm})$$

In particular, it is possible to define $\pi(f)$ for all $f \in L^2(S)$ in a unique manner, extending the usual definition when $f \in C_c(S)$, such that $f \mapsto \pi(f)$ is a unitary isomorphism of $A(\omega)$ with $L^2(\mathscr{H})$, the Hilbert space (under $|||\cdot|||$) of Hilbert–Schmidt operators on $\mathscr{H}$, and $\pi(f) = 0$ for $f \perp A(\check{\omega})$. For any $f \in C_c(S)$, the operator $\pi(f * \tilde{f}) = \pi(f)\pi(f)^\dagger$ is of trace class and let us write

$$\Theta_\omega(f * \tilde{f}) = \operatorname{tr}(\pi(f)\pi(f)^\dagger)$$

Of course $\Theta_\omega(f * \tilde{f}) \geqslant 0$. Let ${}^\circ E$ denote the orthogonal projection

$${}^\circ E : L^2(S) \to {}^\circ L^2(S)$$

Then one has the *discrete Plancherel formula*:

$$\|{}^\circ E f\|^2 = \sum_{\omega \in \hat{S}_{\mathrm{d}}} d(\omega)\Theta_\omega(f * \tilde{f})$$

It is hardly possible to have a more thoroughgoing generalization of the compact theory.

Let us now return to the case of $G = SL(2, \mathbb{R})$. If V is a $(\mathfrak{g}, K)$-module, we shall say that V is *in the discrete series* if there is an $\omega \in \hat{G}_{\mathrm{d}}$ such that the modules of K-finite vectors of the representations in the class ω are isomorphic to V. By Theorem 20 of Chapter 5 we know that ω is then uniquely determined. Let us recall the modules D_m defined in Chapter 5, §5.

Theorem 25. *The D_m are in $\hat{G}_{\mathrm{d}}$ $(m \neq 0)$.*

Proof. The idea behind the proof is very simple. We realize D_m within the appropriate (nonunitary) principal series representation π and suppose for example $m > 0$; then we construct the matrix element f_{m+1} defined by the vector $e_{(m+1)}(f_{m+1}(x) = (\pi(x)e_{m+1} | e_{m+1}))$ which is the vector of lowest weight $m + 1$. The relations

$$H' \cdot e_{m+1} = (m+1)e_{m+1}, \quad Y' \cdot e_{m+1} = 0$$

then lead to *differential equations* for f_{m+1}. The transformation property

$$f_{m+1}(u_{\theta_1} x u_{\theta_2}) = \mathrm{e}^{\mathrm{i}(m+1)(\theta_1 + \theta_2)} f_{m+1}(x)$$

then implies that *$f(a_t)$ satisfies certain ordinary differential equation in t.* These equations can be explicitly solved, so that we obtain an explicit formula for f_{m+1}. From this we can conclude that $f_{m+1} \in L^2(G)$. The closed linear span of the right translates of f_{m+1} in $L^2(G)$ can then be proved to be irreducible, and defines the required ω.

The first step is to express the various differential operators which arise, in polar coordinates. Let $\varphi(K \times A_+ \times K \to G)$ be the map

$$\varphi(u_{\theta_1}, a_t, u_{\theta_2}) = u_{\theta_1} a_t u_{\theta_2} \quad (t > 0)$$

We have already calculated $\mathrm{d}\varphi$ and $\mathrm{d}\varphi^{-1}$ in §2. Since $\mathrm{d}\varphi$ is bijective everywhere on $K \times A_+ \times K$, it follows that any analytic differential operator D on $G^+ = KA_+K$ gives rise to a unique differential operator D^φ on $K \times A_+ \times K$ such that, for any $u \in C^\infty(G^+)$, $(Du)\circ\varphi = D^\varphi(u\circ\varphi)$. For instance, if $D = Z \in \mathfrak{g}$, $D^\varphi = (\mathrm{d}\varphi)^{-1}(Z)$. Using the formulae obtained in §2 we get

$$H'^\varphi = -\mathrm{i}\partial/\partial\theta_2$$

$$2Y'^\varphi = \frac{\mathrm{i}\mathrm{e}^{-2\mathrm{i}\theta_2}}{\sinh 2t}\partial/\partial\theta_1 + \mathrm{e}^{-2\mathrm{i}\theta_2}\partial/\partial t - \mathrm{i}\frac{\cosh 2t}{\sinh 2t}\mathrm{e}^{-2\mathrm{i}\theta_2}\partial/\partial\theta_2$$

$$2X'^\varphi = \frac{-\mathrm{i}\mathrm{e}^{2\mathrm{i}\theta_2}}{\sinh 2t}\partial/\partial\theta_1 + \mathrm{e}^{2\mathrm{i}\theta_2}\partial/\partial t + \mathrm{i}\frac{\cosh 2t}{\sinh 2t}\mathrm{e}^{2\mathrm{i}\theta_2}\partial/\partial\theta_2$$

For higher-order operators we simply use the fact that $D \mapsto D^\varphi$ is a homomorphism. So, as $\Omega^\varphi = (H'^\varphi + 1)^2 + 4Y'^\varphi X'^\varphi$, we have the following.

Lemma 26. *Ω^φ is given by*

$$\Omega^\varphi = \left[\frac{1}{\sinh^2 2t}\left(\frac{\partial^2}{\partial\theta_1^2} + \frac{\partial^2}{\partial\theta_2^2}\right) - 2\frac{\cosh 2t}{\sinh^2 2t}\frac{\partial^2}{\partial\theta_1\partial\theta_2}\right] + \frac{\partial}{\partial t^2} + 2\frac{\cosh 2t}{\sinh 2t}\frac{\partial}{\partial t} + 1$$

Proof. Compute.

We now go back to Theorem 32 of Chapter 5. We suppose $m > 0$ and consider the nonunitary principal series representation $\pi = \pi_{\varepsilon,m}$, ε being of opposite parity to m. Then

$$D_m \cong \langle e_{m+1}, e_{m+3}, \ldots \rangle, \quad D_{-m} \cong \langle \ldots, e_{-m-3}, e_{-m-1} \rangle$$

$$Y' e_{m+1} = 0, \quad X' e_{-m-1} = 0$$

Write

$$f_\pm(x) = (\pi(x)e_{\pm(m+1)} | e_{\pm(m+1)})$$

In what follows we only work with D_m; the details for D_{-m} are similar. If $f(t) = f_+(a_t)$, we have

$$f_+(u_{\theta_1} a_t u_{\theta_2}) = \mathrm{e}^{\mathrm{i}(m+1)(\theta_1+\theta_2)} f(t)$$

and so the equation $Y' f_+ = 0$ becomes, when $\theta_1 = \theta_2 = 0$,

$$\frac{\mathrm{d}f}{\mathrm{d}t} = -(m+1)\frac{\cosh 2t - 1}{\sinh 2t} f = -(m+1)(\tanh t) f$$

Hence

$$f(t) = \text{const.}\,(\cosh t)^{-(m+1)}$$

so that

$$\int_G |f_{m+1}(x)|^2\,\mathrm{d}G = \text{const.}\int_0^\infty (\cosh t)^{-2(m+1)}\sinh 2t\,\mathrm{d}t < \infty$$

since $m \geqslant 1$ and the integrand is $\mathrm{O}(\mathrm{e}^{-2mt})$. On the other hand, as $t \to 0+$, $f(t)$ must tend to 1. Hence

$$f_{m+1}(x) = \mathrm{e}^{\mathrm{i}(m+1)(\theta_1+\theta_2)}(\cosh t)^{-(m+1)} \quad (x = u_{\theta_1}a_t u_{\theta_2})$$

Likewise,

$$f_{-(m+1)}(x) = \mathrm{e}^{-\mathrm{i}(m+1)(\theta_1+\theta_2)}(\cosh t)^{-(m+1)} \quad (x = u_{\theta_1}a_t u_{\theta_2})$$

In particular

$$\bar{f}_{(m+1)} = f_{-(m+1)}$$

We now consider the closed linear span $\mathscr{H}$ of right translates of f_{m+1} in $L^2(G)$. We want to prove that this is irreducible and that the $(\mathfrak{g}, K)$-module of its K-finite vectors is isomorphic to D_m. For this it is sufficient to show that f_{m+1} is an analytic vector for the right regular representation. For, assume that we have shown this. Then $\mathscr{H} = \mathrm{Cl}(U(\mathfrak{g})\cdot f_{m+1})$; and, as $f_{m+1}(x; a) = (\pi(x)\pi(a)e_{m+1}|e_{m+1})$, $\pi(a)e_{m+1} = 0 \Rightarrow af_{m+1} = 0$ and hence $U(\mathfrak{g})f_{m+1}$ is a quotient of D_m, hence isomorphic to D_m because D_m is irreducible. This implies that $\mathscr{H}^0 = U(\mathfrak{g})f_{m+1}$.

We shall check weak analyticity. So we must show that $a(x) = (r(x)f_{m+1}|g)$ is analytic in x, r being the right regular representation and $g \in L^2(G)$. The function a is continuous; and

$$a(x) = \int_G f_{m+1}(yx)\overline{g(y)}\,\mathrm{d}y.$$

We shall view a as a distribution and show that it is an eigendistribution for $(X-Y)^2$ as well as Ω. If E is either of these and $b \in C_c^\infty(G)$,

$$\int aEb\,\mathrm{d}x = \int\int (Eb)(x)f_{m+1}(yx)\overline{g(y)}\,\mathrm{d}y\,\mathrm{d}x$$

The integral is absolutely convergent since

$$\int |f_{m+1}(yx)\overline{g(y)}|\,\mathrm{d}y \leqslant \|r(x)f_{m+1}\|\,\|\bar{g}\| = \|f_{m+1}\|\,\|\bar{g}\|$$

Hence

$$\int aEb\,\mathrm{d}x = \int \overline{g(y)}\,\mathrm{d}y \int (Eb)(x)\, f_{m+1}(yx)\,\mathrm{d}x$$

$$= \int \overline{g(y)}\,\mathrm{d}y \int b(x)\, f_{m+1}(yx{:}E)\,\mathrm{d}x$$

$$= \mu \int\!\!\int b(x)\, f_{m+1}(yx) g(y)\,\mathrm{d}x\,\mathrm{d}y$$

since f_{m+1} is an eigenfunction for E. Hence

$$\int aEb\,\mathrm{d}x = \mu \int ab\,\mathrm{d}x$$

proving that $Ea = \mu a$ as distributions. So a is an eigendistribution for $\Omega + 2(X-Y)^2 = (X-Y)^2 + H^2 + (X+Y)^2 + 1$ which is analytic and *elliptic*. So, by the regularity theorem for elliptic operators, a is analytic. This completes the proof of Theorem 25.

It is of interest to obtain another explicit description of f_{m+1}. Let σ be the two dimensional representation of G and let the unit vector v be such that $\sigma(H')v = -v$, $\sigma(Y')v = 0$. Since

$$H' = \begin{pmatrix} 0 & -\mathrm{i} \\ \mathrm{i} & 0 \end{pmatrix}, \quad Y' = \begin{pmatrix} 1/2 & -\mathrm{i}/2 \\ -\mathrm{i}/2 & -1/2 \end{pmatrix}$$

a simple calculation gives $v = \begin{pmatrix} \mathrm{i}/\sqrt{2} \\ 1/\sqrt{2} \end{pmatrix}$. Hence, if $g(x) = (\sigma(x)v \mid v)$, $g(u_{\theta_1} x u_{\theta_2}) = \mathrm{e}^{-\mathrm{i}(\theta_1+\theta_2)} g(x)$, and

$$g(x) = ((a+d) - \mathrm{i}(b-c))/2, \quad x = \begin{pmatrix} a & b \\ c & d \end{pmatrix}$$

In particular, $g(a_t) = \cosh t$. Thus

$$f_{m+1} = g^{-(m+1)}$$

The explicit formula for f_{m+1} allows us to calculate its formal degree. We have the following.

Theorem 27. *With* dG *as the Haar measure on G, the formal degree of $D_{\pm m}$ is $m/2\pi^2$:*

$$\mathrm{d}(D_{\pm m}) = m/2\pi^2 \quad (m \geqslant 1)$$

Proof. We have

$$\mathrm{d}G = \tfrac{1}{2} \sinh 2t\,\mathrm{d}\theta_1\,\mathrm{d}\theta_2\,\mathrm{d}t$$

Hence writing $f = f_{\pm(m+1)}$ we have

$$\int_G |f(x)|^2\, dG = 4\pi^2 \int_0^\infty (\cosh t)^{-(2m+1)} \sinh t\, dt$$
$$= 4\pi^2 \int_1^\infty u^{-(2m+1)}\, du$$
$$= 2\pi^2/m$$

Remark. If $m \geqslant 2$ we have, exactly from the same type of calculations,

$$\int_G |f_{\pm(m+1)}(x)|\, dG = 4\pi^2/(m-1)$$

It is thus natural to refer to D_m for $m \geqslant 2$ *as integrable representations.* It can be shown that, for these, all K-finite matrix elements are in $L^1(G)$. *$D_{\pm 1}$ are not integrable.* On the other hand, let us consider the modules $D_{0,\pm}$ and the corresponding matrix elements

$$f_{0,\pm}(x) = (\pi(x)e_{\pm 1} | e_{\pm 1})$$

The same method we used in Theorem 25 will now yield

$$f_{0,\ddagger}(x) = g(x)^{-1} = \overline{f_{0,-}(x)}$$

and in particular,

$$|f_{0,\pm}(a_t)| = (\cosh t)^{-1}$$

The modules $D_{0,\pm}$ are *unitary,* for they occur as the two irreducible constituents of the *unitary principal series* $\pi_{1,0}$. But they are *not* in the discrete series since

$$\int_0^\infty (\cosh t)^{-2} \sinh 2t\, dt = \infty$$

However, the integral is only logarithmically divergent, and, for this and other reasons, $D_{0,\pm}$ are often called *mock-discrete series* modules. I shall make a few more comments on them at the end of this section.

Note also that

$$d(D_{\pm m}) = \dim(F_m)/2\pi^2$$

Hence, apart from a normalizing factor, *the formal degree of $D_{\pm m}$ is the same as the dimension of an appropriate irreducible representation of the compact group $SU(2)$.* This and the formulae for the characters of D_m that we shall derive now suggest that the relationship between the representation theory of compact and noncompact real forms of the same complex group goes even deeper than what our development has revealed. I shall not go any further into this.

We now wish to calculate the characters of the D_m. Write

$$\Theta_m = \mathrm{ch}\,(D_m) \quad (\mathrm{ch} = \text{character})$$

We have already seen that

$$\Theta_m^* = \Theta_m + \Theta_{-m}$$

is a locally integrable function on G and

$$\Theta_m^* = T_{\varepsilon,m} - \mathrm{ch}\,(F_m)$$

where $\mathrm{ch}\,(F_m)$ is the character of the finite-dimensional module of dimension m. However, one cannot rule out the presence of mutually compensating singularities in $\Theta_{\pm m}$. So the determination of the character of D_m is more subtle. Observe however that the formal degrees of $D_{\pm m}$ are the same. This means that *in the Plancherel formula* Θ_m *and* Θ_{-m} *will occur only in the combination* $\Theta_m + \Theta_{-m}$, *so that a separate knowledge of* $\Theta_{\pm m}$ *is not needed*. Nevertheless, for a complete theory, the precise description of $\Theta_{\pm m}$ cannot be evaded.

The best way to do this from a conceptual point of view is to use the Harish-Chandra theory of the structure of invariant eigendistributions on G. We shall discuss this in Chapter 7. It follows from that theory that $\Theta_m (m \in \mathbb{Z} \backslash (0))$ is a locally integrable function on G which is (real) analytic on G'; we denote this function also by Θ_m. Moreover, since

$$\Omega \Theta_m = m^2 \Theta_m$$

Harish-Chandra's theory predicts that Θ_m is given on B' and L' by the following formulae:

$$\Theta_m(u_\theta) = \frac{a\mathrm{e}^{im\theta} + b\mathrm{e}^{-im\theta}}{\mathrm{e}^{i\theta} - \mathrm{e}^{-i\theta}} \quad (\theta \not\equiv 0, \pi)$$

$$\Theta_m(a_t) = \frac{a'\mathrm{e}^{mt} + b'\mathrm{e}^{-mt}}{\mathrm{e}^{t} - \mathrm{e}^{-t}} \quad (t > 0)$$

where a, b, a', b' are constants, related by

$$a - b = a' - b'$$

What it does *not* do is to give the actual values of a, b, a', b'. This is as it should be, since that theory is based on the sole fact that Θ_m is an invariant eigendistribution for Ω, and does not use the property that the Θ_m are characters of irreducible representations. So it is a natural, and for us even indispensable question, to determine these constants. We shall now do this, modulo the Harish-Chandra theory. We shall suppress m, write $\Theta = \Theta_m (m > 0)$; let π be the representation corresponding to D_m so that $\Theta = \mathrm{ch}\,(\pi)$. Write

$$\Phi_B(u_\theta) = \Delta_B(u_\theta)\Theta(u_\theta) = a\mathrm{e}^{im\theta} + b\mathrm{e}^{-im\theta}$$

$$\Phi_L(a_t) = \Delta_L(a_t)\Theta(a_t) = a'\mathrm{e}^{mt} + b'\mathrm{e}^{-mt} \quad (t > 0)$$

We follow Harish-Chandra [H5] [H6]. It is a question of proving two things:

$$\sup_{t>1} |\Phi_L(a_t)| < \infty \tag{L}$$

$$\Phi_B(u_\theta) = -e^{im\theta} \tag{B}$$

The first of these will imply that $a' = 0$; and as $a = -1$, $b = 0$ by the second relation, $-1 = a - b = a' - b' = -b'$, so that $b' = 1$. $\Theta(\gamma a_t)$ will then be determined by the rule

$$\Theta(\gamma a_t) = (-1)^{m-1}\Theta(a_t)$$

We take up (L) first. The idea behind proving it is as follows. If we go back to the argument for proving the existence of characters (Theorem 22 of Chapter 5) we see that the character satisfies an estimate of the form

$$|\Theta(f)| \leqslant \text{const.}\, \| af \|_1 \quad (f \in C_c^\infty(G))$$

where a is some differential operator and $\|\cdot\|_1$ is the norm in $L^1(G)$. When we know that the representation π is in the *discrete series*, the *L^1-norm may be replaced by the L^2-norm.* One then uses a little invariant analysis to deduce that Φ_L is *tempered* on $t > 1$, proving *(L)*.

Lemma 28. *Let $\Omega_K = 1 - (X - Y)^2 \in U(\mathfrak{g})$. Then there is a constant $c > 0$ such that, for all $m \in \mathbb{Z}\backslash(0)$,*

$$|\Theta_m(f)| \leqslant c\, \|\Omega_K f\|_2 \quad (\|\cdot\|_2 = L^2\text{-norm})$$

for all $f \in C_c^\infty(G)$.

Proof. We need this lemma now only for fixed m but later in this stranger form. If (e_n) is an orthonormal basis for the Hilbert space of π with $\pi(u_\theta)e_n = e^{ik_n\theta}e_n$ we know that

$$\sum_n (1 + k_n^2)^{-1} = \sum_{p\geqslant 0} (1 + (m + 1 + 2p)^2)^{-1} \leqslant \sum_{r\geqslant 0} (1 + r^2)^{-1} = c_1$$

Let $a_n(x) = (\pi(x)e_n | e_n)$; then $\Omega_K a_n = (1 + k_n^2)a_n$; moreover,

$$\| a_n \|^2 = d(\pi)^{-1} = \text{const.}\frac{1}{|m|} \leqslant c_2$$

Hence, for $f \in C_c^\infty(G)$

$$\begin{aligned}
|\Theta(f)| &\leqslant \sum_n \left| \int a_n(x) f(x)\, \mathrm{d}x \right| \\
&= \sum_n (1 + k_n^2)^{-1} \left| \int (\Omega_K a_n)\cdot f \cdot \mathrm{d}x \right| \\
&= \sum_n (1 + k_n^2)^{-1} \left| \int a_n(\Omega_K f)\, \mathrm{d}x \right| \\
&\leqslant c_1 c_2^{1/2} \| \Omega_K f \|_2
\end{aligned}$$

Let I be the interval $(1, \infty)$ for the variable t. Then the conjugacy map gives rise to a diffeomorphism

$$\varphi:(G/L) \times I \to G_I \quad (xL, t \mapsto x a_t x^{-1})$$

where G_I is open. We choose and fix $u \in C_c^\infty(G/L)$ such that $\int u \, \mathrm{d}(G/L) = 1$. We have a linear map $v \mapsto f_v$ from $C_c^\infty(I)$ to $C_c^\infty(G_I)$ if we put

$$f_v(x a_t x^{-1}) = u(xL) v(t)$$

i.e., $f_v \circ \varphi = u \otimes v$. We want to estimate the norms $\| \Omega_K f_v \|_2$ in terms of the L^2-norms of v and its derivatives. This is the key step; for

$$\int_{G_I} \Theta(x) \, f_v(x) \, \mathrm{d}x = \iint_{G/L \times I} \Theta(a_t) \Delta_L(a_t)^2 u(xL) v(t) \, \mathrm{d}(G/L) \, \mathrm{d}t$$

$$= \int_1^\infty \Phi_L(a_t) v(t) \Delta_L(a_t) \, \mathrm{d}t$$

and the proposed estimates will lead to estimates for

$$\int_1^\infty \Phi_L(a_t) v(t) \Delta_L(a_t) \, \mathrm{d}t$$

Let $\xi(t) = 1 - \mathrm{e}^{-2t} (t > 1)$, $\eta = 1/\xi$. Then $\mathrm{d}\eta/\mathrm{d}t = 2\eta - \eta^2$ so that, if $R = \mathbb{C}[\eta]$, the ring R is closed under $\mathrm{d}/\mathrm{d}t$. Since $\xi \geqslant \frac{1}{2}$ on I, η is bounded on I and so all derivatives of η are bounded on I. So, if $C_{\mathrm{bdd}}^\infty(I)$ is the ring of all C^∞ functions on I which are bounded together with all their derivatives, $R \subset C_{\mathrm{bdd}}^\infty(I)$, and both ξ and η are *units* of $C_{\mathrm{bdd}}^\infty(I)$.

Lemma 29. *If $b \in U(\mathfrak{g})$ and D_b is the differential operator on $G/L \times I$ which corresponds to b via the diffeomorphism φ, then D_b has the following form:*

$$D_b = \sum_{0 \leqslant j \leqslant k} d_j \otimes r_j (\mathrm{d}/\mathrm{d}t)^j$$

where d_j are analytic differential operators on G/L, and the functions r_j belongs to the ring R.

Proof. Note that the form of D_b is preserved under multiplication so that we need only consider $b = A \in \mathfrak{g}$. Also it is enough to prove this only for some open neighbourhood E of $\dot{1}$ (= the coset L). Identifying the tangent space to G/L at $\dot{1}$ with $\mathbb{R} \cdot X + \mathbb{R} \cdot Y$, we have, for $Z' \in \mathbb{R} \cdot X + \mathbb{R} \cdot Y$,

$$(\mathrm{d}\varphi)_{\dot{1},t}(Z') = Z'^{(a_{-t})} - Z, \quad (\mathrm{d}\varphi)_{\dot{1},t}(\mathrm{d}/\mathrm{d}t) = H$$

$$(\mathrm{d}\varphi)_{\dot{1},t}^{-1}: X \mapsto -\eta X, \quad Y \mapsto (\eta - 1) Y, \quad H \mapsto \mathrm{d}/\mathrm{d}t$$

Select a submanifold $Q \subset G$ transversal to L at 1 such that $x \mapsto \dot{x} = xL$

is a diffeomorphism of Q with E. From the commutativity of the diagram

$$\begin{array}{ccc} (\dot{x}, t) & \longrightarrow & x a_t x^{-1} \\ {\scriptstyle l_x}\uparrow & & \uparrow{\scriptstyle \iota_x} \\ (\dot{1}, t) & \longrightarrow & a_t \end{array} \qquad (\mathrm{d}i_x(Z^{x^{-1}}) = Z)$$

we see that

$$(\mathrm{d}\varphi)_{\dot{x},t}^{-1}(Z) = \mathrm{d}l_x(\mathrm{d}\varphi_{\dot{1},t}(Z^{x^{-1}})) \quad (x \in Q)$$

So, if we write $\bar{X}$, $\bar{Y}$ for the vector fields on E defined by $\bar{X}_{\dot{x}} = \mathrm{d}l_x(X)$, $\bar{Y}_{\dot{x}} = \mathrm{d}l_x(Y)$, and if we write $Z^x = a(\dot{x})X + b(\dot{x})Y + c(\dot{x})H$ where a, b, c are analytic on E, we have, on $E \times I$, the formula

$$D_Z = -\eta a\bar{X} + (\eta - 1)b\bar{Y} + c\,\mathrm{d}/\mathrm{d}t$$

which clearly has the required form.

Lemma 30. *Φ_L is bounded on I.*

Proof. We have, for some $a \in U(\mathfrak{g})$,

$$\left|\int_I \Phi_L(a_t)v(t)\Delta_L(a_t)\,\mathrm{d}t\right| = \left|\int_{G_I} \Theta(x)\, f_v(x)\,\mathrm{d}x\right| \leqslant \|af_v\|_2$$

for all $v \in C_c^\infty(I)$. But, under φ, af_v goes over to $D_a(u \otimes v)$ and $\mathrm{d}x$ goes over to $(\mathrm{e}^t - \mathrm{e}^{-t})^2\,\mathrm{d}(G/L)\,\mathrm{d}t$. So, as u is fixed, Lemma 29 gives the following estimate:

$$\left|\int_I \Phi_L(a_t)v(t)(\mathrm{e}^t - \mathrm{e}^{-t})\,\mathrm{d}t\right| \leqslant \text{const.} \sum_{0 \leqslant j \leqslant k} \|(\mathrm{e}^t - \mathrm{e}^{-t})r_j v^{(j)}\|_2$$

for all $v \in C_c^\infty(I)$. If we change v to $\mathrm{e}^{-t}v$ and remember that $\xi, \eta \in C_{\mathrm{bdd}}^\infty(I)$ we get the estimate

$$\left|\int_I \Phi_L(a_t)v(t)\,\mathrm{d}t\right| \leqslant \text{const.} \sum_{0 \leqslant j \leqslant k} \|v^{(j)}\|_2$$

In other words, Φ_L is *tempered* as a distribution on I. *As Φ_L is a linear combination of exponentials*, classical Fourier analysis now shows that Φ_L must be bounded on I.

We take up next the formula for Θ on B'. Let us write (e_n) for the orthonormal basis in the Hilbert space of π such that $\pi(u_\theta)e_n = \mathrm{e}^{\mathrm{i}k_n\theta}e_n$, $\Omega_K = 1 - (X - Y)^2$. Recall (Ch. 5, §5.4, p. 126) the notion of an operator which is summable (with respect to (e_n), which we shall not refer to), and the estimate that, for any such operator A and *any orthonormal basis* (f_m),

$$\sum_m |(Af_m|\, f_m)| \leqslant \sum_{m,n} |(Ae_n|e_m)|$$

Suppose now $h \in C^\infty(B)$ and $\pi_K(h) = \int_B h(u_\theta)\pi(u_\theta)\,d\theta$. Since $(\pi_K(h)e_n|e_m) = \delta_{mn}\hat{h}(k_n)$, π_k is summable and $|\mathrm{tr}\,\pi_k(h)| \leqslant \sum |\hat{h}(k_n)|$ so that $h \mapsto \mathrm{tr}(\pi_K(h))$ is a distribution on B. Let us write Θ_B for it.

Lemma 31. *For all $h \in C_c^\infty(B')$,*

$$\Theta_B(h) = \int_{B'} h(u_\theta) \cdot \frac{-\mathrm{e}^{im\theta}}{\mathrm{e}^{i\theta} - \mathrm{e}^{-i\theta}}\,d\theta$$

Proof. Let r be an auxiliary variable, $0 < r < 1$, which will be allowed to tend to 1. Then, as the characters of B which occur in π are $\mathrm{e}^{\mathrm{i}(m+1+2p)\theta}$ $(p \geqslant 0)$,

$$\begin{aligned}\Theta_B(h) &= \sum_{p \geqslant 0} \int h(u_\theta)\mathrm{e}^{\mathrm{i}(m+1+2p)\theta}\,d\theta \\ &= \lim_{r \to 1-0} \sum_{p \geqslant 0} r^p \int h(u_\theta)\mathrm{e}^{\mathrm{i}(m+1+2p)\theta}\,d\theta \\ &= \lim_{r \to 1-0} \int h(u_\theta)(\mathrm{e}^{\mathrm{i}(m+1)\theta}/(1 - r\mathrm{e}^{2i\theta}))\,d\theta\end{aligned}$$

Since $\mathrm{supp}(h)$ is a compact set not containing $\theta = 0, \pi$,

$$|\mathrm{e}^{\mathrm{i}(m+1)\theta}/(1 - r\mathrm{e}^{2i\theta})| \leqslant \text{const.}\; r \geqslant 1 - \delta, u_\theta \in \mathrm{supp}(h)$$

Hence we can take the limit inside.

Recall from Proposition 12 that, if $F \subset B'$ is compact, the conjugacy map $G \times F \to G$ is proper. In particular, if $E \subset F \subset B'$ and E is open, then, for any $g \in C_c^\infty(E^G)$, the function $x, u_\theta \mapsto g(xu_\theta x^{-1})$ lies in $C_c^\infty(G \times E)$.

Lemma 32. *Let $E \subset B'$ be open such that* Cl(E) *is a compact subset of B'; write $\tilde{G} = G/B$. Then, for any $f \in C_c^\infty(\tilde{G} \times E)$, the operator*

$$T_f = \iint_{\tilde{G} \times E} f(\tilde{x}:u_\theta)\pi(xu_\theta x^{-1})\,d\tilde{G}\,d\theta$$

is summable; and its trace is the distribution in f, given by

$$\mathrm{tr}(T_f) = \iint_{\tilde{G} \times E} f(\tilde{x}:u_\theta)t(u_\theta)\,d\tilde{G}\,d\theta$$

where

$$t(u_\theta) = \frac{-\mathrm{e}^{im\theta}}{\mathrm{e}^{i\theta} - \mathrm{e}^{-i\theta}}$$

Proof. Write

$$f_{mn}(y) = (\pi(y)e_n|e_m), \quad a = \Omega_K$$

Then

$$f_{mn}(a;y;a)=(1+k_m^2)(1+k_n^2)f_{mn}(y)$$

and hence

$$(T_f e_n|e_m)=(1+k_m^2)^{-1}(1+k_n^2)^{-1}\iint f(\tilde{x}:u_\theta)f_{mn}(a;xu_\theta x^{-1};a)\,\mathrm{d}\tilde{G}\,\mathrm{d}\theta$$

Since $(x,u_\theta)\mapsto xu_\theta x^{-1}$ is a diffeomorphism of $\tilde{G}\times E$ with E^G, there is a unique differential operator D_a on $\tilde{G}\times E$ such that, if $v\in C^\infty(E^G)$ and $\tilde{v}(\tilde{x}:u_\theta)=v(xu_\theta x^{-1})$, $v(a;\,xu_\theta x^{-1};\,a)=(D_a\tilde{v})(\tilde{x}:u_\theta)$. So, if D'_a is the transpose of D_a,

$$(T_f e_n|e_m)=(1+k_m^2)^{-1}(1+k_n^2)^{-1}\iint (D'_a f)(\tilde{x}:u_\theta)f_{mn}(xu_\theta x^{-1})\,\mathrm{d}\tilde{G}\,\mathrm{d}\theta$$

As

$$\sum_{m,n}|(T_f e_n|e_m)|\leqslant \text{const.}\iint |D'_a f|\,\mathrm{d}\tilde{G}\,\mathrm{d}\theta$$

the first statement is clear. For computing the trace we may take $f(\tilde{x}:u_\theta)=g(\tilde{x})h(u_\theta)$ where $g\in C_c^\infty(G)$, $h\in C^\infty(E)$. Then,

$$\sum_n (T_f e_n|e_n)=\sum_n\iint g(\tilde{x})h(u_\theta)(\pi(u_\theta)f_n^x|f_n^x)\,\mathrm{d}\tilde{G}\,\mathrm{d}\theta$$

where $f_n^x=\pi(x^{-1})e_n$. Then (f_n^x) is an orthonormal basis also, and

$$\sum_n (T_f e_n|e_n)=\sum_n\int_{\tilde{G}} g(\tilde{x})(\pi_K(h)f_n^x|f_n^x)\,\mathrm{d}G$$

As $\pi_K(h)$ is summable, we know that for all x

$$\sum_n|(\pi_K(h)f_n^x|f_n^x)\leqslant\sum_{m,n}|(\pi_K(h)e_n|e_m)|$$

Hence

$$\begin{aligned}\sum_n (T_f e_n|e_n)&=\int_{\tilde{G}} g(\tilde{x})\sum_n(\pi_K(h)f_n^x|f_n^x)\,\mathrm{d}G\\&=\int_{\tilde{G}} g(\tilde{x})\Theta_B(h)\,\mathrm{d}\tilde{G}\\&=\iint g(\tilde{x})h(u_\theta)t(u_\theta)\,\mathrm{d}\tilde{G}\,\mathrm{d}\theta\end{aligned}$$

Lemma 33. *For $u_\theta\in B'$ we have*

$$\Theta_m(u_\theta)=-\mathrm{e}^{im\theta}/(\mathrm{e}^{i\theta}-\mathrm{e}^{-i\theta})\quad (m\geqslant 1)$$

Proof. With the same notation as in the previous lemma, we have, for

$g \in C_c^\infty(E^G)$ and $f(\tilde{x}:u_\theta) = g(xu_\theta x^{-1})|\Delta_B(u_\theta)|^2$,

$$\pi(g) = \iint f(\tilde{x}:u_\theta)\pi(xu_\theta x^{-1})\,\mathrm{d}\tilde{G}\,\mathrm{d}\theta = T_f$$

Hence

$$\mathrm{tr}\,(\pi(g)) = \mathrm{tr}\,(T_f) = \iint f(\tilde{x}:u_\theta)t(u_\theta)\,\mathrm{d}\tilde{G}\,\mathrm{d}\theta$$

while we also have

$$\mathrm{tr}\,(\pi(g)) = \int_{E^G} \Theta g\,\mathrm{d}G = \iint \Theta(u_\theta)f(\tilde{x}:u_\theta)\,\mathrm{d}\tilde{G}\,\mathrm{d}\theta$$

This finishes the proof.

We can thus formulate the following.

Theorem 34. *For any $m \in \mathbb{Z}$, $m \neq 0$,*

$$\Theta_m(u_\theta) = -\,\mathrm{sgn}\,(m)\frac{\mathrm{e}^{\iota m\theta}}{\mathrm{e}^{\iota\theta} - \mathrm{e}^{-\iota\theta}} \quad (\theta \neq 0, \pi)$$

$$\Theta_m(\gamma^\varepsilon a_t) = (-1)^{(m-1)\varepsilon}\frac{\mathrm{e}^{-|m||t|}}{|\mathrm{e}^t - \mathrm{e}^{-t}|} \quad (t \neq 0, \varepsilon = 0, 1)$$

It is not obvious that Θ_m is locally integrable on G. Let us consider the invariant function $D(x) = \mathrm{tr}\,(x^2 - 2)$. If x has distinct eigenvalues μ, μ^{-1}, $|D(x)| = |\mu - \mu^{-1}|^2$ so that

$$|\Theta_m(x)| \leqslant |D(x)|^{-1/2} \quad (x \in G')$$

I claim that $|D|^{-1/2}$ is locally integrable. It is enough to prove that $|D|^{-1/2}f \in L^1(G)$ for all *nonnegative* $f \in C_c^\infty(G)$. But $|D|^{1/2} = |\Delta_B|$ (resp. $|\Delta_L|$) on B' (resp. L') and so it is enough to check that

$$\int |F_{|D|^{-1/2}f,B}|\,|\Delta_B|\,\mathrm{d}B < \infty, \quad \int |F_{|D|^{-1/2}f,L}|\,|\Delta_L|\,\mathrm{d}L < \infty$$

But these are respectively equal to

$$\int |F_{f,B}|\,\mathrm{d}B, \quad \int |F_{f,L}|\,\mathrm{d}L,$$

which are finite.

I would like to conclude this section with some comments on aspects of the discrete and mock-discrete series which I have not been able to discuss.

Eigendistribution property. For Θ_m we have the differential equation

$$\Omega\Theta_m = m^2\Theta_m$$

Since Θ_m is singular at the identity element one must be careful in interpreting this equation. On G' it is just the classical interpretation, as Θ_m is smooth. However, around 1, the correct meaning is

$$\int_G \Theta_m \Omega f dG = m^2 \int_G \Theta_m f dG$$

It is an instructive exercise to verify this using the integration formula. We have in fact the following.

Proposition 35. *For all $f \in C_c^\infty(G)$,*

$$\int_G \Theta_m f dG = \operatorname{sgn}(m) \int e^{im\theta} F_{f,B}(u_\theta)\, d\theta + \int_0^\infty e^{-|m|t} F_{h,L}(a_t)\, dt$$

where $h = (f + (-1)^{m-1} f_\gamma)$.

Proof. By Proposition 13,

$$\begin{aligned}\int_G \Theta_m f dG = &- \int \Delta_B(u_\theta)\Theta_m(u_\theta) F_{f,B}(u_\theta)\, d\theta \\ &+ \int_0^\infty |e^t - e^{-t}| \Theta_m(a_t) F_{f,L}(a_t)\, dt \\ &+ \int_0^\infty |e^t - e^{-t}| \Theta_m(\gamma a_t) F_{f,L}(\gamma a_t)\, dt \\ = &\operatorname{sgn}(m) \int e^{im\theta} F_{f,B}(u_\theta)\, d\theta + \int_0^\infty e^{-|m|t} F_{h,L}(a_t)\, dt\end{aligned}$$

We now substitute Ωf for f, use the relations

$$F_{\Omega f,B} = -(d/d\theta)^2 F_{f,B}, \quad F_{\Omega f,L} = (d/dt)^2 F_{f,L}$$

and integrate by parts. The point is that the *boundary terms* at $\theta = 0, \pi$, $t = 0$ which one gets *cancel out exactly because of the limit formulae and jump relations.* The reader should check this.

Representations of the covering groups. Essentially the same methods allow one to treat groups which cover $SL(2, \mathbb{R})$ finitely. The representation $D_m (m \in (1/N)\mathbb{Z},\ m \neq 0)$ will now be in the discrete series. The matrix elements $f_{\pm(m+1)}$ have similar formulae and their square integrability is guaranteed as soon as $|m| > 0$. But now all of these may be viewed as representations of the *universal covering group of $SL(2, \mathbb{R})$*; it now makes sense to let $m \to 0\pm$ (through rational values of course). Thus *$D_{0,\pm}$ may actually be viewed as $\lim_{m\to 0\pm} D_m$*. This is the reason why they are also known as *limits of discrete series.* In particular, for their characters, we have

$$\Theta_{0,\pm}(u_\theta) = \mp \frac{1}{e^{i\theta} - e^{i\theta}} \quad (\theta \not\equiv 0, \pi)$$

$$\Theta_{0,\pm}(a_t) = -\Theta_{0,\pm}(\gamma a_t) = \frac{1}{|e^t - e^{-t}|} \quad (t \neq 0)$$

Explicit realizations. Bargmann [Ba] constructed the representations D_m and $D_{0,\pm}$ in Hilbert spaces of analytic functions. This was later on generalized by Harish-Chandra [H11] whose work made it clear that the relationship of the discrete series with complex analysis was a theme to be pursued in depth. The subsequent development of this theme is too long a story to go into here except to mention that the ideas of Langlands, Kostant, Schmid, Blattner and others have turned out to be decisive.

6.6 The Plancherel formula for $SL(2, \mathbb{R})$

Everything is ready for obtaining the Plancherel formula. We have the characters $\Theta_m (m \in \mathbb{Z}\backslash(0))$ and $T_{\varepsilon,\lambda} (\varepsilon = 0, 1, \lambda \in i\mathbb{R})$; $T_{\varepsilon,\lambda} = T_{\varepsilon,-\lambda}$ and all of these are irreducible except $T_{1,0}$; however, this will not matter as $T_{\varepsilon,\lambda}$ enters 'continuously' in the expansion.

Lemma 36. *For any integer $r \geqslant 1$ there is a continuous seminorm μ on $C_c^\infty(G)$ such that*

$$|\Theta_m(f)| \leqslant m^{-2r}\mu(f)$$
$$|T_{\varepsilon,\lambda}(f)| \leqslant (1 + |\lambda|^2)^{-r}\mu(f)$$

for all $f \in C_c^\infty(G)$, all $m \neq 0$, all $\lambda \in i\mathbb{R}$.

Proof. $\Theta_m(f) = m^{-2r}\Theta_m(\Omega^r f)$, and, by Lemma 28, $|\Theta_m(\Omega^r f)| \leqslant$ const. $\|\Omega_K \Omega^r f\|_2$ for all m and f. Further, as $\lambda \in i\mathbb{R}$, $-\lambda^2 = |\lambda|^2$ and so $T_{\varepsilon,\lambda}((1-\Omega)^r f) = (1 + |\lambda|^2)^r T_{\varepsilon,\lambda}(f)$ while $|T_{\varepsilon,\lambda}(f)| = |\hat{F}_{f,L}(\varepsilon:\lambda)| \leqslant \int |F_{f,L}| dL$ which is majorized by a continuous seminorm since we have seen that $f \mapsto F_{f,L}$ is a continuous linear map.

This lemma assures the convergence of the series and integrals that we shall encounter below. The procedure is exactly the same as for the compact and complex groups. We apply the Fourier transform to the limit formula. The first step is to compute the Fourier transform of $F_{f,B}$. Put

$$\hat{F}_{f,B}(m) = \int F_{f,B}(u_\theta) e^{im\theta} \, d\theta$$

(Note the absence of 2π.)

Proposition 37. *For all* $m\in\mathbb{Z}\backslash(0)$, $f\in C_c^\infty(G)$,

$$\hat{F}_{f,B}(m)=\operatorname{sgn}(m)\Theta_m(f)-\frac{1}{m}F_{h,L}(1)-\frac{1}{m}\int_0^\infty \mathrm{e}^{-|m|t}F'_{h,L}(a_t)\,\mathrm{d}t$$

where $h=(f+(-1)^{m-1}f_\gamma)$.

Proof. From Proposition 35 we have

$$\hat{F}_{f,B}(m)=\operatorname{sgn}(m)\Theta_m(f)-\operatorname{sgn}(m)\int_0^\infty \mathrm{e}^{-|m|t}F_{h,L}(a_t)\,\mathrm{d}t$$

We now integrate the last term by parts to get the required result.

Proposition 38. *For all* $f\in C_c^\infty(G)$, $m\in\mathbb{Z}$ *(including* 0), *writing* ′ *for* $\mathrm{d}/\mathrm{d}\theta$ *as well as* $\mathrm{d}/\mathrm{d}t$,

$$(-\mathrm{i}F'_{f,B}(m))\hat{}=-|m|\Theta_m(f)+\int_0^\infty \mathrm{e}^{-|m|t}F'_{h,L}(a_t)\,\mathrm{d}t$$

Proof. We have

$$\begin{aligned}(-\mathrm{i}F'_{f,B}(m))\hat{}&=-\mathrm{i}\int F'_{f,B}(u_\theta)\mathrm{e}^{\mathrm{i}m\theta}\,\mathrm{d}\theta\\&=[-\mathrm{i}F_{f,B}(u_\theta)\mathrm{e}^{\mathrm{i}m\theta}]_{0+}^{\pi-}+[\ldots]_{\pi+}^{\pi-}-m\int F_{f,B}(u_\theta)\mathrm{e}^{\mathrm{i}m\theta}\,\mathrm{d}\theta\\&=-\frac{1}{\mathrm{i}}[F_{f,B}(u_\theta)]_{0-}^{0+}-\frac{1}{\mathrm{i}}[F_{f,B}(u_\theta)\mathrm{e}^{\mathrm{i}m\theta}]_{\pi-}^{\pi+}-m\hat{F}_{f,B}(m)\end{aligned}$$

If we recall that

$$F_{f,B}(u_{\theta+\pi})=-F_{f_\gamma,B}(u_\theta)$$

we get

$$\begin{aligned}(-\mathrm{i}F'_{f,B}(m))\hat{}&=-\frac{1}{\mathrm{i}}[F_{f,B}(u_\theta)]_{0-}^{0+}-\frac{1}{\mathrm{i}}(-1)^{m-1}[F_{f_\gamma,B}(u_\theta)]_{0-}^{0+}-m\hat{F}_{f,B}(m)\\&=-|m|\Theta_m(f)+\int_0^\infty \mathrm{e}^{-|m|t}F'_{h,L}(a_t)\,\mathrm{d}t\\&\quad+\left\{F_{f,L}(1)-\frac{1}{\mathrm{i}}[F_{f,B}(u_\theta)]_{0-}^{0+}\right\}\\&\quad+(-1)^{m-1}\left\{F_{f_\gamma,L}(1)-\frac{1}{\mathrm{i}}[F_{f_\gamma,B}(u_\theta)]_{0-}^{0+}\right\}\end{aligned}$$

by Proposition 37. The expressions within { } are zero by the Harish-Chandra jump relation (Theorem 23). Hence the result.

We can now make the final computations. Since $F_{f,B}$ is piecewise smooth and $F'_{f,B}$ is continuous at $\theta = 0$, we have

$$2\pi\left(\frac{1}{i}F'_{f,B}(u_\theta)\right)_{\theta=0} = \sum_m \left(\frac{1}{i}F'_{f,B}(m)\right)^{\widehat{}}$$

But, by the limit formula (Theorem 22),

$$\frac{1}{i}F'_{f,B}(1) = -\pi f(1)$$

Hence

$$-2\pi^2 f(1) = \sum_m \left(\frac{1}{i}F'_{f,B}(m)\right)^{\widehat{}}$$

$$= \sum_m \left(-|m|\Theta_m(f) + \int_0^\infty e^{-|m|t} F'_{h,L}(a_t)\,dt\right)$$

by Proposition 38. Separating the terms (Lemma 36) we get

$$2\pi^2 f(1) = \sum_m |m|\Theta_m(f) - \sum_m \int_0^\infty e^{-|m|t} F'_{f,L}(a_t)\,dt$$

The sum over m is for nonzero m since the term corresponding to $m = 0$ is zero. We now consider separately the cases $f = \pm f_\gamma$. Note that

$$\Theta_m(f) = 0 \quad (\text{resp. } T_{\varepsilon,\lambda}(f) = 0)$$

if f and $(-1)^{m-1}$ have the opposite parity (resp. f and ε have the opposite parity).

(a) $f = f_\gamma$. Here $f + (-1)^{m-1} f_\gamma = 0$ for m even and $2f$ for m odd. So

$$2\pi^2 f(1) = \sum_{m\,\mathrm{odd}} |m|\Theta_m(f) - 2\sum_{m\,\mathrm{odd}} \int_0^\infty e^{-|m|t} F'_{f,L}(a_t)\,dt$$

$$= \sum_{m\,\mathrm{odd}} |m|\Theta_m(f) - 2\sum_{m\,\mathrm{odd}} \int_0^\infty (te^{-|m|t})\left(\frac{1}{t}F'_{f,L}(a_t)\right)dt$$

where we note that $(1/t)F'_{f,L}(a_t)$ is in $C_c^\infty(\mathbb{R})$ because $F'_{f,L}(a_t)$ is an odd element of $C_c^\infty(\mathbb{R})$. Hence, using Lebesgue's theorem, we get

$$2\pi^2 f(1) = \sum_{m\,\mathrm{odd}} |m|\Theta_m(f) - 2\int_{-\infty}^\infty (t/(e^t - e^{-t}))(F'_{f,L}(a_t)/t)\,dt$$

To the last term we apply Plancherel formula on $\mathbb{R}$. This requires one to know the Fourier transform of $t/(e^t - e^{-t})$. We have the formula

$$(t/(e^t - e^{-t}))^{\sim}(\mu) = (d/d\mu)(\tanh(\pi\mu/2))$$

where $\tilde{u}(\mu) = \int_{-\infty}^\infty u(t)e^{i\mu t}\,dt$. (Start with the formula

$$(\text{p.v.}) \int_{-\infty}^\infty \frac{e^{i\mu t}}{e^t - e^{-t}}\,dt = \frac{\pi}{2}\tanh\left(\frac{\pi\mu}{2}\right)$$

which is established by integrating $e^{i\mu z}/(e^z - e^{-z})$ along the rectangle $-R, R, R + i\pi, -R + i\pi$ and letting $R \to \infty$; then apply $d/d\mu$ to the imaginary part.) Moreover, if $v(t) = F'_{f,L}(a_t)/t$, $F'_{f,L}(a_t) = tv(t)$ and hence $\tilde{F}'_{f,L}(\mu) = -i\tilde{v}'(\mu)$; but, as $f = f_\gamma$,

$$\int_{-\infty}^{\infty} F_{f,L}(a_t)e^{i\mu t}\,dt = \int_{-\infty}^{\infty} F_{f,L}(\gamma a_t)\,e^{i\mu t}\,dt$$
$$= \tfrac{1}{2}\hat{F}_{f,L}(0{:}i\mu)$$

so that $\tilde{v}'(\mu) = i\tilde{F}'_{f,L}(\mu) = (\mu/2)\hat{F}_{f,L}(0{:}i\mu) = (\mu/2)T_{0,i\mu}(f)$. So, finally,

$$\int_{-\infty}^{\infty} (t/e^t - e^{-t})(F'_{f,L}(a_t)/t)\,dt = \frac{1}{2\pi}\int_{-\infty}^{\infty} \frac{\pi}{2}\left(\tanh\left(\frac{\pi\mu}{2}\right)\right)'\tilde{v}(\mu)\,d\mu$$
$$= -\frac{1}{2\pi}\int_{-\infty}^{\infty} \frac{\pi}{2}\tanh\left(\frac{\pi\mu}{2}\right)\frac{\mu}{2}T_{0,i\mu}(f)\,d\mu$$

Hence we get, remembering that $T_{0,i\mu} = T_{0,-i\mu}$,

$$2\pi^2 f(1) = \sum_{m\,\text{odd}} |m|\Theta_m(f) + \frac{1}{2}\int_0^{\infty} \mu\tanh(\pi\mu/2)\cdot T_{0,i\mu}(f)\,d\mu$$

(b) $f = -f_\gamma$. The computations are similar, the sum over m is for even m, and $f + (-1)^{m-1}f_\gamma = 2f$ once again. The formula (established by residue calculus for example)

$$\text{p.v.}\,(1/(1 - e^{-2t}))\tilde{\ }(\mu) = (i\pi/2)\coth(\pi\mu/2)$$

yields

$$(t/(1 - e^{-2t}))\tilde{\ }(\mu) = \frac{\pi}{2}\frac{d}{d\mu}\left(\coth\left(\frac{\pi\mu}{2}\right)\right)$$

But now, as $f = -f_\gamma$, $\tilde{F}_{f,L}(\mu) = \frac{1}{2}\hat{F}_{f,L}(1{:}i\mu) = \frac{1}{2}T_{1,i\mu}(f)$. Hence we obtain

$$2\pi^2 f(1) = \sum_{m\,\text{even}} |m|\Theta_m(f) + \frac{1}{2}\int_0^{\infty} \mu\coth(\pi\mu/2)T_{1,i\mu}(f)\,d\mu$$

Combining both cases we have proved the following.

Theorem 39 (Plancherel formula). *Let the distribution characters be computed with respect to the Riemannian Haar measure* dG. *Then, for all* $f \in C_c^\infty(G)$,

$$2\pi^2 f(1) = \sum_{m\neq 0} |m|\Theta_m(f) + \frac{1}{2}\int_0^{\infty} \mu\tanh(\pi\mu/2)\cdot T_{0,i\mu}(f)\,d\mu$$
$$+ \frac{1}{2}\int_0^{\infty} \mu\coth(\pi\mu/2)\cdot T_{1,i\mu}(f)\,d\mu$$

Remark. We have already calculated the formal degree of D_m (with respect to dG) to be $|m|/2\pi^2$; this formula confirms it.

Problems

1. Let π_m be the discrete series representation corresponding to $D_m (m > 0)$. Let f_{rs} be the matrix element $x \mapsto (\pi_m(x)e_s | e_r)$ where e_r is a unit vector transforming like the character $u_\theta \mapsto e^{ir\theta}$. Obtain an explicit formula for f_{rs}.
2. Prove that if π is an irreducible unitary representation of $G = SL(2, \mathbb{R})$ with character Θ one has an estimate of the form

$$|\Theta(f)| \leqslant \text{const.} \int |af| \,\mathrm{d}x \quad (f \in C_c^\infty(G))$$

where $a \in U(\mathfrak{g})$.
3. Prove, using the previous exercise and the method of Lemma 30, that the corresponding function Φ_L satisfies an estimate of the form

$$|\Phi_L(a_t)| < \text{const.}\, e^t \quad (t > 1)$$

Hence deduce that, if $\pi_{\varepsilon,\lambda}$ is unitary, λ must be real and lie in $[-1, 1]$ (or $\lambda \in i\mathbb{R}$).
4. Let $\mathfrak{h}$ be the upper half-plane on which $G = SL(2, \mathbb{R})$ acts by linear fractional transformations. Write $\gamma(g{:}z) = (cz + d)^{-1}$, $g = \begin{pmatrix} a & b \\ c & d \end{pmatrix} \in G$, $z \in \mathfrak{h}$. Prove that

$$\text{(i)}\ \gamma(g_1 g_2{:}z) = \gamma(g_1{:}g_2[z])\gamma(g_2{:}z)$$
$$\text{(ii)}\ |\gamma(g_1 g_2{:}z)|^2 = \mathrm{Im}\,(g[z])/\mathrm{Im}\,(z)$$

5. Show that

$$\mathrm{d}g[z] = (cz + d)^{-2}\,\mathrm{d}z$$

Hence show that, if $\omega = \mathrm{d}z \wedge \mathrm{d}\bar{z}$,

$$(\mathrm{Im}\,(g[z]))^{-2} g^*\omega = (\mathrm{Im}\,z)^{-2}\omega$$

and hence that

$$\frac{1}{y^2}\,\mathrm{d}x\,\mathrm{d}y$$

is an invariant measure on $\mathfrak{h}$.
6. Let r be an integer > 1 and let H_r be the Hilbert space of holomorphic functions f on $\mathfrak{h}$ such that

$$\|f\|^2 = \iint_{\mathfrak{h}} |f(z)|^2 y^{r-2}\,\mathrm{d}x\,\mathrm{d}y < \infty$$

Prove that the space $H_r \neq 0$ by showing that, if g is a bounded function on $\mathbb{R}$ with $\operatorname{supp}(g) \subset (r, \infty)$ for some $r \in \mathbb{R}$, the Fourier transform $\hat{g}$ is actually well defined on $\mathfrak{h}$ and lies in H_r (Hint: use the Plancherel formula to evaluate $\int_{\mathbb{R}} |g(x + \mathrm{i}y)|^2 \, \mathrm{d}x$). If

$$(L_r(g)f)(z) = \gamma(g^{-1}:z)^r f(g^{-1}[z])$$

show that L_r is a unitary representation in H_r.

7. Prove the following formulae:

$$L_r(H') = \mathrm{i}(1 + z^2)\mathrm{d}/\mathrm{d}z + \mathrm{i}rz$$
$$L_r(X') = ((-2z + \mathrm{i}(z^2 - 1) + (-r + \mathrm{i}rz))/2$$
$$L_r(Y') = ((-2z - \mathrm{i}(z^2 - 1)\mathrm{d}/\mathrm{d}z + (-r - \mathrm{i}rz))/2$$

8. Using the formulae of Problem 7 show that $L_r|_K$ splits as the direct sum of characters $\chi_m(u_\theta \mapsto \mathrm{e}^{\mathrm{i}m\theta})$, $m = -r, -r-2, \ldots$ each with multiplicity 1. Hence verify that L_r corresponds to $D_{-(r-1)}$.

9. Check that the highest-weight vector is

$$\text{const.}\,(z + \mathrm{i})^{-r}$$

Use this to calculate the formal degree.

10. Try to do the theory sketched in the problems above for *real* numbers $r > 1$ and discuss the limiting process when $r \to 1 + 0$ (see [Ba]).

11. If V is the Hilbert space of square integrable functions on $\mathfrak{h}$ with respect to the measure $y^{-2}\,\mathrm{d}x\,\mathrm{d}y$, the same definition as in problem 6 now defines a representation W_r of G. Prove that $W_r = \operatorname{Ind}_K^G \chi_r (\chi_r(u_\theta) = \mathrm{e}^{\mathrm{i}r\theta})$. Prove also that $W_0(\Omega)$ is equal to

$$4y^2(\partial^2/\partial x^2 + \partial^2/\partial y^2) + 1$$

which is essentially the Laplace–Beltrami operator on $\mathfrak{h}$.

7

Invariant eigendistributions

7.1 Invariant eigendistributions and their behaviour on the regular set

The theory of invariant eigendistributions on G and $\mathfrak{g}$ is the indispensable backdrop against which the harmonic analysis on these spaces takes place. I would like to give a brief sketch of this theory of Harish-Chandra by working it out for $SL(2, \mathbb{R})$. Ω is as in Chapter 5, Section 5.

Distributions (resp. differential operators) defined on G and on its invariant open subsets are called *invariant* if they are fixed by the inner automorphisms of G. Differentiation of distributions requires a volume element to begin with: if D is a differential operator on an open subset $V \subset G$ and T a distribution on V,

$$(DT)(f) = T(D^{\mathrm{t}} f) \quad (f \in C_c^\infty(V))$$

Here D^{t} is the differential operator *transpose* to D characterized by

$$\int_V Df \cdot g \, dG = \int_V f \cdot D^{\mathrm{t}} g \, dG \quad (f, g \in C_c^\infty(V))$$

If $D = \partial(a)$, $a \in U(\mathfrak{g})$, then $D^{\mathrm{t}} = \partial(a^{\mathrm{t}})$ where a^{t} is also in $U(\mathfrak{g})$; one has $X^{\mathrm{t}} = -X (X \in \mathfrak{g})$, $(D^{\mathrm{t}})^{\mathrm{t}} = D$, $(D_1 D_2)^{\mathrm{t}} = D_2^{\mathrm{t}} D_1^{\mathrm{t}}$. For distributions on a Cartan subgroup C we use the volume element dC in the above definitions. The distribution T is called an *eigendistribution* if $\Omega T = cT$ for some constant c. More generally it is called *Ω-finite* if the span of $T, \Omega T, \Omega^2 T, \ldots$ is finite-dimensional. For an arbitrary real semisimple Lie group, the term eigendistribution refers to those distributions T such that $zT = \chi(z)T$ for all $z \in \mathfrak{Z}$, $\chi(\mathfrak{Z} \to \mathbb{C})$ being a homomorphism. The significance of this concept lies in the fact that the characters of irreducible unitary representations of a connected semisimple group with finite centre are invariant eigendistributions. More generally, *$\mathfrak{Z}$-finite distributions* are those distributions T such that $\dim(\mathfrak{Z}T) < \infty$. The central result of the theory is the *Harish-Chandra regularity theorem*: if T is an invariant $\mathfrak{Z}$-finite distribution on a connected semisimple group G with finite centre, there is a locally integrable class-function F_T on G, which is analytic on G', such that $T = F_T$, i.e.,

$$T(f) = \int F_T f \, dG \quad (f \in C_c^\infty(G))$$

This is one of the great theorems of the subject; it dominates the entire landscape and one cannot do harmonic analysis without it, or indeed, without the theory of invariant $\mathfrak{Z}$-finite distributions in which it is imbedded at the very core. We have already used it in computing the characters of the D_m in Chapter 6.

There is a companion theory on the Lie algebra $\mathfrak{g}$. Invariance refers to the adjoint action of G and 'eigen' refers to the action of the invariant constant coefficient differential operators. In a sense this is relatively easier, and what Harish-Chandra did was to establish the theory on $\mathfrak{g}$ first and then to carry it over to G, roughly speaking, by the exponential map.

Let me return to $G = SL(2, \mathbb{R})$. The first step is to study the invariant Ω-finite distributions on the regular set. Let C be either B or L. We write x for the coordinate θ on B' $(0 < \theta < \pi$, or $\pi < \theta < 2\pi)$ and the coordinate t on L' $(t > 0$ or $t < 0)$; $\mathrm{d}/\mathrm{d}x$ thus denotes either $\mathrm{d}/\mathrm{d}\theta$ or $\mathrm{d}/\mathrm{d}t$. By an *exponential polynomial* in x we mean a function of the form

$$\mathrm{e}^{\mu_1 x} p_1(x) + \cdots + \mathrm{e}^{\mu_r x} p_r(x)$$

where the μ_j are in $\mathbb{C}$ and the p_j are polynomials. We refer to x as the *canonical* coordinate.

Proposition 1. *Let T be an invariant Ω-finite distribution on G'. Then T is an analytic function, i.e., there is an analytic function Θ on G' such that*

$$T(f) = \int_{G'} \Theta f \, \mathrm{d}G \quad (f \in C_c^\infty(G'))$$

Define the function Φ_C on C' by writing Θ as

$$\Theta(h) = \frac{\Phi_C(h)}{\Delta_C(h)} \quad (h \in C')$$

Then, on any connected component C^+ of C', Φ_C is an exponential polynomial in the canonical coordinate; if T satisfies the differential equation $P(\Omega) \cdot T = 0$ where P is a polynomial in one indeterminate, Φ_C satisfies the differential equation

$$P(\mathrm{d}_C) \cdot \Phi_C = 0$$

where $\mathrm{d}_C = -\mathrm{d}^2/\mathrm{d}\theta^2$ for $C = B$ and $\mathrm{d}^2/\mathrm{d}t^2$ for $C = L$.

Proof. It is obviously enough to prove all of this on the open set $E^G \subset G'$ where $E \subset C'$ is sufficiently small and connected so that the conjugacy map

$$\varphi : (G/C) \times E \to E^G$$

is a diffeomorphism. At first let us assume only that T is defined and invariant on E^G and show that there is a unique distribution σ_T on E

such that

$$T(f) = \sigma_T(F_f) \quad (f \in C_c^\infty(E^G))$$

Observe that, if f corresponds to $u \otimes v$ under φ,

$$F_f = \delta \Delta_C v \cdot \left(\int u \, \mathrm{d}(G/C) \right) \quad (\text{on } E)$$

where $\delta = 1$ for $C = B$ and a constant $= \pm 1$ for $C = L$. So the map $f \mapsto F_f$ is *onto* $C_c^\infty(E^G)$, showing that σ_T is unique. For its existence, transfer T to a distribution $\tilde{T}$ on $(G/C) \times E$; $\tilde{T}$ is invariant under the action $x(yC, h) \mapsto (xyC, h)$ of G. Since this action affects only G/C, and the measure $\mathrm{d}(G/C)$ is the only invariant distribution on G/C, it follows that $\tilde{T}$ has the form $1 \otimes S$ for a distribution S on E:

$$\tilde{T}(u \otimes v) = \int u \, \mathrm{d}(G/C) \cdot S(v) = S(\delta \Delta_C^{-1} F_f)$$

where f corresponds to $u \otimes v$ as above. We thus define σ_T by

$$\sigma_T = \delta^{-1} \Delta_C^{-1} S$$

The mapping $T \mapsto \sigma_T$ takes invariant distributions on E^G to distributions on E. Moreover we have

$$\sigma_{\Omega T} = \mathrm{d}_C \sigma_T$$

since $F_{\Omega F, C} = \mathrm{d}_C F_{f,C}$. Hence the equation $P(\Omega) \cdot T = 0$ gives rise to the equation $P(\mathrm{d}_C)\sigma_T = 0$, which is *an ordinary differential equation in the canonical variable x, with constant coefficients.* Hence

$$\sigma_T = \Phi \, \mathrm{d}C$$

where Φ is an exponential polynomial in x on E; it is of course uniquely determined, when T is given on G', on the entire connected component of C' that contains E. To complete the proof we must show that if Θ is the (unique) invariant function on E^G which is $\varepsilon\Phi/\Delta_C$ on E, where $\varepsilon = -1$ for $C = B$ and $+1$ for $C = L$, then $T = \Theta \, \mathrm{d}G$ on E^G. But for $f \in C_c^\infty(E^G)$

$$T(f) = \sigma_T(F_f) = \int_E (\Phi \Delta_C^{-1}) \Delta_C F_f \, \mathrm{d}C = \int_E (\Theta|_E) \varepsilon \Delta_C F_f \, \mathrm{d}C$$

and, as $\varepsilon \Delta_C = \bar{\Delta}_C$, this is

$$= \int_E (\Theta|_E) |\Delta_C|^2 \int_{G/C} f \, \mathrm{d}(G/C) = \int_{E^G} \Theta f \, \mathrm{d}G$$

B' has two connected components:

$$B^\pm = \{u_\theta | 0 < \pm\theta < \pi\}$$

The set L' has four connected components; we put

$$L^{\pm} = \{a_t | t \gtrless 0\}, \quad L_\gamma^{\pm} = \{\gamma a_t | t \gtrless 0\}$$

and note that it is enough to work with L^+ and L_γ^+ since the other two are obtained by conjugation with respect to $w = \begin{pmatrix} 0 & 1 \\ -1 & 0 \end{pmatrix}$. We write

$$\begin{aligned} \Phi_B(u_\theta) &= (e^{i\theta} - e^{-i\theta})\Theta(u_\theta) \quad (0 < |\theta| < \pi) \\ \Phi_L(a_t) &= (e^t - e^{-t})\Theta(a_t) \quad (t > 0) \\ \Phi_{L,\gamma}(a_t) &= (e^t - e^{-t})\Theta(\gamma a_t) \quad (t > 0) \end{aligned}$$

Then

$$\int \Theta f \, dG = -\int \Phi_B F_{f,B} \, d\theta + \int_0^\infty \Phi_L F_{f,L} \, dt + \int_0^\infty \Phi_{L,\gamma} F_{f_\gamma,L} \, dt$$

Suppose now that

$$\Omega T = \mu^2 T \quad (\mu \neq 0)$$

Then

$$\begin{aligned} &\Phi_B(u_\theta) = a_\pm e^{i\mu\theta} + b_\pm e^{-i\mu\theta} \quad (u_\theta \in B^\pm) \\ &\Phi_L(a_t) = \alpha e^{\mu t} + \beta e^{-\mu t}, \quad \Phi_{L,\gamma}(a_t) = \alpha' e^{\mu t} + \beta' e^{-\mu t} \quad (t > 0) \end{aligned}$$

where $a_\pm, b_\pm, \alpha, \beta, \alpha', \beta'$ are constants. We thus see that character formulae reminiscent of Weyl's emerge already just from the differential equations. Since T *is given only on* G', *nothing more can be said about the constants.* Indeed, given these formulae for *arbitrary* choices of $a_\pm, b_\pm, \ldots$ etc, there is a unique invariant function Θ on G' related as above to Φ_B, Φ_L; and, by our formula for the radial components of Ω, we should have $\Omega\Theta = \mu^2\Theta$. However, if T is a locally integrable function defined and invariant on all of G and satisfying $\Omega T = \mu^2 T$ on all of G, there will be relations between the constants $a_+, b_+, \ldots$ etc. Indeed, we will now have the equation $T(\Omega f) = \mu^2 T(f)$ for all $f \in C_c^\infty(G)$. Calculating $T(\Omega f)$ and $\mu^2 T(f)$ using their expressions in terms of Φ_B and Φ_L and integrating by parts we see that $T(\Omega f) - \mu^2 T(f)$ will come out as the 'boundary terms'. These must then vanish, and this will be equivalent to certain linear relations between these constants or relations between the values of Φ_B, $\Phi'_B, \ldots$ and $\Phi_L, \Phi'_L, \ldots$ at 1 and γ. These are the celebrated matching conditions of Harish-Chandra.

7.2 Matching conditions

Fix an invariant Ω-finite distribution T on G' and let Θ be the invariant analytic function on G' such that $T = \Theta \, dG$ on G'; let Φ_B, Φ_L be as above.

Let D be the invariant analytic function on G such that $D = \Delta_C^2$ on $C = B$ or L; $D(x) = \mathrm{tr}(x^2) - 2$, and $|D|^{1/2} = |\Delta_C|$ on C'. It is now easy to see that, if R is a compact set in G, there is a compact set E in L such that if $y \in B$ or L and y^G meets R, then $y \in B$ or E. So there is a constant $c = c(R) > 0$ such that

$$|\Theta(x)| \leqslant c|D(x)|^{-1/2} \quad (x \in R \cap G')$$

The function Θ is thus locally integrable on G (not just G') and we can define the distribution

$$T^{\#}(f) = \int_G \Theta f \, \mathrm{d}G \quad (f \in C_c^\infty(G))$$

Clearly $T = T^{\#}$ on G' and $T^{\#}$ is the obvious extension of T to all of G. Now $P(\Omega)T = 0$ but one must not assume without further examination that $P(\Omega) \cdot T^{\#} = 0$. (Indeed, if H is the function on $\mathbb{R}$ which is 0 for $x < 0$ and 1 for $x > 0$ and we view H as a distribution on $\mathbb{R}$, then $(\mathrm{d}/\mathrm{d}x)H = \delta$, the Dirac measure at the origin.) Now $\Omega \cdot T$ is an invariant Ω-finite distribution on G' and so it makes sense to speak of $(\Omega T)^{\#}$; this will coincide with $\Omega\Theta \, \mathrm{d}G$ on G' but in general will not coincide on all of G with $\Omega T^{\#}$.

Proposition 2. *We have*

$$\Omega T^{\#} - (\Omega T)^{\#} = J + J^{\gamma}$$

where J and J^{γ} are invariant distributions defined as follows:

$$\begin{aligned} J &= [\Phi_B'(1+) - \mathrm{i}\Phi_L'(1+)]j_+ - [\Phi_B'(1-) - \mathrm{i}\Phi_L'(1+)]j_- \\ &\quad + [\Phi_B(1+) - \Phi_B(1-)](\mathrm{i}\pi\delta) \\ J^{\gamma} &= -[\Phi_B'(\gamma+) - \mathrm{i}\Phi_{L,\gamma}'(1+)](j_+)_{\gamma} + [\Phi_B'(\gamma-) \\ &\quad - \mathrm{i}\Phi_{L,\gamma}'(1+)](j_-)_{\gamma} - [\Phi_B(\gamma+) - \Phi_B(\gamma-)](\mathrm{i}\pi\delta_{\gamma}) \end{aligned}$$

where

$$j_{\pm}(f) = F_{f,B}(1\pm), \quad \delta(f) = f(1)$$

and, for any distribution t on G, t_{γ} is the translate of t by γ.

Proof. We have

$$(\Omega T^{\#})(f) = -\int \Phi_B(-F_{f,B}'')\,\mathrm{d}\theta + \int_0^\infty \Phi_L F_{f,L}''\,\mathrm{d}t + \int_0^\infty \Phi_{L,\gamma} F_{f_\gamma,L}''\,\mathrm{d}t$$

since $F_{\Omega f,B} = -F_{f,B}''$, $F_{\Omega f,L} = F_{f,L}''$. On the other hand, as

$$\Delta_B(\Omega\Theta|_{B'}) = -\Phi_B'', \quad \Delta_L(\Omega\Theta|_{L'}) = \Phi_L''$$

we have

$$(\Omega T)^{\#}(f) = \int f \cdot \Omega\Theta \, \mathrm{d}G$$

$$= -\int(-\Phi_B'')F_{f,B}\,\mathrm{d}\theta + \int_0^\infty \Phi_L'' F_{f,L}\,\mathrm{d}t + \int_0^\infty \Phi_{L,\gamma}'' F_{f_\gamma,L}\,\mathrm{d}t$$

Integration by parts now leads to the result provided we remember that

$$F_{f,B}'(1) = -\mathrm{i}\pi f(1), \quad F_{f,B}(1\pm) = \pm\frac{\mathrm{i}}{4}\int_0^\infty \bar{f}(n_{\pm s})\,\mathrm{d}s$$

$$F_{f,L}(1) = (1/\mathrm{i})(F_{f,B}(1+) - F_{f,B}(1-))$$

J is a linear combination of δ, j_+, and j_-. Let us now take a closer look at $j_\pm$. If we consider

$$n_{\pm 1} = \exp \pm X$$

and define

$$\mathscr{U}_\pm = (n_{\pm 1})^G$$

as the orbits of $n_{\pm 1}$, it is easy to check that they are disjoint, disjoint from (1), (γ), and

$$\mathscr{U} = (1) \coprod \mathscr{U}_+ \coprod \mathscr{U}_-$$

where $\mathscr{U}$ is the variety of *unipotent matrices*. Indeed, as everything is contained in $\mathscr{U} \subset G(\varepsilon)$ for every $\varepsilon > 0$ we may go to the Lie algebra via the inverse of the exponential map; and there the orbits $(\pm X)^G$ are the upper and lower halves of the 'light' cone of nilpotents (Figure 7.1).

If now we wanted to compute the invariant measure $\mu_\pm$ on $\mathscr{U}_\pm$ as a distribution on G, we would have to use the formula

$$\mu_\pm(g) = \int g(n_{\pm 1}^y)\,\mathrm{d}(G/N) = \text{const.} \int\int g(n_{\pm 1}^{u_\theta a_t})\mathrm{e}^{2t}\,\mathrm{d}\theta\,\mathrm{d}t$$

$$= \text{const.} \int \bar{g}(n_{\pm \mathrm{e}^{2t}})\,\mathrm{e}^{2t}\,\mathrm{d}t = \text{const.} \int_0^\infty \bar{g}(n_{\pm s})\,\mathrm{d}s$$

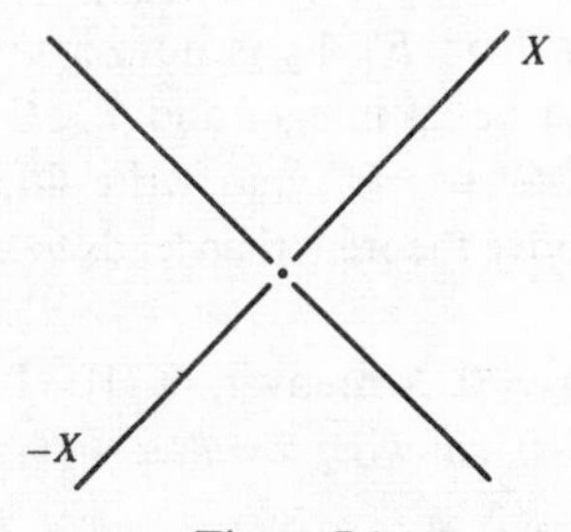

Figure 7.1

We cannot, however, proceed directly like this because, unlike the orbits from B' or L', the orbits of $n_{\pm 1}$ *are not closed in* G, so that, for a $g \in C_c(G)$, the function $yN \mapsto g(n^y_{\pm 1})$ *does not have compact support in* G/N. However, the last expression is obviously convergent and so $\mu_\pm$ are actually Borel measures on G that live on $\mathscr{U}_\pm$. The formula

$$F_{f,B}(1\pm) = j_\pm(f) = \pm(\mathrm{i}/4)\int_0^\infty \bar{f}(n_{\pm s})\,\mathrm{d}s$$

shows that $\mu_\pm$ are mutiples of $j_\pm$. It is obvious from the picture on the Lie algebra that δ, j_+ and j_- are linearly independent, even when restricted to functions with supports in arbitrarily small neighbourhoods of 1. This leads to the following fundamental theorem.

Theorem 3 (Theorem of matching conditions). *Let Θ be an invariant analytic function on G' and let $T^\#$ be the distribution $f \mapsto \int f\Theta\,\mathrm{d}G$ on G. Assume that $\Omega\Theta = \mu^2\Theta$ on $G'(\mu \in \mathbb{C})$ and use the earlier notation. Then $\Omega T^\# = \mu^2 T^\#$ if and only if the following conditions are both satisfied:*

(i) *Φ_B extends to a smooth function on all of B;*
(ii) $-\mathrm{i}\Phi'_B(1) = \Phi'_L(1+)$, $-\mathrm{i}\Phi'_B(\gamma) = \Phi'_{L,\gamma}(1+)$.

If these conditions are satisfied and $\Theta \not\equiv 0$ on B', then μ must be an integer and Φ_B is a linear combination of the characters $\mathrm{e}^{\pm \mathrm{i}\mu\theta}$.

Proof. If (i) and (ii) are satisfied, $J = J^\gamma = 0$ and hence $\Omega T^\# = (\Omega T)^\# = \mu^2 T^\#$. Suppose conversely $\Omega T^\# = \mu^2 T^\#$. Then $\Omega T^\# = (\Omega T)^\#$ and so $J + J^\gamma = 0$. It is then easy to see using functions with small supports around 1 that $J = J^\gamma = 0$ so that

$$\begin{gathered}\Phi'_B(1+) = \Phi'_B(1-) = \mathrm{i}\Phi'_L(1+), \quad \Phi'_B(\gamma+) = \Phi'_B(\gamma-) = \mathrm{i}\Phi'_{L,\gamma}(1+)\\ \Phi_B(1+) - \Phi_B(1-) = 0 = \Phi_B(\gamma+) - \Phi_B(\gamma-)\end{gathered}$$

So Φ_B and Φ'_B extend continuously to B. As $\Phi''_B = -\mu^2\Phi_B$, repeated application yields $\Phi_B \in C^\infty(B)$. The relation (ii) is already contained in the above formulae. If $\Theta \not\equiv 0$ on B', Φ_B is nonzero, C^∞ on B, and satisfies $\Phi''_B = -\mu^2\Phi_B$. So μ must be an integer and $\Phi_B \in \mathbb{C}\cdot\mathrm{e}^{\mathrm{i}\mu\theta} + \mathbb{C}\cdot\mathrm{e}^{-\mathrm{i}\mu\theta}$.

Consider now a $\Theta, \Omega\Theta^\# = \mu^2\Theta^\#$, and write Φ_B, Φ_L in the special form described earlier. The above theorem then leads to a very detailed picture of Θ.

Case 1: $\mu \notin \mathbb{Z}$. Then $\Phi_B = 0$. Moreover, $\Phi'_L(1) = 0$ and $\Phi'_{L,\gamma}(1) = 0$. Hence $\mu(\alpha - \beta) = 0$, $\mu(\alpha' - \beta') = 0$, showing $\alpha = \beta$, $\alpha' = \beta'$. Hence

$$\Theta(a_t) = \alpha\frac{\mathrm{e}^{\mu t} + \mathrm{e}^{-\mu t}}{|\mathrm{e}^t - \mathrm{e}^{-t}|}, \quad \Theta(\gamma a_t) = \alpha'\frac{\mathrm{e}^{\mu t} + \mathrm{e}^{-\mu t}}{|\mathrm{e}^t - \mathrm{e}^{-t}|} \quad (t \neq 0)$$

So the Θ span the same two-dimensional space as the two principal series characters $T_{\varepsilon,\mu}$!

Case 2: $\mu \in \mathbb{Z}\backslash(0)$ We have

$$\Theta(u_\theta) = \frac{a e^{i\mu\theta} + b e^{-i\mu\theta}}{e^{i\theta} - e^{-i\theta}} \quad (\theta \not\equiv 0, \pi)$$

$$\Theta(a_t) = \frac{\alpha e^{\mu t} + \beta e^{-\mu t}}{e^t - e^{-t}} \quad (t > 0)$$

$$\Theta(\gamma a_t) = \frac{\alpha' e^{\mu t} + \beta' e^{-\mu t}}{e^t - e^{-t}} \quad (t > 0)$$

and the conditions are

$$a - b = \alpha - \beta = \alpha' - \beta'$$

These are necessary and sufficient. The eigenspace has dimension 4, and is spanned by Θ_m, Θ_{-m}, F_m, and $T_{\varepsilon,m}$ where ε has the same parity as $m = |\mu|$.

Case 3: $\mu = 0$. We have

$$\Theta(u_\theta) = \frac{a}{e^{i\theta} - e^{-i\theta}} \quad (\theta \not\equiv 0, \pi)$$

and Φ_L, $\Phi_{L,\gamma}$ are *linear* in t. The conditions now give

$$\Theta(a_t) = \frac{\alpha}{e^t - e^{-t}}, \quad \Theta(\gamma a_t) = \frac{\alpha'}{e^t - e^{-t}} \quad (t > 0)$$

and the dimension of the space is 3; it is spanned by $\Theta_{0,+}$, $\Theta_{0,-}$ and $T_{0,0}$.

7.3 The Harish-Chandra regularity theorem

The theorem is as follows.

Theorem 4. *Suppose T is an invariant distribution on G which is Ω-finite. Then $T = T^\#$. In other words there is an invariant locally integrable function Θ on G which is analytic on G' such that*

$$T(f) = \int \Theta f \, dG, \quad f \in C_c^\infty(G)$$

Proof. The space spanned by

$$T, \Omega T, \Omega^2 T, \ldots$$

is finite-dimensional and Ω acts on it. By decomposing it spectrally with respect to this action it is clear that we need prove Theorem 4 only for T such that $(\Omega - \mu^2)^k T = 0$ for some $k \geqslant 1$. We shall do this by induction. Assume the result for $k - 1$ and write $T_1 = (\Omega - \mu^2)T$. Then $(\Omega - \mu^2)^{k-1} T_1 = 0$ so that $T_1 = T_1^\#$. Now $T = \Theta \, dG$ on G' so that $T_1 =$

$(\Omega - \mu^2)\Theta\, dG = \Omega\Theta\, dG - \mu^2\Theta\, dG$ on G'. Hence $T_1 = \Omega\Theta\, dG - \mu^2\Theta\, dG$ on G itself, i.e., $T_1 = (\Omega T)^{\#} - \mu^2 T^{\#}$. But then $T_1 = (\Omega - \mu^2)(T - T^{\#} + T^{\#}) = (\Omega - \mu^2)(T - T^{\#}) + \Omega T - \mu^2 T^{\#}$. We thus obtain the relation

$$(\Omega - \mu^2)(T - T^{\#}) = (\Omega T)^{\#} - \Omega T^{\#}$$

Proposition 2 now implies that

$$(\Omega - \mu^2)(T - T^{\#}) = a_+ j_+ + a_- j_- + b\delta + a'_+(j_+)_\gamma + a'_-(j)_\gamma + b'\delta_\gamma$$

where $a_\pm$, $a'_\pm, b, b'$ are constants. We wish to prove from this that $T - T^{\#} = 0$. It is convenient to separate the various terms on the right. This is done by the following lemma.

Lemma 5. (a) *We can separate $\mathscr{U}$ and $\gamma\mathscr{U}$ by disjoint invariant open sets.*

(*b*) *We can find an invariant open neighbourhood of $\mathscr{U}_+$ (resp. $\mathscr{U}_-$) not meeting* $(1)\coprod\mathscr{U}_- = \mathrm{Cl}(\mathscr{U}_-)$(*resp.*$(1)\coprod\mathscr{U}_+ = \mathrm{Cl}(\mathscr{U}_+)$).

Proof. For $\varepsilon > 0$ sufficiently small, $V = G(\varepsilon)$ and $V_\gamma = \gamma G(\varepsilon)$ are disjoint and contain $\mathscr{U}$ and $\gamma\mathscr{U}$ respectively.

(b) Select an open neighbourhood W_+ of n_{+1} such that $W_+ \cap ((1)\coprod\mathscr{U}_-) = \varnothing$; this is possible since $(1)\coprod\mathscr{U}_- = \mathrm{Cl}(\mathscr{U}_-)$ is closed. Then W_+^x does not meet $(1)\coprod\mathscr{U}_-$ either and so $V_+ = \bigcup_{x\in G} W_+^x$ is an invariant open neighbourhood of $\mathscr{U}_+$ not meeting $(1)\coprod\mathscr{U}_-$. Similarly for $\mathscr{U}_-$

If we denote restriction to $V \cap V_+$ by a suffix, we have

$$(\Omega - \mu^2)((T - T^{\#})_{V\cap V_+}) = a_+ j_+$$

and

$$\mathrm{supp}\,((T - T^{\#})_{V\cap V_+}) = \mathscr{U}_+$$

We now have the following key proposition.

Proposition 5. *Suppose U_+ is an invariant open neighbourhood of $\mathscr{U}_+$ (resp. $\mathscr{U}_-$) and t is an invariant distribution of U_+ such that* $\mathrm{supp}\,(t) \subset \mathscr{U}_+$(*resp. $\mathscr{U}_-$*). *If $(\Omega - \mu^2)t = a_+ j_+$ (resp. $a_- j_-$), then $t = 0$, $a_\pm = 0$.*

This proposition shows immediately that $T - T^{\#}$ is 0 on $V \cap V_+$, and $a_+ = 0$. i.e., $\mathrm{supp}\,(T - T^{\#})$ does not meet $\mathscr{U}_+$, and $a_+ = 0$. Similarly we argue that it does not meet $\mathscr{U}_-$ and $a_- = 0$. Translating by γ we conclude likewise that $\mathrm{supp}\,(T - T^{\#})$ does not meet $\gamma\mathscr{U}_\pm$, and $a'_\pm = 0$. Hence $\mathrm{supp}\,(T - T^{\#}) \subset \{1, \gamma\}$. Localizing once again we see that the proof of the regularity theorem will be complete if we can show that if t is an invariant distribution with $\mathrm{supp}\,(t) \subset (1)$ and $(\Omega - \mu^2)t = b\delta$, then $t = 0$, $b = 0$. But t is necessarily of the form $p(\Omega)\delta$ where p is a uniquely determined polynomial in one indeterminate, and $(\Omega - \mu^2)t = b\delta$ implies $(\Omega - \mu^2)p(\Omega) = b$ which is impossible unless $p = 0$, $b = 0$, i.e., $t = 0$, $b = 0$. This proves the theorem.

It remains to prove Proposition 5. The proof depends on a very interesting property of the space of invariant distributions on $U_\pm$ that are supported by $\mathcal{U}_\pm$, *when this space is viewed as a module for* Ω. Let us write $\mathcal{D}_\pm$ for this module. We consider only $\mathcal{U}_+$ and $\mathcal{D}_+$ as the argument for $\mathcal{U}_-$ and $\mathcal{D}_-$ is similar; we write $\mathcal{D} = \mathcal{D}_+$.

Lemma 6. *Let* $f(x) = \mathrm{tr}(x) - 2$, $x \in G$. *Then* f *is invariant and* $\mathcal{U}_+$ *is exactly the set of zeros of* f *in* U_+. *Moreover,* $\mathrm{d}f$ *is never* 0 *on* $\mathcal{U}_+$.

Proof. It is clear that f is invariant, and $f(x) = 0 \Leftrightarrow x \in \mathcal{U}$. This proves the first assertion since $U_+ \cap \mathcal{U} = U_+$. For the second it is enough to prove that $(\mathrm{d}f)_{n_1} \neq 0$ where $n_1 = \begin{pmatrix} 1 & 1 \\ 0 & 1 \end{pmatrix}$. But

$$(\mathrm{d}f)_{n_1}(Y) = (\mathrm{d}/\mathrm{d}t)_{t=0}\, \mathrm{tr}\,(n_1 \exp tY) = 1$$

It follows from lemma 6 that around any point of $\mathcal{U}_+$ the local structure of distributions supported by $\mathcal{U}_+$ is essentially the same as the structure near 0 of distributions in $\mathbb{R}^3$ supported by $x_3 = 0$. For such a distribution one can define a *local transversal order* with the help of the following lemma.

Lemma 7. *Let* t *be a distribution defined around the origin in* $\mathbb{R}^n (n \geqslant 1)$ *and suppose that* $\mathrm{supp}\,(t) \subset P = \{(x_1, \ldots, x_n) | x_n = 0\}$. *Then there is an integer* $k \geqslant 0$ *such that* $x_n^k t = 0$ *in a neighbourhood of the origin.*

Proof. We can find a number $a > 0$ and an integer $m \geqslant 0$ such that t is defined on the cube $C = \{(x_1, \ldots, x_n) | |x_j| < a \text{ for } 1 \leqslant j \leqslant n\}$ and satisfies the estimate

$$|t(u)| \leqslant A \sum_{|\alpha| \leqslant m} \sup_C |u^{(\alpha)}| \quad (u \in C_c^\infty(C))$$

where $u^{(\alpha)} = (\partial/\partial x_1)^{\alpha_1} \cdots (\partial/\partial x_n)^{\alpha_n}$, $|\alpha| = \alpha_1 + \cdots + \alpha_n$, and $A > 0$ is a constant. Take now u in the form $w \otimes v$ where w depends only on $x_1, \ldots, x_{n-1}$ and v on x_n. For fixed w, $t_w : v \mapsto t(w \otimes v)$ is a distribution in the variable x_n, defined on the interval $(-a, a)$; if v is 0 in a neighbourhood of 0, $w \otimes v$ is zero in a neighbourhood of $P \subset C$ and hence $t(w \otimes v) = 0$ as $\mathrm{supp}\,(t) \subset P \cap C$. So $\mathrm{supp}\,(t_w) \subset (0)$. It is then well known that t_w is a linear combination of the delta function δ at 0 and its derivatives. But (with w fixed still) the estimate for $t(u)$ implies the estimate

$$|t_w(v)| \leqslant A_w \sum_{0 \leqslant r \leqslant m} \sup_{(-a,a)} |v^{(r)}|$$

for all $v \in C_c^\infty((-a, a))$, $A_w > 0$ being a constant. This means that t_w can be a linear combination of only the $\delta^{(r)}, 0 \leqslant r \leqslant m$. In particular, $x_n^{m+1} t_w = 0$.

Hence the distribution $x_n^{m+1}t$ vanishes on all functions of the form $w \otimes v$; it is thus 0.

Let t be a distribution as in the above lemma. We now define the *transversal order of t at* 0 to be the integer $k \geqslant 0$ such that $x_n^{k+1}t = 0$ in some neighbourhood of the origin but $x_n^k t \neq 0$ in any neighbourhood of the origin. It is clear from the above lemma that if t is nonzero in all sufficiently small neighbourhoods of the origin, i.e., if $0 \in \mathrm{supp}(t)$, the transversal order is well defined; we denote it by $o(t)$.

Let us now return to our space $\mathscr{D}$. Since elements of $\mathscr{D}$ are all invariant, their supports are necessarily invariant sets; and, as G acts *transitively* on $\mathscr{U}_+$ which contains the supports of all the members of $\mathscr{D}$, it follows that, for any element $t \in \mathscr{D}$, $t = 0$ or else $\mathrm{supp}(t) \doteq \mathscr{U}_+$. So, if $t \neq 0$, it makes sense in view of Lemmas 6 and 7 to speak of the transversal order $o_n(t)$ of t at any point $n \in \mathscr{U}_+$: $o_n(t) = k$ means $f^{k+1} \cdot t = 0$ in a neighbourhood of n but $f^k \cdot t \neq 0$ (in any neighbourhood of n). By invariance and transitivity it is clear that $o_n(t)$ is independent of n. We write $o(t)$ for this integer and call it the *transversal order of t*. It is defined as soon as $t \in \mathscr{D} \backslash (0)$ and is an integer $k \geqslant 0$; moreover,

$$o(t) = k \Leftrightarrow f^{k+1}t = 0, f^k t \neq 0 \quad (\text{on } U_+)$$

The transversal order gives rise to a *filtration* $(\mathscr{D}_j)$ on $\mathscr{D}$: $t \in \mathscr{D}_j \Leftrightarrow t$ is either 0 or $o(t) \leqslant j$. So

$$\mathscr{D}_0 \subset \mathscr{D}_1 \subset \cdots, \bigcup_j \mathscr{D}_j = \mathscr{D}$$

Now the invariant *measure* j_+ actually *lives on* $\mathscr{U}_+$, i.e., $j_+(g) = 0$ if $g \in C_c^\infty(U_+)$ and g is 0 on $\mathscr{U}_+$. Hence $f \cdot j_+ = 0$, i.e.,

$$o(j_+) = 0$$

The remarkable property of $\mathscr{D}$ as an Ω-module is now contained in the following lemma.

Lemma 8. *Suppose* $0 \neq t \in \mathscr{D}$ *and* $o(t) = k$. *Then* $o(\Omega t) = k + 1$. *In other words,* $\Omega \mathscr{D}_j \subset \mathscr{D}_{j+1}$ *for all* $j \geqslant 0$, *and the map induced by* Ω *on* $\mathscr{D}^{\mathrm{gr}} = \bigoplus_{j \geqslant 0} (\mathscr{D}_j / \mathscr{D}_{j-1})(\mathscr{D}_{-1} = 0)$ *is injective from* $\mathscr{D}_j / \mathscr{D}_{j-1}$ *to* $\mathscr{D}_{j+1} / \mathscr{D}_j$ *for all* $j \geqslant 0$.

Proof. We are given that

$$f^{k+1}t = 0, \quad f^k t \neq 0$$

Let $s = \Omega t$. We must prove that

$$f^{k+2}s = 0, \quad f^{k+1}s \neq 0$$

We do this by establishing the following formula that calculates the functions $\Omega(f^{r+1})$ $r \geqslant 0$:

$$\Omega(f^{r+1}) = (2r+3)(2r+2)f^r + (r+2)^2 f^{r+1} \qquad (*)$$

Let us assume that (*) has been proved. Then, locally around a point u_0 of $\mathscr{U}_+$, let us take x_1, x_2 and $x = f$ as a coordinate system, and express Ω in the form

$$A + B\partial/\partial x + C\partial^2/\partial x^2 + E_1\partial/\partial x_1 + E_2\partial/\partial x_2$$

where E_1 and E_2 are differential operators of degree $\leqslant 1$. We now claim that

(i) $A = 1$, $B = 3(x+2)$, $C = x^2 + 4x$,
(ii) if $E = E_1\partial/\partial x_1 + E_2\partial/\partial x_2$, $Et' = 0$ for any invariant distribution t'(around u_0).

The formulae (i) follow at once from (*) since $\Omega 1 = 1$, $\Omega x = 4x + 6$ and $\Omega x^2 = 9x^2 + 20x$. To prove (ii), note that the planes $x = \text{const.}$ describe the orbits of G locally around u_0 and hence the vector fields $\partial/\partial x_1$ and $\partial/\partial x_2$ must be in the module (over C^∞ functions) generated by the vector fields $\tilde{Z}$ induced by the elements $Z \in \mathfrak{g}$. Since $\tilde{Z}t' = 0$, we must have $\partial/\partial x_j t' = 0$, hence $Et' = 0$. So,

$$\Omega t = t + 3(x+2)\mathrm{D}t + x(x+4)\mathrm{D}^2 t \quad (\mathrm{D} = \partial/\partial x).$$

The key point is that the *coefficient of* $\mathrm{D}^2 t$ *already has* x *as a factor*. Hence *all* terms above have transversal order $\leqslant k+1$, showing $x^{k+2}\Omega t = 0$. Let us now calculate $x^{k+1}\Omega t$. We have

$$x^{k+1}\Omega t = 6x^{k+1}\mathrm{D}t + 4x^{k+2}\mathrm{D}^2 t$$

But

$$x^{k+1}\circ \mathrm{D} = \mathrm{D}\circ x^{k+1} - (k+1)x^k$$
$$x^{k+2}\circ \mathrm{D}^2 = \mathrm{D}^2\circ x^{k+2} - 2(k+2)\mathrm{D}\circ x^{k+1} + (k+2)(k+1)x^k$$

Hence

$$\begin{aligned} x^{k+1}\Omega t &= -6(k+1)x^k t + 4(k+2)(k+1)x^k t \\ &= (2k+2)(2k+1)x^k t \neq 0 \end{aligned}$$

So it remains to prove (*). Since we are dealing with analytic functions it is enough to prove (*) on A'. We do this using our formula for radial components. For $x = \mathrm{diag}(\mathrm{e}^t, \mathrm{e}^{-t})$, we have

$$f(x) = \mathrm{e}^t + \mathrm{e}^{-t} - 2$$
$$\Delta(x)^2 = f(x)^2 + 4f(x)$$

If we denote the restrictions to A' by the suffix A, then, for any invariant (analytic) function g,

$$(\Omega g)_A = \Delta^{-1}(\mathrm{d}^2/\mathrm{d}t^2)(g_A\Delta)$$

We take $g = f^{r+1}$. The verification of (*) is a simple but tedious calculation;

indeed,

$$(d^2/dt^2)(\Delta f^{r+1}) = (r+1)rf^{r-1}\Delta^3 + 3(r+1)f^r\Delta(e^t + e^{-t}) + f^{r+1}\Delta$$

so that, as $\Delta^2 = f^2 + 4f$,

$$\begin{aligned}\Delta^{-1}(d^2/dt^2)(\Delta f^{r+1}) &= f^{r-1}\{(r+1)r\Delta^2 + 3(r+1)f(f+2) + f^2\} \\ &= f^r\{(r+1)r(f+4) + 3(r+1)(f+2) + f\} \\ &= f^r\{(r+2)^2 f + (2r+3)(2r+2)\}\end{aligned}$$

which is $(*)$. This proves the lemma.

***Proof of Proposition** 5.* Suppose now $t \in \mathscr{D}, t \neq 0$, and $(\Omega - \mu^2)t = a_+ j_+$. Let $k = o(t)$; it is then immediate from lemma 8 that $\Omega t - \mu^2 t$ has transversal order $k+1$. As $k \geqslant 0$ and as j_+ has transversal order 0, we have a contradiction. This proves the proposition, and hence completes the proof of the regularity theorem.

If one want to extends all of this to *general* semisimple groups, it is clear what one must do. The orbital integrals, radial components, and distributions with unipotent supports must all be studied in complete generality. That however will take anyone along Harish-Chandra's epic journey.

Problems

1. Carry out the analogous development of the structure of invariant $\partial(I)$-finite distributions on $\mathfrak{g}$ where I is the algebra of G-invariants in the symmetric algebra $S(\mathfrak{g})$, G being $SL(2, \mathbb{R})$.
2. Prove that if U_+ is an invariant open neighbourhood of $\mathscr{U}_+$ not meeting $(0) \coprod \mathscr{U}_-$, and $\mathscr{D}_+$ is as in the text, the distributions

$$\Omega_n j_+ \quad (n \geqslant 0)$$

form a basis for $\mathscr{D}_+$.
3. Prove the regularity theorem for $G = SL(2, \mathbb{C})$.
4. Prove that, if Θ is an invariant eigendistribution on a compact Lie group, it must be multiple of the character of some irreducible representation (hint: use harmonic analysis on the group).

8

Harmonic analysis of the Schwartz space

8.1 The point of view of differential equations and eigenfunction expansions. Plancherel measure and its relation to eigenfunction asymptotics

In the general theory of locally compact separable groups the decomposition of the regular representation is one of the basic objectives, and the abstract Plancherel theorem is formulated in that context. Although we have obtained the Plancherel formula, we have worked only over $C_c^\infty(G)$. For a deeper and more complete understanding of the decomposition theory and harmonic analysis of $L^2(G)$ it is essential to work in a space of rapidly decreasing functions and to develop a theory of Fourier transforms for such functions. For such functions f, the corresponding operators $\pi(f)$ ($\pi = \pi_m$, the discrete series, or $\pi = \pi_{\varepsilon, i\mu}$, the unitary principal series) will be represented by matrices (relative to the basis in which K acts diagonally) which are actually functions of m and μ. For a given f the associated matrix function may be thought of as the *operator Fourier transform* of f, and the fundamental goal of this chapter is to prove that *the operator Fourier transform gives an isomorphism of the convolution algebra of rapidly decreasing functions on the group with an algebra (under pointwise multiplication) of rapidly decreasing matrix functions of m and μ.*

For $SL(2, \mathbb{R})$ this was originally done by Ehrenpreis and Mautner [EM] who introduced the Schwartz space and carried out its harmonic analysis. However, it was only after Harish-Chandra introduced the Schwartz space $\mathscr{C}(G)$ of a general semisimple group with finite centre and carried out its harmonic analysis [H6][H7] that its fundamental role and importance became clear. There is no question that the theory of the Schwartz space provides the correct framework for all questions of L^2 harmonic analysis on semisimple groups.

A clue to the correct definition of the Schwartz space may be obtained by taking the point of view of differential equations that goes back to Bargmann [Ba]. Let $\mathscr{H} = L^2(G)$ and $\mathscr{H}_{mn}$ be the subspace of functions f in $L^2(G)$ *of type* (m, n), i.e., satisfying

$$f(u_{\theta_1} x u_{\theta_2}) = e^{i(m\theta_1 + n\theta_2)} f(x)$$

The space $\mathscr{H}_{00}$ is the space of *spherical functions.* One can view Ω as an operator in $\mathscr{H}_{mn}$ and ask for its spectral resolution. Roughly speaking this is the problem of expanding an arbitrary function in $\mathscr{H}_{mn}$ in terms of eigenfunctions of Ω, i.e., solutions to the differential equation

$$\Omega f = -\mu^2 f$$

The transformation property defining $\mathscr{H}_{mn}$ allows one to use polar coordinates and reduce the problem to a corresponding problem on the half-line

$$\mathbb{R}_+ = \{t \mid t \in \mathbb{R}, t > 0\}$$

Since

$$dG = \tfrac{1}{2} \sinh 2t \, d\theta_1 \, dt \, d\theta_2$$

it is natural to introduce

$$\tilde{f}(t) = (\sinh 2t)^{1/2} f(a_t).$$

Ω goes to an operator D_{mn},

$$D_{mn} = d^2/dt^2 - q_{mn}$$

$$q_{mn}(t) = \frac{m^2 + n^2 - 2mn \cosh 2t - 1}{\sinh^2 2t}$$

The associated equation is

$$D_{mn} \tilde{f} = -\mu^2 \tilde{f}$$

D_{mn} is a special instance of the type of singular differential operator whose spectral theory was worked out by Weyl, Kodaira, Titchmarsh, Stone and others in a series of papers going all the way back to the celebrated work of Hermann Weyl (1908–10, [W4]). In the case of spherical functions, i.e., $\mathscr{H}_{00}$, we will have a continuous spectrum and nothing else; for general m and n there will be a *finite* discrete spectrum in addition to the continuous spectrum.

This spectral point of view is related to harmonic analysis on G because unitary representation theory enables one to construct globally defined eigenfunctions on G for the operator Ω. If π is any irreducible unitary representation of G with infinitesimal character χ_π, and (e_n) is an orthonormal basis of the space of π such that $\pi(u_\theta)e_n = e^{in\theta} e_n$, and if $\chi_\pi(\Omega) = -\mu^2$, then the functions

$$f_{mn}(\mu : x) = (\pi(x)e_n \mid e_m)$$

satisfy

$$\Omega f_{mn}(\mu : x) = -\mu^2 f_{mn}(\mu : x)$$

Moreover, as a general rule, there are no other eigenfunctions apart from the ones provided by the irreducible representations (see Theorem 3). Hence the spectral resolution of Ω is essentially the same as the problem of expanding an arbitrary function in $\mathscr{H}_{mn}$ in terms of the matrix elements of type (m, n) of irreducible unitary representations. The discrete part of the spectrum obviously corresponds to the discrete series. It will turn out that the continuous spectrum will correspond to the unitary principal series.

It should be noted that an eigenfunction which is a matrix element of an irreducible unitary representation is bounded, but in general one cannot say anything more – consider just the constant 1. However, for the L^2 spectral theory of D_{mn} the appropriate eigenfunctions will be of moderate growth in t which translates into the growth condition

$$(\sinh 2t)^{1/2} f(a_t) = \mathrm{O}(t^q) \quad (t \to +\infty)$$

for f, which gives a much sharper decay at infinity (like e^{-t}). One may then expect to find irreducible unitary representations not connected with $L^2(G)$ such as the trivial representation whose matrix elements do not go to zero so fast. It is natural to call the representations whose matrix elements satisfy the above growth (decay) condition *tempered*; the extra irreducible unitary representations will then be nontempered or *exceptional.* The trivial representation and in fact all representations of the supplementary series are exceptional; their matrix elements satisfy

$$f(a_t) \sim \text{const.}\, \mathrm{e}^{-\beta t} \quad (0 \leqslant \beta < 1)$$

The discovery and elucidation of temperedness was one of the great contributions of Harish-Chandra. It lies at the very heart of L^2 harmonic analysis on G.

Let us now take a closer look at the spectral theory of $D_{mn} = \mathrm{d}^2/\mathrm{d}t^2 - q_{mn}$. Proceeding naively, we may think of D_{mn} as a perturbation of $\mathrm{d}^2/\mathrm{d}t^2$ which is analytic in e^{-2t}; for, q_{mn} has a convergent expansion of the form

$$c_1\mathrm{e}^{-2t} + c_2\mathrm{e}^{-4t} + \cdots$$

So one should expect the eigenfunctions $\tilde{f}_{mn}$ to behave asymptotically like $\mathrm{e}^{i\mu t}$:

$$\tilde{f}_{mn}(\mu\!:\!t) \sim c^+_{mn}(\mu)\mathrm{e}^{i\mu t} + c^-_{mn}(\mu)\,\mathrm{e}^{-i\mu t}$$

where $\sim$ means the difference is $\mathrm{O}(\mathrm{e}^{-2t})$. In analogy with classical Fourier analysis we may now form the 'wave packets' of the eigenfunctions f_{mn}, namely

$$g_b(x) = \int f_{mn}(\mu\!:\!x) b(\mu)\, \mathrm{d}\mu$$

where the auxiliary function b is in the Schwartz space of μ. For these wave packets one would expect the asymptotic behaviour

$$\mathrm{e}^t g_{mn}(a_t) = \mathrm{O}(t^{-r}) \quad (\text{for every } r \geqslant 0)$$

These heuristic considerations suggest that the Schwartz space of G should consist of all smooth functions g such that each derivative of g satisfies a decay estimate of the above form. If B_T is the 'ball' defined by

$$B_T = \{u_{\theta_1} a_t u_{\theta_2} | 0 \leqslant t \leqslant T\}$$

then

$$\mathrm{vol}(B_T) \sim \mathrm{const.}\, \mathrm{e}^{2T}$$

so that the definition of the Schwartz space is entirely compatible with this. The elements of the discrete series will already be in Schwartz space because they are $O(\mathrm{e}^{-2t})$. For the principal series they will be *oscillatory around* e^{-t} *and so will fail to be in Schwartz space, but just barely; their wave packets formed with rapidly decreasing functions will be in Schwartz space.*

With reference to the formation of wave packets it should be observed that m, n and μ are parameters of the eigenvalue problem and *all estimates depend at most polynomially on these.* It follows that the integrals representing the wave packets will converge at infinity in μ; and moreover, if the dependence of the auxiliary function b in m and n is rapidly decreasing, one can even sum over m and n. However, it will turn out that $\mu = 0$ is a possible pole of the functions $c_{mn}^{\pm}$ and so it is necessary to compensate for this when choosing the function b. It is not surprising that $\mu = 0$ plays a special role; the character of the eigenvalue problem changes when μ passes through zero because the solutions change from $\mathrm{e}^{\pm i\mu t}$ to 1, t.

The formation of wave packets is just one half of the problem of harmonic analysis, for the map $b \mapsto g_b$ is *essentially the inverse of the Fourier transform map.* I say essentially because one cannot be sure that the transform of g_b will be exactly equal to, or at least a constant multiple of b. In particular the L^2- norm of g_b may not be proportional to the L^2-norm of b. To understand this a little more, let us introduce, for any f in the Schwartz space of type (m, n), its *Harish-Chandra transform* $\mathscr{H} f$ defined by

$$\mathscr{H} f(\mu) = \int f(x) \overline{f_{mn}(\mu : x)}\, \mathrm{d}G$$

The decay properties of f and the asymptotic growth of f_{mn} match each other nicely so that $\mathscr{H} f$ will be a well-defined function in the Schwartz space of μ. However, it will *not* be an arbitrary Schwartz function of μ. To see what additional restrictions it should satisfy recall the fact that the representations $\pi_{\varepsilon, i\mu}$ and $\pi_{\varepsilon, -i\mu}$ are equivalent (Theorem 35, Chapter 5). Let $U(\mu)$ be a unitary operator of $L^2(K)$ such that

$$U(\mu)^{-1} \pi_{\varepsilon, i\mu} U(\mu) = \pi_{\varepsilon, -i\mu}$$

Of course $\mu \neq 0$ here, and $U(\mu)$ is determined up to a scalar of absolute value 1 since $\pi_{\varepsilon,\pm i\mu}$ are irreducible. We have $U(\mu)e_n = t_n(\mu)e_n$, and one can normalize $U(\mu)$ so that

$$\overline{t_n(\mu)} = t_n(-\mu) = t_n(\mu)^{-1}$$

The ratios $t_n(\mu)/t_m(\mu)$ are unambiguous, and one has

$$f_{mn}(-\mu:x) = \frac{t_n(\mu)}{t_m(\mu)} f_{mn}(\mu:x)$$

Consequently,

$$(\mathscr{H}f)(-\mu) = \frac{t_m(\mu)}{t_n(\mu)}(\mathscr{H}f)(\mu)$$

We refer to these as the functional equations of type (m, n).

The Harish-Chandra transform thus maps the type (m, n) subspace of the Schwartz space into the subspace $\mathscr{S}_{mn}(\mathbb{R})$ of the Schwartz space $\mathscr{S}(\mathbb{R})$ of functions of μ satisfying the functional equation of type (m, n). However, one should not rush to conjecture that it is an isomorphism because there may well be matrix elements of the *discrete series* of type (m, n) and $\mathscr{H}$ will send these to zero. Let ${}^\circ\mathscr{C}_{mn}$ be the linear span of all matrix coefficients of the representations $D_k (k \neq 0)$ which are of type (m, n). Then ${}^\circ\mathscr{C}_{mn}$ is *a finite-dimensional space*; if $\mathscr{C}^{\#}_{mn}$ *is its orthogonal complement within* $\mathscr{C}_{mn}$, $\mathscr{C}^{\#}_{mn}$ is a closed space of $\mathscr{C}_{mn}$

$$\mathscr{C}_{mn} = {}^\circ\mathscr{C}_{mn} + \mathscr{C}^{\#}_{mn},$$

and the sum is direct and smooth, namely the projections are continuous on $\mathscr{C}$. The main result of the theory is that $\mathscr{H}$ is an isomorphism $\mathscr{C}^{\#}_{mn} \cong \mathscr{S}_{mn}(\mathbb{R})$ and $\mathscr{H}^{-1}$ sends b to the *normalized or exact wave packet*

$$f_b(x) = \int f_{mn}(\mu:x)\beta_{mn}(\mu)\,\mathrm{d}\mu$$

where β_{mn} will be a nonnegative, even, analytic function whose derivatives will all have at most polynomial growth; and

$$\int |f|^2\,\mathrm{d}G = \int |\mathscr{H}f|^2 \beta_{mn}\,\mathrm{d}\mu$$

The measure $\beta_{mn}\,\mathrm{d}\mu$ will be called the *Plancherel measure.* For a definitive theory it will be essential to describe it completely.

The central formula of the theory is the one that computes β_{mn} in terms of the asymptotic data of the eigenfunctions $f_{mn}(\mu:\cdot)$. The eigenfunctions $f_{mn}(\mu:\cdot)$ satisfy the asymptotic relation

$$\mathrm{e}^t f_{mn}(\mu:a_t) \sim c^+_{mn}(\mu)\mathrm{e}^{\mathrm{i}\mu t} + c^-_{mn}(\mu)\mathrm{e}^{-\mathrm{i}\mu t}$$

It turns out that for the coefficients $c^{\pm}_{mn}(\mu)$ we have

$$|c^+_{mn}(\mu)| = |c^-_{mn}(\mu)|$$

and β_{mn} is obtained from the formula

$$\beta_{mn}(\mu) = \text{const.}\, |c^{\pm}_{mn}(\mu)|^{-2}$$

In particular this carries with it the assertion that $c^{\pm}_{mn}$ will not have any zero for real μ. We have already mentioned that they will also be free of poles for real μ except possibly for $\mu = 0$. Although we thus have apparently as many Plancherel measures as there are types (m, n), in fact there are only two: β_{mn} depends only on the parity of $m - n$, and coincides with the Plancherel measures for the corresponding principal series, to wit, with $\mu \tanh(\pi\mu/2)$ for even parity and $\mu \coth(\pi\mu/2)$ for the odd one. The eigenfunction expansion point of view is thus fully compatible with the character point of view. The most elegant way to see this is from the point of view of the operator Fourier transform mentioned at the very beginning; the two kinds of expansions – types and characters – are then special cases obtained by going over to a specific type or to the trace.

The formula

$$\beta_{mn} = |c^{\pm}_{mn}|^{-2}$$

for $SL(2, \mathbb{R})$ can be obtained directly from Weyl's theory. However, Harish-Chandra established it in complete generality; moreover, he proved that it survives without any change even when we go over to the p-adic groups. His discovery of the relationship of the Plancherel measure to the asymptotic data coming from the matrix elements is perhaps the most beautiful result in the theory of the continuous spectrum of semisimple groups.

Finally, let me remark that the differential equation for f_{mn} in polar coordinates becomes the *hypergeometric equation* in suitable exponential coordinates. So one can resort to the hathayoga of special functions to do everything done here. But this will be an exercise in futility for it will tell us almost nothing of what is likely to happen in the general case.

8.2 Definition and the elementary theory of the Schwartz space

The remarks in the previous section are already precise enough so far as the definition of the Schwartz space is concerned. However, Harish-Chandra replaced the function e^{-t} by a function Ξ defined on all of G with essentially the same decay but with very good formal properties. We shall also do the same.

The functions Ξ and σ. We put

$$\Xi(x) = f_{00}(0:x) = (\pi_{0,0}(x)e_0 | e_0)$$

From the explicit definition of $\pi_{0,0}$ (see Chapter 5, Section 5), we get

$$\Xi(x) = \frac{1}{2\pi}\int \mathrm{e}^{-t(x^{-1}u_\theta)}\,\mathrm{d}\theta$$

Since e_0 is fixed by K, Ξ is *spherical*, i.e.,

$$\Xi(x) = \Xi(u_{\theta_1} x u_{\theta_2})$$

Since a_t and a_{-t} are always conjugate under K, x and x^{-1} always belong to the same (K, K) double coset, i.e.,

$$\Xi(x) = \Xi(x^{-1})$$

Now $\pi_{0,0}$ is unitary while $t(x^{-1}u_\theta)$ is real. Hence

$$0 < \Xi(x) \leqslant 1, \quad \Xi(1) = 1$$

More generally, the function of θ given by

$$t(a_r^{-1}u_\theta) = \tfrac{1}{2}\log(\mathrm{e}^{-2r}\cos^2\theta + \mathrm{e}^{2r}\sin^2\theta)$$

has critical points precisely when

$$\theta = 0, \frac{\pi}{2}, \quad -\frac{\pi}{2}, \quad \pi$$

where the second derivative has values respectively

$$>0, <0, <0, >0.$$

Hence

$$-r \leqslant t(a_r^{-1}u_\theta) \leqslant r \quad (r > 0)$$

It follows that

$$\Xi(a_t) \geqslant \mathrm{e}^{-t} \quad (t \geqslant 0)$$

The next proposition shows that Ξ is quite regular in its behaviour.

Proposition 1. *If $R \subset G$ is any compact set, there is a constant $c = c(R) > 0$ such that*

$$\Xi(yxy') \leqslant c\,\Xi(x) \quad (x \in G,\ y, y' \in R)$$

If $b, b' \in U(\mathfrak{g})$, there is a constant $c = c(b, b') > 0$ such that

$$|\Xi(b; x; b')| \leqslant c\,\Xi(x) \quad (x \in G)$$

Proof. For $u, v \in C(K)$,

$$(\pi_{0,0}(x)u|v) = \frac{1}{2\pi}\int \mathrm{e}^{-t(x^{-1}u_\theta)} u(x^{-1}[u_\theta])\overline{v(u_\theta)}\,\mathrm{d}\theta$$

so that

$$|(\pi_{0,0}(x)u|v)| \leqslant \|u\|_\infty \|v\|_\infty \Xi(x)$$

If $u=\pi_{0,0}(b')\cdot 1$, $v=\pi_{0,0}(b^{\dagger})\cdot 1$, then u and v are in $C(K)$ and the left side is $\Xi(b;x;b')$, proving the second assertion. If $u=\pi_{0,0}(y')\cdot 1$, $v=\pi_{0,0}(y^{-1})\cdot 1$, $y,y'\in R$, $\|u\|_{\infty}$ and $\|v\|_{\infty}$ are bounded when y and y' vary in R. This proves the first assertion.

We shall now prove the basic result concerning the asymptotic behaviour of Ξ.

Theorem 2. *There are constants c_1, c_2 such that as $t\to+\infty$*

$$\Xi(a_t)e^t = c_1 t + c_2 + O(t^4 e^{-3t})$$

In particular, there is a constant $c>0$ such that

$$e^{-t}\leqslant \Xi(a_t)\leqslant ce^{-t}(1+t) \quad (t\geqslant 0)$$

Proof. We shall obtain this using the differential equation for Ξ, namely,

$$\Omega\Xi = 0$$

If

$$g(t)=(\sinh 2t)^{1/2}\Xi(a_t)$$

then

$$(d^2/dt^2)-q)g=0, \quad q(t)=-1/\sinh^2 2t$$

For Ξ we have the initial estimates, for $t>0$,

$$0<\Xi(a_t)\leqslant 1, \quad |(d/dt)\Xi(a_t)| = |\Xi(a_t;H)|\leqslant \text{const.}$$

Hence, for $t\geqslant 1$,

$$|g(t)|\leqslant ce^t, \quad |g'(t)|\leqslant ce^t$$

Let us pass to the vector

$$V(t)=\begin{pmatrix} g(t) \\ g'(t)\end{pmatrix}$$

Then

$$dV/dt = XV+PV, \quad X=\begin{pmatrix} 0 & 1 \\ 0 & 0\end{pmatrix}, \quad P=\begin{pmatrix} 0 & 0 \\ q & 0\end{pmatrix}$$

From now on we use c as a generic symbol for a constant >0, and all estimates are for $t\geqslant 1$. Then

$$\|V(t)\|\leqslant ce^t, \quad \|P(t)\|\leqslant ce^{-4t}$$

so that the operator $V\mapsto P\cdot V$ is a very small perturbation when $t\to+\infty$. Let

$$W=e^{-tX}V=(1-tX)V$$

Then

$$dW/dt = QW, \quad Q=e^{-tX}Pe^{tX}$$

Obviously

$$\|Q(t)\| \leqslant c(1+t)^2 e^{-4t}, \quad \|W(t)\| \leqslant c(1+t)e^t$$

Hence $\|dW/dt\| = O(t^3 e^{-3t})$ for large t, proving that $\int_1^\infty \|dW/dt\| dt < \infty$. So,

$$W(\infty) = \lim_{t \to +\infty} W(t)$$

exists, and

$$W(t) = W(\infty) - \int_t^\infty (dW/ds)\, ds$$

This means

$$\begin{aligned} \|W(t) - W(\infty)\| &\leqslant \int_t^\infty \|dW/ds\|\, ds \\ &\leqslant \int_t^\infty \|Q(s)\|\,\|W(s)\|\, ds \\ &\leqslant c\int_t^\infty (1+s)^3 e^{-3s}\, ds \\ &\leqslant c\int_0^\infty (1+t+x)^3 e^{-3(t+x)}\, dx \\ &\leqslant c(1+t)^3 e^{-3t} \end{aligned}$$

As $V(t) = e^{tX}W = (1+tX)W$, we have

$$\|V(t) - (1+tX)W(\infty)\| \leqslant c(1+t)^4 e^{-3t}$$

We get Theorem 2 from the first component of this vector estimate.

For fixed $m, n \in \mathbb{Z}$ and $\mu \in \mathbb{C}$ let $F_{mn}(\mu)$ be the space of C^∞ functions f on G such that

(i) $f(u_{\theta_1} x u_{\theta_2}) = e^{i(m\theta_1 + n\theta_2)} f(x)$
(ii) $\Omega f = -\mu^2 f$.

It must be noted that, if m and n have opposite parity, 0 is the only function satisfying (i) because, if we take $\theta_1 = \theta_2 = \pi$ in (i), we get $f(x) = (-1)^{m+n} f(x)$.

Theorem 3. *$F_{mm}(\mu) = \mathbb{C} \cdot f_{mm}(\mu{:}\cdot)$. In particular $f_{mm}(\mu{:}\cdot)$ is the unique element of $F_{mm}(\mu)$ which is 1 at 1; Ξ is the unique spherical C^∞ function on G such that $\Xi(1) = 1$ and $\Omega\Xi = 0$; and*

$$\frac{1}{2\pi}\int \Xi(x u_\theta y)\, d\theta = \Xi(x)\Xi(y) \quad (x, y \in G)$$

Proof. For the last assertion note that, for fixed x, the integral is a function $g_x \in C^\infty(G)$ which is spherical and satisfies $\Omega g_x = 0$, $g_x(1) = \Xi(x)$, and so must be $\Xi(x)\cdot\Xi$. To prove the first statement let $f \in F_{mm}(\mu)$ and $f(1) = 0$; we shall prove that $f = 0$. Now f is an eigenfunction for $X - Y$ and Ω and hence also for $2(X - Y)^2 + \Omega = (X - Y)^2 + H^2 + (X + Y)^2$ which is elliptic. So f is analytic and it suffices to prove that $f(1; a) = 0$ for all $a \in U(\mathfrak{g})$. Since $X - Y$ and Ω generate the centralizer of B in $U(\mathfrak{g})$, we have $f(1; a) = 0$ whenever $a^{u_\theta} = a$ for all θ. If $a^{u_\theta} = e^{ik\theta} a$ for all θ where $k \neq 0$, $f(1; a) = f(1; a^{u_\theta})e^{-ik\theta}$ while $f(1; a^{u_\theta}) = e^{i(m-m)\theta} f(1; a) = f(1; a)$, so $f(1; a) = f(1; a)e^{-ik\theta}$ giving $f(1; a) = 0$. Since k was arbitrary, $\neq 0$, we are through.

We introduce next the function σ on G. It is the unique spherical function on G such that $\sigma(a_t) = t, t \geqslant 0$:

$$\sigma(u_{\theta_1} a_t u_{\theta_2}) = |t|$$

It is not difficult to show that $\sigma(x)$ is the Riemannian distance $d(K, xK)$ between K and xK in the space G/K (identify G/K with the Poincaré half-plane $\mathrm{Re}(z) > 0$ with metric $ds^2 = (dx^2 + dy^2)/y^2$ and calculate distance from i to ie^{2t} along the imaginary axis):

$$\sigma(x) = d(K, xK)$$

Then we have the *triangle inequality for* σ:

$$\sigma(xy) \leqslant \sigma(x) + \sigma(y) \quad (x, y \in G)$$

which follows from $d(K, xyK) \leqslant d(K, xK) + d(xK, xyK)$, since $d(xK, xyK) = d(K, yK)$ by the G-invariance of d. It is also obvious that

$$\sigma(x) = \sigma(x^{-1}) \quad (x \in G)$$

Although σ is continuous, it is not differentiable. It is easy to show that σ^2 is analytic.

Definition. A function f on G with values in some finite dimensional normed vector space is said to satisfy the *weak inequality if*, for some constants $c > 0$, $r \geqslant 0$,

$$|f(x)| \leqslant c\Xi(x)(1 + \sigma(x))^r \quad (x \in G)$$

It is said to satisfy the *strong inequality* if for every $r \geqslant 0$ there is a constant $c_r > 0$ such that

$$|f(x)| \leqslant c_r \Xi(x)(1 + \sigma(x))^{-r} \quad (x \in G)$$

The *Schwartz space* $\mathscr{C}(G:V)$ of V-valued functions is the space of all $f \in C^\infty(G:V)$ such that, for each $a, b \in U(\mathfrak{g})$, the function $afb: x \mapsto f(b; x; a)$ satisfies the strong inequality. If $V = \mathbb{C}$ we write $\mathscr{C}(G)$ for $\mathscr{C}(G:\mathbb{C})$ and refer to it simply as *the Schwartz space of* G. It is closed under translations.

The quantitative link between Ξ and $L^2(G)$ is given by the following.

Proposition 4. *We have, for all* $q > 3$

$$\int \Xi(x)^2(1 + \sigma(x))^{-q}\,\mathrm{d}G < \infty$$

Proof. By the estimate of Theorem 2, the integrand is $O(e^{-2t}(1+t)^{-(q-2)})$ for large t when $x = a_t$. The result follows by going over to polar coordinates.

We shall regard $\mathscr{C}(G)$ as a Fréchet space with respect to the seminorms

$$\mu_{a,b:r}(f) = \sup_G(|f(b;x;a)|\Xi(x)^{-1}(1+\sigma(x))^r)$$

Theorem 5. *We have*

$$C_c^\infty(G) \subset \mathscr{C}(G) \subset L^2(G)$$

the natural inclusions being continuous with dense images.

Proof. The inclusions are obvious in view of Proposition 4. The only thing not immediately obvious is the density of the first inclusion. Let $B_t(t > 0)$ be the 'ball' where $\sigma(x) \leqslant t$; write $f_t = 1_{B_t}$, the characteristic function of B_t. Choose $\alpha \in C_c^\infty(G)$ such that $\alpha \geqslant 0$, $\int \alpha\,\mathrm{d}G = 1$, $\mathrm{supp}(\alpha) \subset B_{1/2}$, and let $h_t = \alpha * f_t * \alpha$, $*$ being convolution. The triangle inequality for σ shows that $h_t = 1$ on $B_{t-1}\,(t > 1)$, $\mathrm{supp}(h_t) \subset B_{t+1}$, and h_t is C^∞. Moreover, if $a, b \in U(\mathfrak{g})$, it follows from the usual arguments that $a(\alpha * f_t * \alpha)b = \alpha b * f_t * a\alpha$. Hence, as $\|u * v\|_\infty \leqslant \|u\|_1 \|v\|_\infty$, we have

$$|h_t(a;x;b)| \leqslant \|\alpha b\|_1 \|a\alpha\|_1$$

as $0 \leqslant f_t \leqslant 1$. If now $f \in \mathscr{C}(G)$, we have

$$\lim_{t \to +\infty} (h_t f) = f$$

This is quite easy to show. Indeed, $h_t f = f$ on B_{t-1} and so for estimating the difference we may restrict to x with $\sigma(x) > t - 1$; but, for such x, both $(h_t f)(x)$ and $f(x)$ are $O(e^{-t}(1+t)^{-r})$ so that $h_t f - f = O(e^{-t}(1+t)^{-r})$. The argument is similar for the derivatives.

We consider convolutions.

Theorem 6. *$\mathscr{C}(G)$ is closed under convolution and becomes a Fréchet algebra.*

Proof. The most elegant way to prove this is as follows. Define, for any real number s, the spherical function Ξ_s by

$$\Xi_s(x) = \Xi(x)(1 + \sigma(x))^{-s}$$

Then, if $r_0 \geqslant 0$ and

$$c_0 = \int \Xi^2(1 + \sigma)^{-r_0}\, dG < \infty$$

and $s_1, s_2 \in \mathbb{R}$ with $s_1 > |s_2| + r_0$, the integrals defining $\Xi_{s_1} * \Xi_{s_2}$ converge, uniformly when x varies in compact subsets of G, and

$$\Xi_{s_1} * \Xi_{s_2}(x) \leqslant c_0 \Xi_{s_2}(x)$$

To prove this, we have

$$\Xi_{s_1} * \Xi_{s_2}(x) = \int \Xi_{s_1}(y)\Xi_{s_2}(y^{-1}x)\, dG(y)$$

We now use the estimate

$$\frac{1 + \sigma(y)}{1 + \sigma(x)} \leqslant 1 + \sigma(x^{-1}y) \leqslant (1 + \sigma(y))(1 + \sigma(x))$$

which gives, when $x \in R$, a compact set in G,

$$\Xi_{s_1}(y)\Xi_{s_2}(y^{-1}x) \leqslant \text{const.}\, \Xi(y^2)(1 + \sigma(y))^{-(s_1 - |s_2|)}(1 + \sigma(x))^{|s_2|}$$

So the uniform convergence of

$$\int \Xi_{s_1}(y)\Xi_{s_2}(y^{-1}x)\, dG(y)$$

is clear, for $x \in R$. For getting an estimate of the integral, we average over K. We have

$$\begin{aligned}\Xi_{s_1} * \Xi_{s_2}(x) &= \frac{1}{2\pi} \iint \Xi_{s_1}(y)\Xi_{s_2}(y^{-1}u_\theta x)\, dG(y)\, d\theta \\ &\leqslant \frac{1}{2\pi} \iint \Xi_{s_1}(y)\Xi(y^{-1}u_\theta x)(1 + \sigma(y))^{|s_2|}(1 + \sigma(x))^{-s_2}\, dG\, d\theta \\ &\leqslant c_0 \Xi_{s_2}(x) \quad \text{(Theorem 3)}\end{aligned}$$

This done, we can prove Theorem 5 quite easily. The above convergence result shows at once that, for $f, g \in \mathscr{C}(G)$, the convolution $f * g$ is well defined everywhere, and in fact the integrals

$$\int |f(y)||g(y^{-1}x)|\, dG = \int |f(xy^{-1})||g(y)|\, dG$$

converge uniformly when x varies over compact subsets of G. Since we can also use derivatives of f and g here, it is clear that $f * g$ is C^∞ and its derivatives can be calculated formally. The estimate $\Xi_{s_1} * \Xi_{s_2} \leqslant \text{const.}\, \Xi_{s_2}$

for $s_1 > |s_2| + r_0$ then implies easily that $f * g \in \mathscr{C}(G)$ and the map $f, g \mapsto f * g$ is continuous.

8.3 Asymptotic behaviour of matrix elements

We shall now develop the programme hinted at in the introduction to this chapter. The first step is to determine the asymptotic behaviour of the matrix elements. We shall use the same method as we did for Ξ. The idea is to treat the differential operator D_{mn} as an exponentially small perturbation of d^2/dt^2 and to keep track of the dependence in m, n and μ. The initial estimates are provided by the estimate for Ξ.

Recall that, if $\chi_q(u_\theta) = e^{iq\theta}$,

$$f_{mn}(\mu:x) = \frac{1}{2\pi}\int e^{-(i\mu+1)t(x^{-1}u_\theta)}\chi_{-n}(x^{-1}[u_\theta])\chi_m(u_\theta)\, d\theta$$

It is clear that this is defined for all $\mu \in \mathscr{C}$ and is holomorphic in μ. Write

$$\tilde{f}_{mn}(\mu:t) = (2\sinh 2t)^{1/2} f_{mn}(\mu; a_t) \quad (t > 0)$$

Then

$$(d^2/dt^2 - q_{mn})\tilde{f}_{mn} = -\mu^2 \tilde{f}_{mn}$$
$$q_{mn}(t) = (m^2 + n^2 - 2mn\cosh 2t - 1)/\sinh^2 2t$$

All estimates will be for $t \geqslant 1$. Even though we are eventually interested only for $\mu \in \mathbb{R}$, it will be essential to work in the complex domain for μ, at least in small strips $|Im\mu| < \delta$. We shall use the symbol c generically for a constant > 0 that is independent of μ, m, n, t; it may depend on other variables and then we shall write c_s, $c_{s,\delta}$ etc. It will be our intention to keep track of the dependence of the estimates on μ, m, n only to the extent of assuring that they depend *polynomially* on these variables. Thus we abbreviate $1 + |\mu|$ by $[\mu]$, $(1 + |\mu|)(1 + |m| + |n|)$ by $[\mu, m, n]$ etc. Finally note that $(2\sinh 2t)^{1/2} \sim e^t$ in the sense that $(2\sinh 2t)^{1/2} = e^t h(t)$ where $0 < a \leqslant h(t) \leqslant b < \infty$ for $t \geqslant 1$, and, as $t \to +\infty$,

$$h(t) = 1 + O(e^{-4t}), \quad h^{(s)}(t) = O(e^{-4t}) \quad (s \geqslant 1)$$

So in our estimates we may go back and forth between $(2\sinh 2t)^{1/2}$ and e^t. Put $(' = d/dt)$

$$V_{mn}(\mu:t) = \begin{pmatrix} \tilde{f}_{mn}(\mu:t) \\ \tilde{f}'_{mn}(\mu:t) \end{pmatrix}$$

Then

$$V'_{mn} = \begin{pmatrix} 0 & 1 \\ q_{mn} - \mu^2 & 0 \end{pmatrix} V_{mn}$$

which can be written as

$$(\mathrm{d}/\mathrm{d}t)V_{mn} = M(\mu)V_{mn} + P_{mn}V_{mn}$$

with

$$M(\mu) = \begin{pmatrix} 0 & 1 \\ -\mu^2 & 0 \end{pmatrix} \quad P_{mn}(t) = \begin{pmatrix} 0 & 0 \\ q_{mn}(t) & 0 \end{pmatrix}$$

Initial estimates. We begin with the following.

Lemma 7. *For* $\exp tM(\mu)$ *and the derivatives* $P_{mn}^{(s)}$ *the estimates are*

$$\| P_{mn}^{(s)}(t) \| \leqslant c_s[m,n]^2 \mathrm{e}^{-2t}$$
$$\| \exp tM(\mu) \| \leqslant c\mathrm{e}^{t|\mathrm{Im}\,\mu|}[\mu]^2(1+t)$$

Proof. It is clear that

$$q_{mn}(t) = (m^2+n^2-1)\sum_{k\geqslant 1} a_k \mathrm{e}^{-4kt} + 2mn \sum_{k\geqslant 1} b_k \mathrm{e}^{-2kt}$$

where $|a_k| + |b_k| = \mathrm{O}(k)$. Hence

$$|q_{mn}^{(s)}(t)| \leqslant c[m,n]^2 \mathrm{e}^{-2t}$$

Giving the first estimate. As $|\sin z/z| \leqslant 2\mathrm{e}^{|\mathrm{Im}\,z|}$, $z \in \mathbb{C}$, and

$$\exp tM(\mu) = (\cos \mu t)\cdot 1 + (\sin \mu t/\mu)M(\mu)$$

we get the second.

Lemma 8 (Estimate for f_{mn} and its derivatives). *We have*

$$|f_{mn}(\mu{:}x)| \leqslant \mathrm{e}^{\sigma(x)|\mathrm{Im}\,\mu|}\Xi(x) \quad (x \in G)$$

If $u, v \in U(\mathfrak{g})$ *have degrees* $\leqslant r$, $\leqslant s$, *respectively,*

$$f_{mn}(\mu{:}v;x;u) \leqslant c_{uv}[\mu,m]^s[\mu,n]^r \cdot \mathrm{e}^{\sigma(x)|\mathrm{Im}\,\mu|}\Xi(x)$$

Moreover, for any integer $k \geqslant 0$,

$$|(\mathrm{d}/\mathrm{d}\mu)^k f_{mn}(\mu{:}x)| \leqslant \mathrm{e}^{\sigma(x)|\mathrm{Im}\,\mu|}\Xi(x)(1+\sigma(x))^k$$

and, for $u, v \in U(\mathfrak{g})$ *as before,*

$$|(\mathrm{d}/\mathrm{d}\mu)^k f_{mn}(\mu{:}v;x;u)| \leqslant c_{k,u,v}[\mu,m]^s[\mu,n]^r \mathrm{e}^{\sigma(x)|\mathrm{Im}\,\mu|}\Xi(x)(1+\sigma(x))^k$$

Proof. Since $-r \leqslant t(a_r^{-1}u_\theta) \leqslant r$ for $r > 0$,

$$|f_{mn}(\mu{:}a_r)| \leqslant \mathrm{e}^{r|\mathrm{Im}\,\mu|} \frac{1}{2\pi}\int \mathrm{e}^{-t(a_r^{-1}u_\theta)}\,\mathrm{d}\theta$$
$$\leqslant \mathrm{e}^{r|\mathrm{Im}\,\mu|}\Xi(a_r)$$

For the derivatives we may assume $u = H'^a X'^b Y'^c$, $v = H'^p X'^q Y'^t$, $a+b+c \leqslant r, p+q+t \leqslant s$, and use Proposition 31, Chapter 5. Notice that, for $\mu \in \mathbb{R}$,

$$(\pi(v)\pi(x)\pi(u)e_n | e_m) = (\pi(x)\pi(u)e_n | \pi(v^\dagger)e_m)$$

where $v^\dagger = (-1)^{q+t} X'' Y'^q H'^p$ and we are writing π for $\pi_{\varepsilon, i\mu}$, this is because π is unitary. Then by the result mentioned above there are polynomials $p_r(\mu{:}n)$ and $p_s(\mu{:}m)$ of degrees $\leqslant r, s$ respectively such that the left side reduces to

$$p_r(\mu{:}n) p_s(\mu{:}m) f_{m'n'}(\mu{:}x)$$

where $m' = m - 2q + 2t$, $n' = n - 2c + 2b$. By holomorphy this is then true for all $\mu \in \mathbb{C}$. Hence

$$f_{mn}(\mu{:}v; x; u) = p_r(\mu{:}n) p_s(\mu{:}m) f_{m'n'}(\mu{:}x)$$

and the second estimate follows from the first. For the last we have

$$(\mathrm{d}/\mathrm{d}\mu)^{k'} f_{m'n'}(\mu{:}x) = \frac{1}{2\pi} \int (-\mathrm{i}t(x^{-1}u_\theta))^{k'} \,\mathrm{e}^{-(\mathrm{i}\mu+1)t(x^{-1}u_\theta)} \chi_{-n'}(x^{-1}[u_\theta]) \chi_{m'}(u_\theta) \,\mathrm{d}\theta$$

so that when $x = a_r (r > 0)$

$$|(\mathrm{d}/\mathrm{d}\mu)^{k'} f_{m'n'}(\mu{:}a_r)| \leqslant r^{k'} \mathrm{e}^{r|\mathrm{Im}\,\mu|} \Xi(a_r)$$

So if we observe that $r = \sigma(a_r)$ the third set of estimates follows by differentiating the earlier identity relating $u f_{mn}(\mu{:}\cdot)v$ and $f_{m'n'}(\mu{:}\cdot)$.

Lemma 9. *We have, for any $s \geqslant 0$,*

$$\| V_{mn}^{(s)}(\mu{:}t) \| \leqslant c_s[\mu, n]^{s+1} \mathrm{e}^{t|\mathrm{Im}\,\mu|}(1+t)$$

Proof. Note that, for any function g, $g(a_t; H^s) = (\mathrm{d}/\mathrm{d}t)^s g(a_t)$. Now replace $(\sinh 2t)^{1/2}$ by e^t (cf. remark made earlier), observe that V_{mn} involves terms only up to $f_{mn}^{(s+1)}$, that only right derivatives occur, and use Theorem 2.

Write

$$W_{mn}(\mu{:}t) = \mathrm{e}^{-tM(\mu)} V_{mn}(\mu{:}t)$$

Then

$$W'_{mn} = Q_{mn} W_{mn}, \quad Q_{mn} = \mathrm{e}^{-tM} P_{mn} \mathrm{e}^{tM}$$

Lemma 10. *We have, for $s \geqslant 1$,*

$$\| Q_{mn}(\mu{:}t) \| \leqslant c[\mu, m, n]^4 \mathrm{e}^{2t|\mathrm{Im}\,\mu| - 2t}(1+t)^2$$
$$\| W_{mn}(\mu{:}t) \| \leqslant c[\mu, m, n]^3 \mathrm{e}^{2t|\mathrm{Im}\,\mu|}(1+t)^2$$
$$\| W_{mn}^{(s)}(\mu{:}t) \| \leqslant c[\mu, m, n]^{2s+3} \mathrm{e}^{2t|\mathrm{Im}\,\mu| - 2t}(1+t)^2$$

Proof. Use Lemmas 7 and 9 for the first two. For the last, start with $W'_{mn} = \mathrm{e}^{-tM} P_{mn} V_{mn}$, differentiate $s - 1$ times, and again use Lemmas 7 and 9.

Lemma 11. *The limit*

$$W_{mn}(\mu:\infty) = \lim_{t \to \infty} W_{mn}(\mu:t)$$

exists for all μ, $|\mathrm{Im}\,\mu| < 1$, *and is holomorphic there, with*

$$|W^{(s)}_{mn}(\mu:\infty)| \leqslant c_s\,[\mu, m, n]^3$$

for $|\mathrm{Im}\,\mu| \leqslant 1 - \delta (0 < \delta < 1)$. *Moreover,*

$$W_{mn}(\mu:t) = W_{mn}(\mu:\infty) - \int_t^\infty W'_{mn}(\mu:y)\,\mathrm{d}y$$

Proof. From the estimate for W'_{mn} provided by Lemma 10 we find that, for $|\mathrm{Im}\,\mu| < 1$, the integral

$$\int_1^\infty \| W'_{mn}(\mu:t) \|\,\mathrm{d}t$$

converges, uniformly when $|\mathrm{Im}\,\mu| \leqslant 1 - \delta$. So the first assertion is clear and

$$W_{mn}(\mu:\infty) = W_{mn}(\mu:1) + \int_1^\infty W'_{mn}(\mu:t)\,\mathrm{d}t$$

From the uniform convergence it is obvious that $W_{mn}(\mu:\infty)$ is holomorphic; the above representation leads to the estimate for $W_{mn}(\mu:\infty)$ when $|\mathrm{Im}\,\mu| \leqslant 1 - \delta$. To estimate $(\mathrm{d}/\mathrm{d}\mu)^s W_{mn}(\mu:\infty)$ use Cauchy's formula. The last formula is trivial.

Proposition 12. *We have*

$$V_{mn}(\mu:t) = \mathrm{e}^{tM(\mu)} W_{mn}(\mu:\infty) - \mathrm{e}^{tM(\mu)} \int_t^\infty W'_{mn}(\mu:y)\,\mathrm{d}y$$

Moreover, for any $s \geqslant 0$, *and* $|\mathrm{Im}\,\mu| \leqslant \frac{1}{6}$,

$$\| V^{(s)}_{mn}(\mu:t) - \mathrm{e}^{tM(\mu)} M(\mu)^s W_{mn}(\mu:\infty) \| \leqslant c_s[\mu, m, n]^{2s+5}(1+t)^3 \mathrm{e}^{-3t/2}$$

Proof. The formula of Lemma 11 for W_{mn} leads to the present formula for V_{mn}, since $V = \mathrm{e}^{tM} W$. For the estimate it comes down to estimating terms of the form

$$\mathrm{e}^{tM(\mu)} M(\mu)^s \int_t^\infty W'_{mn}(\mu:y)\,\mathrm{d}y, \mathrm{e}^{tM(\mu)} M(\mu)^j W^{s-j}_{mn}(\mu:t) \quad (s - j \geqslant 1)$$

We now use Lemmas 7 and 10.

Remark. Note that the dominant term approximating V_{mn} is at worst $O(\mathrm{e}^{t/6})$ while the error is $\mathrm{e}^{-3t/2}$.

To proceed further we must diagonalize $M(\mu)$. Although M is holomorphic in μ, it changes from regular semisimple to nilpotent when μ passes through 0. So the spectral projections will have a pole at $\mu = 0$. The eigenvalues are $\pm i\mu$; and for $\mu \neq 0$ the projections $E^{\pm}(\mu)$ on the eigenspaces corresponding to $\pm i\mu$ are computed to be

$$E^{\pm}(\mu) = \begin{pmatrix} \frac{1}{2} & \pm 1/2i\mu \\ \pm i\mu/2 & \frac{1}{2} \end{pmatrix}$$

We have

$$e^{tM(\mu)} = e^{i\mu t}E^{+}(\mu) + e^{-\mu t}E(-\mu)$$
$$M(\mu)^s = (i\mu)^s E^{+}(\mu) + (-i\mu)^s E^{-}(\mu)$$
$$\| \mu E^{+}(\mu) \| \leqslant c[\mu]^2$$

Define

$$c^{\pm}_{mn}(\mu) = (E^{\pm}(\mu) W_{mn}(\mu : \infty))_1$$

where the suffix indicates first component.

Proposition 13*. The functions $\mu c^{\pm}_{mn}(\mu)$ are holomorphic for $|\mathrm{Im}\,\mu| < 1$; for any $\delta > 0$ we have, for $|\mathrm{Im}\,\mu| \leqslant 1 - \delta$,*

$$|(d/d\mu)^s(\mu c^{\pm}_{mn}(\mu))| \leqslant c_s[\mu, m, n]^5$$

Moreover,

$$(E^{\pm}(\mu) W_{mn}(\mu : \infty))_2 = \pm i\mu c^{\pm}_{mn}(\mu)$$

Proof. Only the last statement needs proof. In view of holomorphy, we need only prove it for each fixed real $\mu \neq 0$, m, n. Let $d^{\pm}_{mn}(\mu) = (E^{\pm}(\mu) W_{mn}(\mu : \infty))_2$. We now observe that

$$e^{tM(\mu)} M(\mu)^s W_{mn}(\mu : \infty) = e^{it\mu}(i\mu)^s \begin{pmatrix} c^{+}_{mn}(\mu) \\ d^{+}_{mn}(\mu) \end{pmatrix} + e^{-it\mu}(-i\mu)^s \begin{pmatrix} c^{-}_{mn}(\mu) \\ d^{-}_{mn}(\mu) \end{pmatrix}$$

The second component of the estimate of Proposition 12 for $s = 0$ gives

$$(d/dt)\tilde{f}_{mn}(\mu : t) = e^{i\mu t} d^{+}_{mn}(\mu) + e^{-i\mu t} d^{-}_{mn}(\mu) + O(e^{-t})$$

Similarly the first component of the same estimate for $s = 1$ gives

$$(d/dt)\tilde{f}_{mn}(\mu : t) = e^{i\mu t}(i\mu) c^{+}_{mn}(\mu) - e^{-i\mu t}(i\mu) c^{-}_{mn}(\mu) + O(e^{-t})$$

Hence, writing $a^{\pm}_{mn} = d^{\pm}_{mn}(\mu) - (\pm i\mu) c^{\pm}_{mn}(\mu)$, we have:

$$\lim_{t \to \infty} (e^{i\mu t} a^{+}_{mn} + e^{-i\mu t} a^{-}_{mn}) = 0$$

We may therefore conclude that $a^{\pm}_{mn} = 0$. This follows from the general principle that, if $a_1, \ldots, a_r$ are *distinct* real numbers and $c_1, \ldots, c_r$ any complex numbers,

$$\frac{1}{T}\int_0^T |c_1 e^{ita_1} + \cdots + c_r e^{ita_r}|^2 dt \to |c_1|^2 + \cdots + |c_r|^2 \quad (T \to +\infty)$$

showing that

$$\lim_{t\to\infty} (c_1 e^{ita_1} + \cdots + c_r e^{ita_r}) = 0 \Rightarrow c_1 = \cdots = c_r = 0$$

From Proposition 12 we get the following; we replace $(2\sinh 2t)^{1/2}$ by e^t.

Proposition 14. *For all* $\mu \neq 0$, $|\mathrm{Im}\,\mu| \leqslant \frac{1}{6}$, $t \geqslant 1, m, n \in \mathbb{Z}$,

$$|(d/dt)^s(e^t f_{mn}(\mu : a_t)) - (e^{it\mu}(i\mu)^s c^+_{mn}(\mu) + e^{-it\mu}(-i\mu)^s c^-_{mn}(\mu))|$$
$$\leqslant c_s[\mu, m, n]^{2s+5}(1+t)^3 e^{-3t/2}$$

The basic result on asymptotics of f_{mn} is now almost in our hands, except that we must also control the derivatives. Now, one can differentiate on either side and use any element of $U(\mathfrak{g})$ for that purpose; so a little more work is necessary before we can reach our final goal, and so we take care of it with the following two lemmas.

Lemma 15. *If* $h \in C^\infty(G)$ *and* $h(u_\theta x) = e^{im\theta} h(x)$, *then*

$$h((X-Y)^r H^s X^q; a_t; v) = (im)^r e^{-2qt} h(a_t; H^s X^q v)$$

Proof. We have

$$h((X-Y)^r H^s X^q; a_t; v) = (im)^r h(H^s X^q; a_t; v) = (im)^r e^{-2qt} h(a_t; H^s X^q v)$$

Here we use the fact that

$$(H^s X^q)^{(a_t)^{-1}} = e^{-2qt} H^s X^q$$

The next lemma deals with reduction of derivatives from $U(\mathfrak{g})$ to the $(d/dt)^r$. Let R be the complex algebra of functions on $(0, \infty)$ generated by $(\sinh 2t)^{-1}$ and $\cosh 2t \cdot (\sinh 2t)^{-1}$. Since the derivatives of the generators are in R, $(d/dt)R \subset R$. Given any $u \in U(\mathfrak{g})$ of degree $\leqslant r$, we can write $u = \gamma H^r + \cdots$ where $\cdots$ refers to terms involving $(X-Y)^p H^q (X+Y)^k$ where $p+q+k \leqslant r$ but $q \leqslant r-1$.

Lemma 16. *If* $u \in U(\mathfrak{g})$ *is of degree* $\leqslant r$ *and* γ *is as above, we can find* $f_0, \ldots, f_{r-1} \in \mathbb{C}[m, n] \otimes R$ *such that for any* $h \in C^\infty(G)$ *of type* (m, n) *we have, for all* (m, n),

$$h(a_t; u) = h(a_t; D_u)$$

where

$$D_u = \gamma(\mathrm{d}/\mathrm{d}t)^r + \sum_{0\leqslant j\leqslant r-1} f_j(\mathrm{d}^j/\mathrm{d}t^j)$$

Proof. We use the formulae for the inverse of the differential of the polar decomposition map (Chapter 6, Section 2) to transfer u to a differential operator $\tilde{u}$ in θ_1, θ_2, t. The coefficients of $\tilde{u}$ are in the algebra generated by R and $\cos 2\theta_1, \cos 2\theta_2, \sin 2\theta_1, \sin 2\theta_2$. On the space of functions h of type (m, n), $\partial/\partial\theta_1 = \mathrm{i}m$, $\partial/\partial\theta_2 = \mathrm{i}n$. So $\tilde{u}$ may be viewed when acting on these functions as an operator in t with coefficients in $\mathbb{C}[m, n] \otimes R$. It only remains to find out the coefficient of $\mathrm{d}^r/\mathrm{d}t^r$ in this operator. In $\tilde{u}$ the coefficient of $\mathrm{d}^r\mathrm{d}t^r$ is $\gamma(\cos 2\theta_2)^r + \cdots$ where $\cdots$ refers to terms $(\cos 2\theta_2)^q(\sin 2\theta_2)^{q'}$ with $q' \geqslant 1$, so that when $\theta_1 = \theta_2 = 0$ this reduces to γ.

The main result of this section is the following.

Theorem 17. *There is a constant $c > 0$ such that, for all $\mu \neq 0$ with $|\mathrm{Im}\,\mu| \leqslant \frac{1}{6}, m, n, \in \mathbb{Z}, t \geqslant 1$,*

$$|\mathrm{e}^t f_{mn}(\mu{:}a_t) - (\mathrm{e}^{\mathrm{i}\mu t}c^+_{mn}(\mu) + \mathrm{e}^{-\mathrm{i}\mu t}c^-_{mn}(\mu)| \leqslant c[\mu, m, n]^5(1+t)^5\mathrm{e}^{-3t/2}$$

Furthermore, if $u, v \in U(\mathfrak{g})$, we can find an integer $q = q_{u,v} \geqslant 0$ and polynomials $p^{\pm} = p^{\pm}_{u,v}$ in μ, m, n, such that, for some constant $c_{u,v} > 0$,

$$\begin{aligned}|\mathrm{e}^t f_{mn}(\mu; u; a_t; v) - (p^+(\mu{:}m, n)c^+_{mn}(\mu)\mathrm{e}^{\mathrm{i}\mu t} + p^-(\mu; m, n)c^-_{mn}(\mu))\mathrm{e}^{-\mathrm{i}\mu t}| \\ \leqslant c_{u,v}[\mu, m, n, t]^q\mathrm{e}^{-3t/2}\end{aligned}$$

for μ, m, n, t as before.

Proof. Only the estimates for the derivatives need proof. It is done in stages, each stage reducing the complexity of the derivatives involved till we are left with only derivatives with respect to t. We use Lemmas 15 and 16 for this purpose.

Stage 1 (*reduction to one-sided derivatives*). By Lemma 15,

$$f_{mn}((X-Y)^r H^s X^q; a_t; v) = (\mathrm{i}m)^r \mathrm{e}^{-2qt} f_{mn}(\mu{:}a_t; H^s X^q v)$$

If $q \geqslant 1$, we use the estimate in Lemma 8 to conclude that $\mathrm{e}^t f_{mn}(\mu{:}a_t; H^s X^q v)$ is of the order $\mathrm{e}^{t|\mathrm{Im}\,\mu|} \leqslant \mathrm{e}^{t/6}$ so that the e^{-2qt} in front makes this of greater decay than $\mathrm{e}^{-3t/2}$. So we just absorb this in the error term. For $q = 0$, this is to be treated as part of Stage 2.

Stage 2 (*reduction to derivatives with respect to t, hence to Proposition 12*). By Lemma 16, given $u \in U(\mathfrak{g})$ of degree r,

$$\mathrm{e}^t f_{mn}(\mu{:}a_t; u) = \gamma \mathrm{e}^t(\mathrm{d}/\mathrm{d}t)^r f_{mn}(\mu{:}a_t) + \sum_{0\leqslant j\leqslant r-1} f_j(m, n; t)\mathrm{e}^t(\mathrm{d}/\mathrm{d}t)^j f_{mn}(\mu{:}a_t)$$

We now observe that, for any function r in the ring R of Lemma 16, we have $r(t) = r(\infty) + O(e^{-2t})$ when $r(\infty)$ is a constant (check this for generators). So we can rewrite this as

$$e^t f_{mn}(\mu:a_t;u) = \gamma(d/dt)^r(e^t f_{mn}(\mu:a_t)) + \sum_{0\leqslant j\leqslant r-1} r_j(m,n)(d/dt)^j(e^t f_{mn}(\mu:a_t))$$
$$+ \sum_{0\leqslant j\leqslant r-1} g_j(m,n:t)(d/dt)^j(e^t f_{mn}(\mu:a_t))$$

We estimate the terms in the first two groups above by Proposition 14. The last term is to be absorbed with the error, and Lemma 8 is used to estimate

$$(d/dt)^j(e^t f_{mn}(\mu:a_t)) = e^t(d/dt+1)^j f_{mn}(\mu:a_t)$$

This finishes the proof.

Remark. The estimates for *the derivatives with respect to* μ are already contained in Theorem 17. This is because the estimates are valid in a strip $|\mathrm{Im}\,\mu| < \delta$ and hence one can use Cauchy's theorem.

8.4 The wave packet theorem

There will have to be some rewards after all this work. For any $b\in\mathscr{S}(\mathbb{R})$ put

$$g_{b:mn}(x) = \int_{\mathbb{R}} \mu f_{mn}(\mu:x)b(\mu)\,d\mu$$

We shall think of g_b as a (provisional) *wave packet* defined by the auxiliary function b. Since, by Lemma 8, f_{mn} and its derivatives in μ are $O([\mu]^s\Xi(x))$ it is clear that g_b is a well-defined C^∞ function on G and its derivatives are majorized by const. Ξ. But we can do much better. Write $\mathscr{C}$ for $\mathscr{C}(G)$, and $\mathscr{C}_{mn}$ for the subspace of $\mathscr{C}$ of functions of type (m,n).

Theorem 18 (Wave packet theorem). *For any $b\in\mathscr{S}(\mathbb{R})$, $g_{b;mn}\in\mathscr{C}_{mn}$. Given $r\geqslant 0$ and $u,v\in U(\mathfrak{g})$, we can find a continuous seminorm $\nu_{u,v:r}$ on $\mathscr{S}(\mathbb{R})$ and an integer $q\geqslant 0$ such that, for all $b\in\mathscr{S}(\mathbb{R})$ and all m,n,*

$$\mu_{u,v:r}(g_{b:mn}) \leqslant [m,n]^q \nu_{u,v:r}(b)$$

Proof. Here $\mu_{u,v:r}$ refers to the defining seminorms on $\mathscr{C}$. First consider the case where there are no derivatives, i.e., $u=v=1$. Write

$$\mu e^t f_{mn}(\mu:a_t) = e^{i\mu t}(\mu c^+_{mn}(\mu)) + e^{-i\mu t}(\mu c^-_{mn}(\mu)) + E_{mn}(\mu:t)$$

where the 'error' E_{mn} is to be estimated by Theorem 17. Let $d^+_{mn}(\mu) = \mu c^+_{mn}(\mu)$, $d^-_{mn}(\mu) = -\mu c^-_{mn}(-\mu)$. Then ($b^0(\mu) = b(-\mu)$)

$$e^t g_{b:mn}(a_t) = \widehat{bd^+_{mn}}(t) + \widehat{b^0 d^-_{mn}}(t) + \int E_{mn}(\mu:t)\mu b(\mu)\,d\mu$$

The functions $d^{\pm}_{mn}$ and their derivatives have at most polynomial growth by Proposition 13 so that $bd^{\pm}_{mn}\in\mathscr{S}(\mathbb{R})$, while $|E_{mn}|\leqslant ce^{-3t/2}(1+t)^5[\mu,m,n]^5$ by Theorem 17. It is now obvious that the left side goes to zero faster than any power t^{-q}. Moreover, Proposition 13 implies that, if v is a continuous seminorm on $\mathscr{S}(\mathbb{R})$, there is a continuous seminorm v' on $\mathscr{S}(\mathbb{R})$ and an integer $q\geqslant 0$ such that

$$v(bd^{\pm}_{mn})\leqslant [m,n]^q v'(b)$$

for all $b\in\mathscr{S}(\mathbb{R}), m, n$. It follows easily from this that for any $r\geqslant 0$ we can find a continuous seminorm v_r on $\mathscr{S}(\mathbb{R})$ and an integer $q=q_r\geqslant 0$ such that

$$\sup_{\sigma(x)\geqslant 1}|\Xi^{-1}(1+\sigma)^r g_{b:mn}|\leqslant [m,n]^q v_r(b)$$

for all b, m, n. In the general case, let $u, v\in U(\mathfrak{g})$; we first note the existence of $u_j, v_j\in U(\mathfrak{g})$ and finite Fourier series c_j on $K\times K$ $(1\leqslant j\leqslant p)$ such that, for *any* $h\in C^\infty(G)$,

$$h(u; u_{\theta_1}xu_{\theta_2}; v)=\sum_{1\leqslant j\leqslant p} c_j(u_{\theta_1}, u_{\theta_2})({}_{\theta_1}h_{\theta_2})(u_j; x; v_j)$$

where $({}_{\theta_1}h_{\theta_2})(y)=h(u_{\theta_1}yu_{\theta_2})$. Hence, for h of type (m,n),

$$\sup_{\sigma(x)\geqslant 1}\Xi^{-1}(1+\sigma)^r|vhu|\leqslant c\sup_{t\geqslant 1, j} e^t(1+t)^{r+1}|h(u_j; a_t; v_j)|$$

where c is a constant independent of m and n. If we now take $h=g_{b:mn}$ we may proceed as in the previous case, this time using the derivatives part of Theorem 17. Finally, the estimating of

$$\sup_{\sigma(x)\leqslant 1}\Xi^{-1}(1+\sigma)^r|vg_{b:mn}u|$$

is trivial, for we are dealing with a compact set in G. This finishes the proof of the theorem.

8.5 The Harish-Chandra transform on the spaces $\mathscr{C}_{mn}$

For the moment we shall study the $\mathscr{C}_{mn}$ for each m, n fixed. First we introduce the appropriate subspace of $C^\infty(G)$ that is dual to $\mathscr{C}$. Let $\mathscr{A}=\mathscr{A}(G)$ denote the space of all $f\in C^\infty(G)$ such that for each $u, v\in U(\mathfrak{g})$ there are an integer $r=r_{u,v}\geqslant 0$ and a constant $c=c_{u,v}>0$ such that

$$|f(u; x; v)|\leqslant c\Xi(x)(1+\sigma(x))^r \quad (x\in G)$$

In our terminology this is just saying that each vfu satisfies the weak inequality. It is natural to introduce this space, for all the $f_{mn}(\mu:\cdot)$ are in it. Also $u\mathscr{A}v\subset\mathscr{A}$ for $u, v\in U(\mathfrak{g})$.

Proposition 19. *If f satisfies the weak inequality, then*

$$\int fg\,\mathrm{d}G \quad (g\in\mathscr{C})$$

converges absolutely and is a continuous linear functional on $\mathscr{C}$. *If* $f \in \mathscr{A}$ *and* $u, v \in U(\mathfrak{g})$, *then*

$$\int ufv \cdot g \, dG = \int f \cdot u^t g v^t \, dG$$

Proof. Let $r_0 \geqslant 0$ be such that

$$c_0 = \int \Xi^2 (1+\sigma)^{-r_0} \, dG < \infty$$

If $f = O(\Xi(1+\sigma)^r)$, say $|f| \leqslant c\Xi(1+\sigma)^r$,

$$|fg| \leqslant c\mu_{1,1:r+r_0}(g) \cdot \Xi^2(1+\sigma)^{-r_0}$$

and the first statement is clear. For the second use the density of $C_c^\infty(G)$ in $\mathscr{C}$.

This proposition allows us to introduce the *Harish-Chandra transform* $\mathscr{H}$ on the $\mathscr{C}_{mn}$: for $f \in \mathscr{C}_{mn}$,

$$(\mathscr{H} f)(\mu) = \int_G f(x) \overline{f_{mn}(\mu : x)} \, dG$$

Theorem 20. $\mathscr{H}$ *is a continuous linear map of* $\mathscr{C}_{mn}$ *into* $\mathscr{S}(\mathbb{R})$. *More precisely if* ν *is any continuous seminorm on* $\mathscr{S}(\mathbb{R})$, *there is a continuous seminorm* ν' *on* $\mathscr{C}$ *such that, for all* $f \in \mathscr{C}_{mn}$, *for all* m, n,

$$\nu(\mathscr{H} f) \leqslant \nu'(f)$$

Proof. That $\mathscr{H}$ is well defined is immediate from Proposition 19. If $\Box = 1 - \Omega$, then

$$\Box^k f_{mn} = (1+\mu^2)^k f_{mn}, \; \Box^k \bar{g} = \overline{\Box^k g}$$

Hence

$$(1+\mu^2)^k (\mathscr{H} f)(\mu) = \left(\int f(x) f_{mn}(\mu : x; \Box^k) \, dG \right)^{\text{conj}}$$

$$= \left(\int \Box^k f \cdot f_{mn}(\mu : \cdot) \, dG \right)^{\text{conj}}$$

So, by Lemma 8,

$$(1+\mu^2)^k |(\mathscr{H} f)(\mu)| \leqslant \int |\Box^k f| \Xi \, dG = \nu_0(\Box^k f)$$

and the right side is a continuous seminorm on $\mathscr{C}$ by Proposition 19. For the derivatives, first note that, for any $l \geqslant 0$, the integral

$$\int f(x) (d^l / d\mu^l) f_{mn}(\mu : x) \, dG$$

converges uniformly when μ varies in compact subsets of $\mathbb{R}$, in view of the estimates of Lemma 8 and Proposition 19. So, $\mathscr{H}f \in C^\infty(\mathbb{R})$ and

$$(\mathscr{H}f)^{(l)}(\mu) = \int f(x)(\mathrm{d}^l/\mathrm{d}\mu^l) f_{mn}(\mu : x)\,\mathrm{d}G$$

We now apply Lemma 8 to get

$$|(\mathscr{H}f)^{(l)}(\mu)| \leqslant \int |f|\Xi(1+\sigma)^l\,\mathrm{d}G = v_l(f)$$

To get rapid decrease we proceed as before although the argument is a little more delicate. We start by noting that, if $\partial = \mathrm{d}/\mathrm{d}\mu$,

$$(1+\mu^2)\partial^l = \partial^l \circ (1+\mu^2) - 2l\mu\partial^{l-1} - l(l-1)\partial^{l-2}$$

where the last term is omitted for $l = 1$. So,

$$(1+\mu^2)(\mathscr{H}f)^{(l)}(\mu) = (\mathscr{H}\cdot\Box f)^{(l)}(\mu) - 2l\mu(\mathscr{H}f)^{(l-1)}(\mu) - l(l-1)(\mathscr{H}f)^{(l-2)}(\mu)$$

from which we get the following: if

$$\beta_{k,l}(g) = \sup_\mu [\mu]^k |g^{(l)}(\mu)| \quad (g \in \mathscr{S}(\mathbb{R}))$$

$$\beta_{k+1,l}(\mathscr{H}f) \leqslant 2\beta_{k-1,l}(\mathscr{H}\cdot\Box f) + 4l^2(\beta_{k,l-1}(\mathscr{H}f) + \beta_{k,l-2}(\mathscr{H}f))$$

The assertion of the theorem for $v = \beta_{k,l}$ now follows by a double induction on l and k; the case $l = 0$ has already been taken care of.

To determine the range of $\mathscr{H}$ we must determine the functional equations.

Lemma 21. *Let Γ be the classical gamma function and define*

$$t_n(\mu) = \frac{\Gamma((n+1+\mathrm{i}\mu)/2)}{\Gamma((n+1-\mathrm{i}\mu)/2)} \quad (n \in \mathbb{Z})$$

Then the t_n are holomorphic and zero free if $|\mathrm{Im}\,\mu| < 1$ and are of absolute value 1 for real μ. Moreover,

$$\frac{t_n(\mu)}{t_{n+2}(\mu)} = \frac{n+1-\mathrm{i}\mu}{n+1+\mathrm{i}\mu} \quad (n \in \mathbb{Z}, |\mathrm{Im}\,\mu| < 1)$$

$$\overline{t_n(\mu)} = t_n(-\mu) = t_n(\mu)^{-1} \quad (n \in \mathbb{Z}, \mu \in \mathbb{R})$$

In particular, if $n - m$ is even, t_n/t_m has at most polynomial growth on $|\mathrm{Im}\,\mu| \leqslant 1 - \delta$ $(0 < \delta < 1)$; and multiplication by it is a continuous endomorphism of $\mathscr{S}(\mathbb{R})$.

Proof. Elementary.

Proposition 22. *If $\mu \in \mathbb{R}$ there is a unitary operator $U(\mu)$ of $L^2(K)$ such that*

$U(\mu)e_n = t_n(\mu)e_n$; and

$$U(\mu)^{-1}\pi_{\varepsilon,i\mu}U(\mu) = \pi_{\varepsilon,-i\mu}$$

Proof. Obvious (cf. Theorem 35, Chapter 5, Section 5).

Theorem 23 (functional equations). *We have*

$$f_{mn}(-\mu:x) = \frac{t_n(\mu)}{t_m(\mu)} f_{mn}(\mu:x)$$

for all $m, n \in \mathbb{Z}$, $x \in G$, $\mu \in \mathbb{R}$. Moreover, for any $f \in \mathscr{C}_{mn}(G)$,

$$(\mathscr{H} f)(-\mu) = \frac{t_m(\mu)}{t_n(\mu)}(\mathscr{H} f)(\mu)$$

Proof. The first assertion is immediate from the intertwining property; the second follows from the first if we multiply by f and integrate and use the relation $\overline{t_n(\mu)} = t_n(\mu)^{-1}$.

The asymptotic theory of §3 now gives the following.

Theorem 24. *We have*

$$c^{\pm}_{mn}(-\mu) = \frac{t_n(\mu)}{t_m(\mu)} c^{\mp}_{mn}(\mu)$$

for all $m, n \in \mathbb{Z}$, $\mu \in \mathbb{R}\backslash(0)$.

Proof. We know that, for each $\mu \neq 0, m, n \in \mathbb{Z}$,

$$e^t f_{mn}(\mu:a_t) \sim e^{i\mu t} c^+_{mn}(\mu) + e^{-i\mu t} c^-_{mn}(\mu)$$

where $\sim$ means that the difference goes to zero exponentially when $t \to +\infty$. We now change μ to $-\mu$, use the functional equations, and argue as in the proof of Proposition 13.

At this stage we have maps going in both directions: $\mathscr{H}$, from $\mathscr{C}_{mn}$ to $\mathscr{S}$ and the provisional wave packet map, from $\mathscr{S}$ to $\mathscr{C}_{mn}$. They will *not* be inverse to each other, and we must determine the measure that should be used in the definition of the wave packet to make it inverse to $\mathscr{H}$. This is however quite involved and will have to be done in stages. Right now we shall only prove the following.

Proposition 25. *There is a unique continuous function $\beta = \beta_{mn}$ on $\mathbb{R}$ such* *that, writing g_b for $g_{b:mn}$,*

$$\mu\mathscr{H}g_b(\mu) = \beta(\mu)\left\{b(\mu) - \frac{t_n(\mu)}{t_m(\mu)}b(-\mu)\right\}$$

for all $b\in\mathscr{S}(\mathbb{R})$. Moreover β is even, $\geqslant 0$, and

$$\int |g_b|^2 \, dG = \frac{1}{2}\int \left| b - \frac{t_n}{t_m}b^0 \right|^2 \beta \, d\mu$$

(where $b^0(\mu) = b(-\mu)$) for all $b\in C_c^\infty(\mathbb{R})$.

Proof. The equality of norms is stated only for $b\in C_c^\infty(\mathbb{R})$ because nothing is known at this stage of the growth of β at infinity. It is clear by choosing b to have small support in $(0,\infty)$ and $(-\infty,0)$ that β is unique on $\mathbb{R}\backslash(0)$, hence on $\mathbb{R}$ by continuity. For existence, we begin with the formulae

$$\mathscr{H}\Omega f = -\mu^2\mathscr{H}f, \quad \Omega g_b = g_{-\mu^2 b}$$

where μ is the coordinate function on $\mathbb{R}$, both of which follow from $\Omega f_{mn}(\mu{:}\cdot) = -\mu^2 f_{mn}(\mu{:}\cdot)$. For fixed $\mu_0 \neq 0$, $J{:}\, b \mapsto (\mu\mathscr{H}g_b)(\mu_0)$ is a tempered distribution on $\mathbb{R}$, and the above formulae imply easily that $(\mu^2 - \mu_0^2)J = 0$. Hence $\operatorname{supp} J \subset \{-\mu_0, \mu_0\}$, showing that for all $b\in\mathscr{S}(\mathbb{R})$

$$\mu_0(\mathscr{H}g_b)(\mu_0) = \beta_+(\mu_0)b(\mu_0) + \beta_-(\mu_0)b(-\mu_0)$$

In other words, there are well-defined complex functions $\beta_\pm$ on $\mathbb{R}\backslash(0)$ such that for all $b\in\mathscr{S}(\mathbb{R})$

$$\mu\mathscr{H}g_b = \beta_+ b + \beta_- b^0 \quad \text{on } \mathbb{R}\backslash(0)$$

It is obvious from this that $\beta_\pm$ are smooth on $\mathbb{R}\backslash(0)$. We now change μ to $-\mu$ and use the functional equations to get

$$-(t_m/t_n)(\beta_+ b + \beta_- b^0) = \beta_-^0 b + \beta_+^0 b^0$$

which leads to

$$\beta_- = -(t_n/t_m)\beta_+^0$$

On the other hand, from the definition of g_b, we have $g_{b^0} = g_{-(t_n/t_m)b}$ so that we also have

$$\beta_+ b^0 + \beta_- b = \beta_+(-t_n/t_m)b + \beta_-(-t_m/t_n)b^0$$

This gives

$$\beta_- = -\beta_+(t_n/t_m)$$

We thus see that β_+ is even and

$$\mu\mathscr{H}g_b = \beta_+(b - (t_n/t_m)b^0)$$

on $\mathbb{R}\backslash(0)$, valid for all $b\in\mathscr{S}(\mathbb{R})$; we put $\beta = \beta_+$.

We should now examine what happens at $\mu = 0$. From the definition of t_n we find easily that

$$t_n(0) = \begin{cases} 1 & n \geqslant 0 \\ (-1)^n & n < 0 \end{cases}$$

In particular, $t_n(0)/t_m(0) = \pm 1$. *Case 1*: $t_n(0)/t_m(0) = 1$. Select $b_1 \in \mathscr{S}(\mathbb{R})$, $b_1(0) = \frac{1}{2}$, and let $b = \mu b_1$. Then $\mu \mathscr{H} g_b = \beta \cdot \mu b_2$ where $b_2(0) = 1$. Hence $\beta = \mathscr{H} g_b \cdot b_2^{-1}$ near 0, proving it extends continuously across $\mu = 0$.

Case 2: $t_n(0)/t_m(0) = -1$. Select b such that $b = b^0$, $b(0) = \frac{1}{2}$. Then $\mu \mathscr{H} g_b = \beta b_1, b_1(0) = 1$, and $\beta = \mu \mathscr{H} g_b \cdot b^{-1}$ near $\mu = 0$, again proving that β extends continuously across $\mu = 0$. The first assertion is thus fully proved.

We shall now determine the L^2-norm of the wave packet. For any $g \in \mathscr{C}_{mn}$

$$\int g_b \bar{g}\, \mathrm{d}G = \int_G \left(\int_{\mathbb{R}} \mu b(\mu) f_{mn}(\mu : x)\, \mathrm{d}\mu \right) \bar{g}\, \mathrm{d}G$$

Since

$$|f_{mn}(\mu : x)| \bar{g}| |\mu b(\mu)| \leqslant |\mu| |b(\mu)| \Xi(x) |g(x)|$$

the integrand is in L^1 $(G \times \mathbb{R})$ and hence, by Fubini,

$$\int g_b \bar{g}\, \mathrm{d}G = \int \mu b(\mu) \overline{\mathscr{H} g(\mu)}\, \mathrm{d}\mu$$

Put now $g = g_b$ and substitute for $\mu \mathscr{H} g_b$, and for safety assume $b \in C_c^\infty(\mathbb{R})$. Then

$$\int |g_b|^2\, \mathrm{d}G = \int b(\bar{b} - (t_m/t_n)\bar{b}^0)\bar{\beta}\, \mathrm{d}\mu$$

We now remark that the function $(b + (t_n/t_m)b^0)(\bar{b} - (t_m/t_n)\bar{b}^0)$ is odd and hence its integral will be 0. Hence

$$\int |g_b|^2\, \mathrm{d}G = \frac{1}{2} \int |b - (t_n/t_m)b^0|^2 \bar{\beta}\, \mathrm{d}\mu$$

This shows that β has to be real and $\geqslant 0$, and finishes the proof.

8.6 The relation $\beta(\mu) = 2\pi^3 \mu^2 |c_{mn}^{\pm}(\mu)|^2$ and the determination of $c_{mn}^{\pm}$

There are at least two methods of proving this, and both have far-reaching generalizations to arbitrary semisimple groups. The first is the so-called method of *Maass–Selberg relations* and is the more 'L^2' of the two; the other is based on computing $\mathscr{H} g_b$ directly by integrating along N. We work with the first method. Throughout this calculation we shall keep m, n fixed and suppress all references to them. Thus we write $c^{\pm}$ for $c_{mn}^{\pm}$,

D for $d^2/dt^2 - q$ where $q(t) = (m^2 + n^2 - 2mn\cosh t - 1)/\sinh^2 2t$. We write $f_\mu(x)$ for $f_{mn}(\mu{:}x)$.

The idea behind the computation of

$$\int g_a \bar{g}_b \, dG$$

is as follows. We substitute for g_a and g_b their expressions as wave packets and reverse the order of integrations. Since the f_μ are not in $L^2(G)$, this would not work, but we can write

$$\int g_a \bar{g}_b \, dG = \lim_{T \to +\infty} \int_{B_T} g_a \bar{g}_b \, dG$$

where

$$B_T = \{x \in G \mid \sigma(x) \leqslant T\},$$

compute the scalar product over B_T by this method, and let $T \to +\infty$. The integral $\int_{B_T} f_\lambda \bar{f}_\mu \, dG$ reduces to an integral over $(0, T)$, and, because f_λ and f_μ are eigenfunctions for D, it reduces in the usual way *to a boundary term* (see definitions below)

$$-\frac{[\tilde{f}_\lambda, \tilde{f}_\mu](T)}{\lambda^2 - \mu^2}$$

where $\tilde{f}_\lambda(t) = f_\lambda(a_t)\ (2\sinh 2t)^{1/2}$. We now replace $\tilde{f}_\lambda$ and $\tilde{f}_\mu$ by the dominant terms in their approximations, and find that, when $T \to +\infty$,

$$-\iint a(\lambda)\overline{b(\mu)}([\tilde{f}_\lambda, \tilde{f}_\mu](T)/(\lambda^2 - \mu^2)) \, d\lambda \, d\mu$$

converges to

$$\text{const.} \int |c^+|^2 \mu a \mu \bar{b} \, d\mu$$

But for this it is necessary to know that

$$|c^+| = |c^-| \quad \text{on } \mathbb{R}\backslash(0)$$

which are the celebrated *Maass–Selberg relations.* If we compute the functions $c^\pm$ explicitly, this would be an easy verification; this is the case here, but we give a direct proof not requiring any calculations. This proof will go over to the general context.

Let us write, for C^∞ functions u, v on $(0, \infty)$,

$$[u, v] = u'\bar{v} - u\bar{v}'$$

As before write $\tilde{f}_\lambda(t) = f_\lambda(a_t)\ (2\sinh 2t)^{1/2}$ so that

$$\Omega f_\lambda = -\lambda^2 f_\lambda, \quad \Omega f_\mu = -\mu^2 f_\mu, \quad D\tilde{f}_\lambda - q^2\tilde{f}_\lambda = -\lambda^2\tilde{f}_\lambda$$

Writing $f_\lambda(t) = f_\lambda(a_t)$ etc. we have

$$[\tilde{f}_\lambda, \tilde{f}_\mu](t) = 2\sinh 2t[f_\lambda, f_\mu](t)$$

so that $[\tilde{f}_\lambda, \tilde{f}_\mu]$ is C^∞ across $t = 0$ and vanishes there. Now, for any $T > 0$,

$$\begin{aligned}
-(\lambda^2 - \mu^2)\int_{B_T} f_\lambda \bar{f}_\mu \, \mathrm{d}G &= -\pi^2(\lambda^2 - \mu^2)\int_0^T \tilde{f}_\lambda \bar{\tilde{f}}_\mu \, \mathrm{d}t \\
&= \pi^2 \int_0^T (D\tilde{f}_\lambda \cdot \bar{\tilde{f}}_\mu - \tilde{f}_\lambda D\bar{\tilde{f}}_\mu)\, \mathrm{d}t \\
&= \pi^2 \int_0^T (\tilde{f}''_\lambda \bar{\tilde{f}}_\mu - \tilde{f}_\mu \bar{\tilde{f}}''_\mu)\, \mathrm{d}t \\
&= \pi^2[\tilde{f}_\lambda, \tilde{f}_\mu](T)
\end{aligned}$$

An easy calculation based on $f_\lambda(0) = f_\mu(0) = \delta_{mn}$ shows the contribution at 0 is zero. Hence

$$(\lambda^2 - \mu^2)\int_{B_T} f_\lambda \bar{f}_\mu \, \mathrm{d}G = -\pi^2[\tilde{f}_\lambda, \tilde{f}_\mu](T)$$

We also write

$$f_\lambda^\infty(t) = c^+(\lambda)\mathrm{e}^{\mathrm{i}\lambda t} + c^-(\lambda)\mathrm{e}^{-\mathrm{i}\lambda t}$$
$$\tilde{f}_\lambda = f_\lambda^\infty + \theta_\lambda$$

Then

$$[\tilde{f}_\lambda, \tilde{f}_\mu] = [f_\lambda^\infty, f_\mu^\infty] + \theta_{\lambda\mu}$$

while

$$\begin{aligned}
[f_\lambda^\infty, f_\mu^\infty](T) = {} & \mathrm{i}(\lambda - \mu)(c^+(\lambda)\overline{c^-(\mu)}\,\mathrm{e}^{\mathrm{i}(\lambda+\mu)T} - c^-(\lambda)\overline{c^+(\mu)}\,\mathrm{e}^{-\mathrm{i}(\lambda+\mu)T}) \\
& + \mathrm{i}(\lambda + \mu)(c^+(\lambda)\overline{c^+(\mu)}\,\mathrm{e}^{\mathrm{i}(\lambda-\mu)T} - c^-(\lambda)\overline{c^-(\mu)}\,\mathrm{e}^{-\mathrm{i}(\lambda-\mu)T})
\end{aligned}$$

Theorem 26 (Maass–Selberg relations). *For $\mu \neq 0$ in $\mathbb{R}$, we have*

$$|c_{mn}^+(\mu)| = |c_{mn}^-(\mu)|$$

We also have

$$|c_{mn}^+(\mu)| = |c_{mn}^-(-\mu)|$$

Proof. The second follows from the functional equations of Theorem 24. So we need only prove the first.

Taking $\lambda = \mu$ in the formula for $\int_{B_T} f_\lambda \bar{f}_\mu \, \mathrm{d}G$ we get

$$[\tilde{f}_\mu, \tilde{f}_\mu](T) = 0 \quad (T > 0)$$

Hence

$$[f_\mu^\infty, f_\mu^\infty](T) = -\theta_{\mu\mu}(T)$$

But as $[\cdot,\cdot]$ is bilinear, we have

$$\theta_{\mu\mu} = [f_\mu^\infty, \theta_\mu] + [\theta_\mu, f_\mu^\infty] + [\theta_\mu, \theta_\mu]$$

From Theorem 17 we know, when $T\to+\infty$, $\theta_\mu(T)\to 0$, $\theta'_\mu(T)\to 0$ exponentially fast. Hence $\theta_{\mu\mu}(T)\to 0$ when $T\to+\infty$, showing that $[f^\infty_\mu, f^\infty_\mu](T)\to 0$. But

$$[f^\infty_\mu, f^\infty_\mu](T)=2i\mu(|c^+(\mu)|^2-|c^-(\mu)|^2))$$

identically, proving the theorem.

Corollary 27. $[f^\infty_\mu, f^\infty_\mu]=-\theta_{\mu\mu}=0$.

We are now ready to calculate the scalar products of wave packets. It is enough to calculate this for $a,b\in C^\infty_c((0,\infty))$. Starting from the identity for $\int_{B_T} f_\lambda \bar{f}_\mu \mathrm{d}G$ we have

$$\int_{B_T} g_a\bar{g}_b\,\mathrm{d}G=-\pi^2\iint \lambda a(\lambda)\mu\overline{b(\mu)}([\tilde{f}_\lambda,\tilde{f}_\mu](T)/(\lambda^2-\mu^2))\mathrm{d}\lambda\,\mathrm{d}\mu$$

We must remember that

$$\frac{[\tilde{f}_\lambda,\tilde{f}_\mu](T)}{\lambda^2-\mu^2}$$

has no singularity anywhere since it equals

$$\int_{B_T} f_\lambda\bar{f}_\mu\,\mathrm{d}G$$

To study the above integral we go back to

$$[\tilde{f}_\lambda,\tilde{f}_\mu]=[f^\infty_\lambda,f^\infty_\mu]+\theta_{\lambda\mu}$$

We claim that both terms on the right, when divided by $\lambda^2-\mu^2$, have no singularity anywhere on $\operatorname{supp}(a)\times\operatorname{supp}(b)$. It is enough to check this for $[f^\infty_\lambda,f^\infty_\mu]$. If we look at the expression for it and observe that $\lambda+\mu$ is bounded away from 0 when $(\lambda,\mu)\in\operatorname{supp}(a)\times\operatorname{supp}(b)$, it becomes a question of verifying that

$$(\lambda-\mu)^{-1}(c^+(\lambda)\overline{c^+(\mu)}\mathrm{e}^{\mathrm{i}(\lambda-\mu)T}-c^-(\lambda)\overline{c^-(\mu)}\mathrm{e}^{-\mathrm{i}(\lambda-\mu)T})$$

has no singularity. But now, by the Maass–Selberg relations, $c^+(\lambda)\overline{c^+(\mu)}-c^-(\lambda)\overline{c^-(\mu)}$ vanishes for $\lambda=\mu$ and hence can be written as

$$c^+(\lambda)\overline{c^+(\mu)}-c^-(\lambda)\overline{c^-(\mu)}=(\lambda-\mu)d(\lambda:\mu)$$

where $d\in C^\infty((0,\infty)\times(0,\infty))$. Hence the expression under discussion reduces to

$$d(\lambda;\mu)\cos(\lambda-\mu)T+\frac{\mathrm{i}\sin(\lambda-\mu)T}{(\lambda-\mu)}(c^+(\lambda)\overline{c^+(\mu)}+c^-(\lambda)\overline{c^-(\mu)})$$

and the regularity of this for $\lambda=\mu$ is clear. Our claim is thus verified.

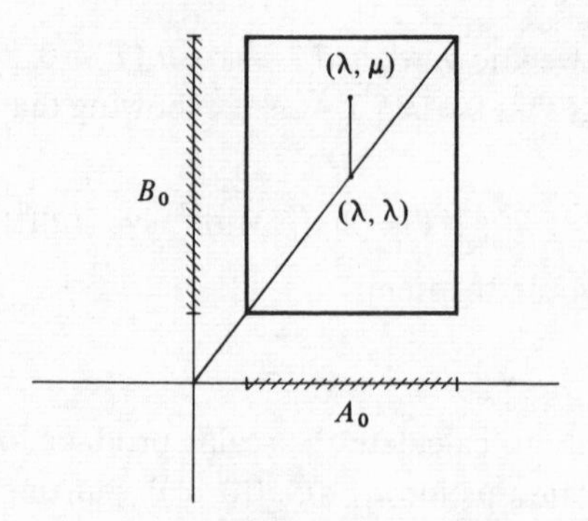

Figure 8.1

Write now

$$J_1(T) = -\pi^2 \iint \lambda a(\lambda)\mu\overline{b(\mu)}\frac{[f_\lambda^\infty, f_\mu^\infty]}{\lambda^2-\mu^2}\,\mathrm{d}\lambda\,\mathrm{d}\mu$$

$$J_2(T) = -\pi^2 \iint \lambda a(\lambda)\mu\overline{b(\mu)}\frac{\theta_{\lambda\mu}(T)}{\lambda^2-\mu^2}\,\mathrm{d}\lambda\,\mathrm{d}\mu$$

Lemma 28. *$J_2(T) \to 0$ as $T \to +\infty$.*

Proof. See Figure 8.1. Let A_0 (resp. B_0) be a closed interval in $(0, \infty)$ containing $\operatorname{supp}(a)$ (resp. $\operatorname{supp}(b)$). It is enough to find an estimate for $\theta_{\lambda\mu}(T)$ of the form

$$|\theta_{\lambda\mu}(T)| \leqslant C|\lambda-\mu|\mathrm{e}^{-\alpha T} \quad (\lambda\in A_0, \mu\in B_0)$$

for all $T \geqslant 1$, where $C > 0$, $\alpha > 0$ are constants; as $\lambda + \mu \geqslant \gamma > 0$ for all $\lambda\in A_0, \mu\in B_0$, we would then have

$$\left|\frac{\theta_{\lambda\mu}(T)}{\lambda^2-\mu^2}\right| \leqslant C\gamma^{-1}\mathrm{e}^{-\alpha T} \quad (T \geqslant 1)$$

and the desired conclusion would be immediate. Let us write $\theta(\lambda:\mu:T) = \theta_{\lambda\mu}(T)$. Then by Theorem 17 and the definition of $\theta_{\lambda\mu}(T)$ it is clear that we have estimates of the form (see remark following Theorem 17)

$$|\theta(\lambda:\mu:T)| + \left|\frac{\partial}{\partial\lambda}\theta(\lambda:\mu:T)\right| + \left|\frac{\partial}{\partial\mu}\theta(\lambda:\mu:T)\right| \leqslant C\mathrm{e}^{-\alpha T}$$

for all $T \geqslant 1$, $(\lambda, \mu)\in A_0 \times B_0$. We now note that $\theta(\lambda:\lambda T) = 0$ *identically* by

Corollary 27. Hence (if $\mu \geqslant \lambda$)

$$\begin{aligned}\theta(\lambda:\mu:T) &= |\theta(\lambda:\mu:T) - \theta(\lambda:\lambda:T)| \\ &\leqslant |\lambda - \mu| \sup_{A_0 \times B_0}\left(\left|\frac{\partial}{\partial\lambda}\theta\right| + \left|\frac{\partial}{\partial\mu}\theta\right|\right) \\ &\leqslant C|\lambda - \mu|\mathrm{e}^{-\alpha T}\end{aligned}$$

We next study J_1.

Lemma 29. *We have*

$$\lim_{T\to\infty} J_1(T) = 2\pi^3\int |c^+(\mu)|^2\mu a(\mu)\overline{\mu b(\mu)}\,\mathrm{d}\mu$$

Proof. This is just classical Fourier analysis since only exponential functions are involved. We substitute for $[f_\lambda^\infty, f_\mu^\infty]$ and write it in the following form (cf. remarks before Lemma 28):

$$\frac{[f_\lambda^\infty, f_\mu^\infty]}{\lambda^2 - \mu^2} = \mathrm{i}(\lambda + \mu)^{-1}(c^+(\lambda)\overline{c^-(\mu)}\mathrm{e}^{\mathrm{i}(\lambda+\mu)T} - c^-(\lambda)\overline{c^+(\mu)}\mathrm{e}^{-\mathrm{i}(\lambda+\mu)T})$$

$$+ \mathrm{i}d(\lambda:\mu)\cos(\lambda - \mu)T - \frac{\sin(\lambda-\mu)T}{\lambda - \mu}(c^+(\lambda)\overline{c^+(\mu)} + c^-(\lambda)\overline{c^-(\mu)})$$

Since $\lambda + \mu \geqslant \gamma > 0$ for $(\lambda, \mu) \in A_0 \times B_0$, the contribution to $J_1(T)$ from the first term tends to zero when $T \to \infty$; similarly the contribution from the second term also goes to zero. The last term gives the contribution

$$\pi^2\iint \frac{\sin(\lambda-\mu)T}{\lambda-\mu}(c^+(\lambda)\overline{c^+(\mu)} + c^-(\lambda)\overline{c^-(\mu)})\lambda a(\lambda)\overline{\mu b(\mu)}\,\mathrm{d}\lambda\,\mathrm{d}\mu$$

which tends to

$$2\pi^3\int \mu a(\mu)\overline{\mu b(\mu)}|c^+(\mu)|^2\,\mathrm{d}\mu$$

by the classical theorem of Dirichlet.

Theorem 30. *We have*

$$\beta_{mn}(\mu) = 2\pi^3\mu^2|c_{mn}^{\pm}(\mu)|^2$$

In particular, for all $b \in \mathscr{S}(\mathbb{R})$,

$$\int |g_b|^2\,\mathrm{d}G = \pi^3\int |b - (t_n/t_m)b^0|^2\mu^2|c_{mn}^{\pm}|^2\,\mathrm{d}\mu$$

Proof. If we take b to be in $C_c^\infty((0, \infty))$ in Proposition 25, we have

$$\int |g_b|^2 \, dG = \int |b|^2 \beta \, d\mu$$

On the other hand, Lemmas 28 and 29 show that

$$\int |g_b|^2 \, dG = 2\pi^3 \int |b(\mu)|^2 \mu^2 |c^+(\mu)|^2 \, d\mu$$

Hence, as b is arbitrary,

$$\beta(\mu) = 2\pi^3 \mu^2 |c^+(\mu)|^2 \quad (\mu > 0)$$

Since both sides are continuous even functions the first statement follows immediately. For the second we note that the formula is true for $b \in C_c^\infty(\mathbb{R})$. As $\mu^2 |c_{mn}^\pm|^2$ has at most polynomial growth by Proposition 13, the formula is true for $b \in \mathscr{S}(\mathbb{R})$.

We shall now compute $c_{mn}^\pm$ explicitly.

Theorem 31. (a) *On* $|\text{Im}\, \mu| < 1, c_{mn}^\pm$ *are zero free for all* m, n, *pole free for odd* m, n, *and have* $\mu = 0$ *as a simple pole for even* (m, n) *(and no other poles)*

(b) $c_{mn}^+(\mu) = i^n \theta_*(\mu) t_n(\mu)^{-1}$, $c_{mn}^-(\mu) = i^n \theta_*(-\mu) t_m(\mu)$ *where*

$$\theta_{\text{odd}}(\mu) = \frac{-i}{\sqrt{\pi}} \frac{\Gamma(\frac{1}{2} + \frac{1}{2}(i\mu))}{\Gamma(1 - \frac{1}{2}(i\mu))}, \quad \theta_{\text{even}}(\mu) = \frac{1}{\sqrt{\pi}} \frac{\Gamma(\frac{1}{2}(i\mu))}{\Gamma(\frac{1}{2} - \frac{1}{2}(i\mu))}$$

(c) *If* $\mu \in \mathbb{R} \backslash (0)$,

$$c_{mn}^+(\mu)^{\text{conj}} = (-1)^n i^{n-m} c_{nm}^-(\mu)$$
$$|c_{mn}^\pm(\mu)|^{-2} = (\mu\pi/2) \tanh(\mu\pi/2) \quad (m, n \textit{ even})$$
$$|c_{mn}^\pm(\mu)|^{-2} = (\mu\pi/2) \coth(\mu\pi/2) \quad (m, n \textit{ odd})$$

Proof. If $\pi = \pi_{\varepsilon, i\mu}$, the formulae (Chapter 5, Proposition 31) for $\pi(X')$ and $\pi(Y')$ give (as $X'^\dagger = -Y'$),

$$f_{m,n}(\mu{:}x; X') = ((n + 1 + i\mu)/2) f_{m,n+2}(\mu{:}x)$$
$$f_{m,n}(\mu{:}X'; x) = ((m - 1 + i\mu)/2) f_{m-2,n}(\mu{:}x)$$

We now use polar coordinates and find (see Chapter 6, Section 5) that, for any function g of type (m, n) with bounded derivatives on G,

$$g(a_t; X') \sim \tfrac{1}{2} d/dt - (n/2)$$
$$g(X'; a_t) \sim \tfrac{1}{2} d/dt + (n/2)$$

where $\sim$ means neglecting differential operators whose coefficients decay exponentially when $t \to +\infty$ (for the second of these relations write $X' = H/2 + \frac{1}{2}i(2X - (X - Y))$ and note that $g(X; a_t) = e^{-2t} g(a_t; X) \sim 0$). Hence

$$\tilde{f}_{m,n+2}(\mu:a_t) \sim (n+1+\mathrm{i}\mu)^{-1}[(-n+\mathrm{i}\mu-1)c^+_{mn}(\mu)\mathrm{e}^{\mathrm{i}\mu t} + (-n-\mathrm{i}\mu-1)c^-_{mn}(\mu)\mathrm{e}^{-\mathrm{i}\mu t}]$$

$$\tilde{f}_{m-2,n}(\mu:a_t) \sim (m-1+\mathrm{i}\mu)^{-1}[(m+\mathrm{i}\mu-1)c^+_{mn}(\mu)\mathrm{e}^{\mathrm{i}\mu t} + (m-\mathrm{i}\mu-1)c^-_{mn}(\mu)\mathrm{e}^{-\mathrm{i}\mu t}]$$

From these we derive

$$c^+_{m,n+2}(\mu) = -c^+_{mn}(\mu)\frac{n+1-\mathrm{i}\mu}{n+1+\mathrm{i}\mu}, \quad c^-_{m,n+2}(\mu) = -c^-_{mn}(\mu)$$

$$c^+_{m-2,n}(\mu) = c^+_{mn}(\mu), \quad c^-_{m-2,n}(\mu) = \frac{m-1-\mathrm{i}\mu}{m-1+\mathrm{i}\mu}c^-_{mn}(\mu)$$

It follows already that $|c^{\pm}_{mn}(\mu)|$ depends only on the parity of m (or n). Let us now calculate $c_{00}(\mu)$, $c_{11}(\mu)$.

(a) $c^{\pm}_{11}(\mu)$. Take μ with $\mathrm{Im}\,\mu > 0$ *but small*. Then $\mathrm{e}^{\mathrm{i}\mu t}$ has exponential decay so that

$$\mathrm{e}^t f_{11}(\mu:a_t) \sim c^-_{11}(\mu)\mathrm{e}^{-\mathrm{i}\mu t}$$

On the other hand, for $r > 0$,

$$f_{11}(\mu:a_r) = \frac{1}{2\pi}\int \mathrm{e}^{-(\mathrm{i}\mu+1)t(a_r^{-1}u_\theta)}\chi_{-1}(a_r^{-1}[u_\theta])\chi_1(u_\theta)\,\mathrm{d}\theta$$

Now,

$$t(a_r u) = \tfrac{1}{2}\log(\mathrm{e}^{-2r}\cos^2\theta + \mathrm{e}^{2r}\sin^2\theta)$$

$$X_{-1}(a_r^{-1}[u_\theta]) = \frac{\cos\theta\cdot\mathrm{e}^{-r} - (\mathrm{i}\sin\theta)\mathrm{e}^r}{(\mathrm{e}^{-2r}\cos^2\theta + \mathrm{e}^{2r}\sin^2\theta)^{1/2}}$$

Writing

$$t(a_r^{-1}u_\theta) = r + \tfrac{1}{2}\log(\sin^2\theta + \mathrm{e}^{-2r}\cos^2\theta)$$

we have

$$f_{11}(\mu:a_r)\mathrm{e}^{(\mathrm{i}\mu+1)r} = \frac{1}{2\pi}\int g(\mu:r:u_\theta)\,\mathrm{d}\theta$$

where

$$g(\mu:r:u_\theta) = \exp(-\tfrac{1}{2}(\mathrm{i}\mu+1)\log(\sin^2\theta + \mathrm{e}^{-2r}\cos^2\theta))\chi_{-1}(a_r^{-1}[u_\theta])\chi_1(u_\theta)$$

$$|g(\mu:r:u_\theta)| \leqslant |\sin\theta|^{1-\varepsilon} \quad \varepsilon = |\mathrm{Im}\,\mu|$$

while for $r \to +\infty$

$$g(\mu:r:u_\theta) \to |\sin\theta|^{-(1+\mathrm{i}\mu)}(-\mathrm{i}\,\mathrm{sgn}(\sin\theta))\mathrm{e}^{\mathrm{i}\theta}$$

Hence, as $t \to +\infty$,

$$\mathrm{e}^{\mathrm{i}\mu+t}f_{11}(\mu:a_t) \to \frac{-\mathrm{i}}{2\pi}\int |\sin\theta|^{-(1+\mathrm{i}\mu)}\,\mathrm{sgn}(\sin\theta)\mathrm{e}^{\mathrm{i}\theta}\,\mathrm{d}\theta$$

This proves that

$$c_{11}^{-}(\mu)=\frac{-\mathrm{i}}{2\pi}\int|\sin\theta|^{-(1+\mathrm{i}\mu)}\,\mathrm{sgn}\,(\sin\theta)\mathrm{e}^{\mathrm{i}\theta}\,\mathrm{d}\theta$$

$$=\left(\frac{2}{\pi}\right)\int_{0}^{\pi/2}|\sin\theta|^{-\mathrm{i}\mu}\,\mathrm{d}\theta$$

$$=\frac{1}{\sqrt{\pi}}\frac{\Gamma(\frac{1}{2}-\frac{1}{2}(\mathrm{i}\mu))}{\Gamma(1-\frac{1}{2}(\mathrm{i}\mu))}$$

where Γ is the classical gamma function. By Theorem 24, we now obtain

$$c_{11}^{\pm}(\mu)=\frac{1}{\sqrt{\pi}}\frac{\Gamma(\frac{1}{2}\pm\frac{1}{2}(\mathrm{i}\mu))}{\Gamma(1\pm\frac{1}{2}(\mathrm{i}\mu))}$$

(b) $c_{00}^{\pm}(\mu)$. The calculation is similar. For $\mathrm{Im}\,\mu>0$ and small θ

$$c_{00}^{-}(\mu)=\frac{1}{2\pi}\int|\sin\theta|^{-(1+\mathrm{i}\mu)}\,\mathrm{d}\theta$$

Thus

$$c_{00}^{\pm}(\mu)=\frac{1}{\sqrt{\pi}}\frac{\Gamma(\pm\frac{1}{2}(\mathrm{i}\mu))}{\Gamma(\frac{1}{2}\pm\frac{1}{2}(\mathrm{i}\mu))}$$

Now we have

$$|c_{00}^{\pm}(\mu)|^{-2}=\pi\frac{\Gamma(\frac{1}{2}+\frac{1}{2}(\mathrm{i}\mu))\Gamma(\frac{1}{2}-\frac{1}{2}(\mathrm{i}\mu))}{\Gamma(\frac{1}{2}(\mathrm{i}\mu))\Gamma(-\frac{1}{2}(\mathrm{i}\mu))}=(\mu\pi/2)\tanh\mu\pi/2$$

and similarly we get

$$|c_{11}^{\pm}(\mu)|^{-2}=(\mu\pi/2)\coth\mu\pi/2$$

In particular, within $|\mathrm{Im}\,\mu|<1$, $c_{00}^{\pm}$, $c_{11}^{\pm}$, hence all of $c_{mn}^{\pm}$ are zero free; if m,n are even, $c_{mn}^{\pm}$ have a simple pole at $\mu=0$; for m,n odd, $c_{mn}^{\pm}$ is also pole free.

Finally, let us fix a parity class of m and n; then from the relations linking $c_{m,n+2}^{\pm}$ to $c_{m,n}^{\pm}$ we see that $c_{mn}^{+}(\mu)t_n(\mu)\mathrm{i}^{-n}$ and $c_{mn}^{-}(\mu)\mathrm{i}^{-n}$ are independent of n; using Theorem 24 they are respectively $\theta_m(\mu)$ and $\theta_m(-\mu)t_m(\mu)$. From the other set relations we find $c_{mn}^{-}(\mu)=\varphi_n(\mu)t_m(\mu)$, $c_{mn}^{+}(\mu)=\varphi_n(-\mu)/t_n(\mu)$. Hence

$$c_{mn}^{+}(\mu)=\mathrm{i}^{n}\theta(\mu)t_n(\mu)^{-1},\quad c_{mn}^{-}(\mu)=\mathrm{i}^{n}\theta(-\mu)t_m(\mu)$$

where θ is independent of m and n; of course it will depend on the parity class. From the definition of t_n and the formulae for $c_{11}^{\pm}$ and $c_{00}^{\pm}$, θ_{odd} and θ_{even} are seen to have the expressions indicated. This proves the theorem.

The second method of proving Theorem 23 depends on the theory of orbital integrals of Schwartz functions on G. We shall not take it up in this book.

8.7 Exact wave packets. The Plancherel measure revisited

Let us recall (Lemma 21) that multiplication by t_n/t_m is a well-defined continuous endomorphism of $\mathscr{S}(\mathbb{R})$. For any $b \in \mathscr{S}(\mathbb{R})$ we now define the *exact or normalized wave packet* associated to b by

$$f_b(x) = \int b(\mu)|c_{mn}^{+}(\mu)|^{-2} f_{mn}(\mu : x)\, \mathrm{d}\mu$$

Put

$$b_{mn}^{\pm}(\mu) = \begin{cases} \mu c_{mn}^{\pm}(\mu) & m, n \text{ even} \\ c_{mn}^{\pm}(\mu) & m, n \text{ odd} \end{cases}$$

Lemma 32. *$(b_{mn}^{\pm})^{-1}$ and $b_{mn}^{\pm}$ are holomorphic on $|\mathrm{Im}\, \mu| < 1$; and on $|\mathrm{Im}\, \mu| \leqslant 1 - \delta$ there is an integer $q \geqslant 0$ such that*

$$|(b_{mn}^{\pm})^{\pm 1}| \leqslant c_\delta[\mu, m, n]^q$$

Proof. Only the second statement needs a proof. Actually for $b_{mn}^{\pm}$ one can get the estimate from Proposition 13. But for $(b_{mn}^{\pm})^{-1}$ it appears necessary to use the explicit formulae. It is a question of proving such estimates for $(t_n^{\pm})^{-1}$, $\theta_{\mathrm{odd}}^{\pm 1}$, $(\mu \cdot \theta_{\mathrm{even}})^{\pm 1}$. This comes down to estimates of the form

$$\left|\frac{\Gamma(a + x + \mathrm{i}y)}{\Gamma(b - x - \mathrm{i}y)}\right| \leqslant c \cdot [a, y]^q$$

where $c > 0$, $q \geqslant 0$ are constants, a, b, x, y real with $|x| \leqslant \frac{1}{2} - \delta$ $(0 < \delta < \frac{1}{2})$, $a, b \geqslant \frac{1}{2}$ or $\leqslant -\frac{1}{2}$, and $|a - b| \leqslant \alpha$. We may assume $a, b \geqslant \frac{1}{2}$ by using the formula $\Gamma(z)\Gamma(-z) = \pi/(-z)\sin z\pi$. The estimate then follows from Stirling's formula.

We now define, for any $b \in \mathscr{S}(\mathbb{R})$ *the exact or normalized wave packet associated to* b to be the function f_b given by

$$f_{b:mn}(x) = f_b(x) = \int b(\mu)|c_{mn}^{+}(\mu)|^{-2} f_{mn}(\mu : x)\, \mathrm{d}\mu$$

Theorem 33. *For any $b \in \mathscr{S}(\mathbb{R})$, $f_b \in \mathscr{C}_{mn}$ and the map $b \mapsto f_b$ is continuous. More precisely, given $r \geqslant 0$, and $u, v \in U(\mathfrak{g})$ we can find a continuous seminorm $\nu_{u,v:r}$ on $\mathscr{S}(\mathbb{R})$ such that for all $m, n \in \mathbb{Z}$, $b \in \mathscr{S}(\mathbb{R})$*

$$\mu_{u,v:r}(f_{b:mn}) \leqslant [m, n]^q \nu_{u,v:r}(b)$$

Moreover we have

$$\mathscr{H} f_{b:mn} = 2\pi^3(b + (t_n/t_m)b^0)$$

$$\int |f_{b:mn}|^2 \, dG = \pi^3 \int |b + (t_n/t_m)b^0|^2 |c_{mn}^+|^{-2} \, d\mu$$

for all $b \in \mathscr{S}(\mathbb{R})$.

Proof. We consider the even and odd parities separately.
(a) m, n *even.* Then

$$|c_{mn}^+(\mu)|^{-2} = \mu^2 |b_{mn}^+(\mu)|^{-2}$$

Now, by Theorem 31,

$$|b_{mn}^+(\mu)|^{-2} = \varepsilon b_{mn}^+(\mu)^{-1} b_{nm}(\mu)^{-1}$$

where ε is a constant of absolute value 1 and hence, by Lemma 31, multiplication by $\mu |b_{mn}^+|^{-2}$ is a continuous endomorphism of $\mathscr{S}(\mathbb{R})$; and in fact, if ν is a continuous seminorm on $\mathscr{S}(\mathbb{R})$, there is a continuous seminorm ν' on $\mathscr{S}(\mathbb{R})$ so that

$$\nu(|b_{mn}^+|^{-2} \cdot \mu \cdot h) \leqslant [m, n]^q \nu'(h)$$

for all h, m, n. Clearly, if

$$b_1 = \mu |b_{mn}^+|^{-2} b \quad (b \in \mathscr{S}(\mathbb{R}))$$

then

$$f_{b:mn} = g_{b_1}$$

and we are through, by Theorems 18 and 30 and Proposition 25.
(b) m, n *odd.* Here $c_{mn}^{\pm}$ is regular at $\mu = 0$ and nonzero. The (wave packet) theorem, 18, does not apply; however, the factor μ was introduced in the proof of Theorem 18 *solely* for controlling the poles; when there are no poles, there is no need for that factor. The proof of Theorem 18 goes through without any change. The formulae for $\mathscr{H} f_b$ and $\| f_b \|^2$ can now be deduced by continuity arguments from Proposition 25 and Theorem 30. First, if $\mu_0 \neq 0$, we argue as before that $\mathscr{H} f_b(\mu_0)$ is equal to $\alpha b(\mu_0) + \beta b(-\mu_0)$ for all $b \in \mathscr{S}(\mathbb{R})$; α, β can then be found out by using $b \in C_c^\infty(\mathbb{R})$ with supports localized so that $0 \notin \operatorname{supp}(b)$. Then $b = \mu b_1, b_1 \in C_c^\infty(\mathbb{R})$, and $f_b = g_{b_2}$ where $b_2 = b_1 |c^+|^{-2}$; and Proposition 25 and Theorem 30 give $\alpha = 2\pi^3$, $\beta = 2\pi^3 t_n(\mu_0)/t_m(\mu_0)$. The final result for $\mathscr{H} f_b$ is valid even for $\mu_0 = 0$ by continuity. Moreover, the argument in Proposition 25 for calculating $(g_b | g)$ goes through without any change to give, for $b \in \mathscr{S}(\mathbb{R})$,

$$\int f_b \bar{f} \, dG = \int \overline{\mathscr{H} f(\mu)} \, b(\mu) |c_{mn}^+(\mu)|^{-2} \, d\mu$$

If we take $f = f_b$, the right side becomes

$$2\pi^3 \int b(\overline{\bar{b} + (t_n/t_m)\bar{b}^0})|c_{mn}^+|^{-2}\,\mathrm{d}\mu$$

We now check that $(b - (t_n/t_m)b^0)(\bar{b} + \overline{(t_n/t_m)\bar{b}^0})$ is odd and hence will have integral 0. Hence

$$\int |f_b|^2\,\mathrm{d}G = \pi^3 \int |b + (t_n/t_m)b^0|^2 |c_{mn}^+|^{-2}\,\mathrm{d}\mu$$

This proves the theorem.

Let me reformulate Theorem 33 so as to focus more sharply on the mutually reciprocal relationship between $\mathscr{H}$ and the wave packet map W_{mn} defined by

$$W_{mn} = W: b \mapsto \frac{1}{4\pi^3} \int b|c_{mn}^+|^{-2}\,\mathrm{d}\mu$$

on $\mathscr{S}_{mn}$. Let us recall $\mathscr{S}_{mn} = \mathscr{S}_{mn}(\mathbb{R})$, the closed subspace of all b in $\mathscr{S}(\mathbb{R})$ such that $b = (t_n/t_m)b^0$. Then $\mathscr{H}$ maps $\mathscr{C}_{mn}$ into $\mathscr{S}_{mn}$; and, by Theorem 33, $\mathscr{H}Wb = b$. Thus

$$\mathscr{H}W = \text{id} \quad (\text{on } \mathscr{S}_{mn})$$

and

$$\| Wb \|^2 = \frac{1}{4\pi^3} \| b \|_{mn}^2$$

where the left side is with respect to $L^2(\mathrm{d}G)$ and the right side with respect to $|c_{mn}^+|^{-2}\,\mathrm{d}\mu$. Let us write

$$\mathscr{C}_{mn}^\# = \{Wb \mid b \in \mathscr{S}_{mn}\}$$

Since $\mathscr{H}$ and W are continuous, it is then immediate that $\mathscr{C}_{mn}^\#$ is a *closed* subspace of $\mathscr{C}_{mn}$, and that

$$W: \mathscr{S}_{mn} \to \mathscr{C}_{mn}^\#$$

is a topological linear isomorphism that is unitary up to a multiplicative constant. Let us now define the map $E^\#$ on $\mathscr{C}_{mn}$ by

$$E^\# = W\mathscr{H}$$

Clearly $E^\#$ is a projection operator in $\mathscr{C}_{mn}$ and its range is $\mathscr{C}_{mn}^\#$. Let

$${}^0\mathscr{C}_{mn} = \text{kernel}\,(E^\#)$$

Lemma 34. *${}^0\mathscr{C}_{mn}$ is the orthogonal complement of $\mathscr{C}_{mn}^\#$ in $\mathscr{C}_{mn}$. In particular,*

$$\mathscr{C}_{mn} = {}^0\mathscr{C}_{mn} \oplus \mathscr{C}_{mn}^\#$$

and the two projections are continuous in $\mathscr{C}_{mn}$.

Proof. It is a question of showing that for any $f \in \mathscr{C}_{mn}$

$$E^{\#} f = 0 \Leftrightarrow (Wb | f) = 0 \quad \text{for all } b \in \mathscr{S}_{mn}$$

But we have seen earlier that

$$(Wb | f) = \frac{1}{4\pi^3} (b | \mathscr{H} f)_{mn}$$

where the suffix mn on the right means we take the scalar product in $L^2(|c_{mn}^+|^{-2}\, d\mu)$. As $E^{\#} f = W\mathscr{H} f$, $E^{\#} f = 0 \Leftrightarrow \mathscr{H} f = 0 \Leftrightarrow (Wb | f) = 0$ for all $b \in \mathscr{S}_{mn}$.

What is ${}^0\mathscr{C}_{mn}$? Since there is only the discrete spectrum to be taken into account it is a fair guess that ${}^0\mathscr{C}_{mn}$ is spanned by the discrete series matrix elements. For any integer $k \geqslant 1$ let $D_{\pm k}$ be the corresponding discrete series modules. The linear span of functions

$$x \mapsto (x \cdot u | v) \quad (u, v \text{ are } K\text{-finite}, \in D_{\pm k}, k \geqslant 1)$$

which are of type (m, n) will be denoted by ${}^{00}\mathscr{C}_{mn}$. Note that u (resp. v) must then transform according to $e^{in\theta}$ (resp. $e^{im\theta}$) under u. This means that

$$k + 1 \leqslant \min(m, n) (\text{for } D_k) \quad \text{or} \quad k + 1 \leqslant \min(-m, -n) (\text{for } D_{-k})$$

Hence

$$\dim({}^{00}\mathscr{C}_{mn}) < \infty$$

Lemma 35. ${}^{00}\mathscr{C}_{mn} \subset {}^0\mathscr{C}_{mn}$

Proof. It is enough to prove that ${}^{00}\mathscr{C}_{mn} \subset \mathscr{C}_{mn}$. For suppose this were done. Let $h \in \mathscr{C}_{mn}$ be a matrix element of $D_{\pm k}$. Then $\Omega \bar{h} = k^2 \bar{h}$ while $\Omega f_{mn}(\mu : \cdot) = -\mu^2 f_{mn}(\mu : \cdot)$; as $k \geqslant 1$, $k^2 \neq -\mu^2$ and hence

$$\int \bar{h} f_{mn}(\mu : \cdot)\, dG = 0$$

in view of Proposition 19. Multiplying by $b|c_{mn}^+|^{-2}$ $(b \in \mathscr{S}_{mn})$ and integrating we get by Fubini, as

$$b|c_{mn}^+|^{-2} \bar{h} f_{mn}(\mu : \cdot) \in L^1(G \times \mathbb{R})$$

$$\int \bar{h} Wb\, dG = 0$$

Thus $h \perp \mathscr{C}_{mn}^{\#}$, proving that it is in ${}^0\mathscr{C}_{mn}$.

To prove that these are in Schwartz space it is enough to consider the matrix elements $f_{\pm(k+1)}$ defined by the highest- or lowest-weight vectors (see Chapter 6, Section 5); for the others are obtained by differentiating these. We know that

$$f_{\pm(k+1)}(a_t) = (\cosh t)^{-(k+1)}$$

It is then clear that for any $r \geqslant 0$, as $t \to +\infty$

$$((\mathrm{d}^r/\mathrm{d}t^r) f_{\pm(k+1)})(a_t) = \mathrm{O}(\mathrm{e}^{-(k+1)t})$$

Arguing as we did in Theorem 17 we conclude that, for any u, $v \in U(\mathfrak{g})$,

$$f_{\pm(k+1)}(u; a_t; v) = \mathrm{O}(\mathrm{e}^{-(k+1)t})$$

It is an easy consequence now that vfu satisfies the strong inequality for all u, v. This proves the lemma. Summarizing, we have the following result.

Theorem 36. *The wave packet map $W = W_{mn}$ defined by*

$$Wb = \frac{1}{4\pi^3} \int b |c^+_{mn}|^{-2} \, \mathrm{d}\mu \quad (b \in \mathscr{S}_{mn})$$

is a linear topological isomorphism of $\mathscr{S}_{mn}$ with a closed subspace $\mathscr{C}^\#_{mn}$ of $\mathscr{C}_{mn}$; and $\mathscr{H}$, when restricted to $\mathscr{C}^\#_{mn}$, inverts W. $\mathscr{C}_{mn}$ is the algebraic direct sum of $\mathscr{C}^\#_{mn}$ and the kernel ${}^0\mathscr{C}_{mn}$ of $\mathscr{H}$, and the two projections are continuous in $\mathscr{C}_{mn}$; moreover the projection $E^\#(\mathscr{C}_{mn} \to \mathscr{C}^\#_{mn})$ is just $W\mathscr{H}$. The space ${}^0\mathscr{C}_{mn}$ contains all the matrix elements of type (m, n) of the discrete series representations (which span a finite-dimensional space). Finally $2\pi^{3/2}\, W$ is unitary, and for all $f \in \mathscr{C}_{mn}$

$$2\pi^2 \int |E^\# f|^2 \, \mathrm{d}G = \frac{1}{2} \int_0^\infty |\mathscr{H} f|^2 \gamma \, \mathrm{d}\mu$$

where

$$\gamma(\mu) = \begin{cases} \mu \tanh(\mu\pi/2) & (m, n \text{ even}) \\ \mu \coth(\mu\pi/2) & (m, n \text{ odd}) \end{cases}$$

It will be clear to the reader that this formula gives the contribution to the Plancherel formula coming from the principal series. So we should expect that ${}^{00}\mathscr{C}_{mn} = {}^0\mathscr{C}_{mn}$, and that, for any $h \in {}^0\mathscr{C}_{mn}$, $\| h \|^2$ can be computed in terms of the discrete series. We cannot get this immediately from the Plancherel formula of Chapter 6 because that result is in $C_c^\infty(G)$; and even if we start with $f \in C_c^\infty(G)$, of type (m, n), the projections $E^\# f$ and $f - E^\# f$ *will not be in* $C_c^\infty(G)$. (Why?) Hence for completeness in this sense it is necessary to extend the Plancherel formula for Chapter 6 to the Schwartz space. To do this we must be able to view characters, matrix elements, and conjugacy classes as distributions defined on $\mathscr{C}(G)$, not just on $C_c^\infty(G)$, i.e., as *tempered distributions.* I turn now to this aspect of harmonic analysis.

8.8 Tempered distributions. Temperedness of matrix elements, characters, and conjugacy classes

A distribution T on G is said to be *tempered* (according to Harish-Chandra) if it is the restriction to $C_c^\infty(G)$ of a continuous linear functional on $\mathscr{C}(G)$; since $C_c^\infty(G)$ is densely imbedded in $\mathscr{C}(G)$, there is at most one such, and we shall denote it also by T. We have already seen in Proposition 19 that if a function g satisfies the weak inequality, i.e., $g = O(\Xi(1+\sigma)^r)$ for some $r \geqslant 0$, then g defines a tempered distribution, namely the functional $f \mapsto \int gf \, dG$ $(f \in \mathscr{C}(G))$. Similarly, if p is a *nonnegative* Borel measure and

$$\int (1+\sigma)^{-r} \, dp < \infty$$

for some $r \geqslant 0$, then p is tempered; the corresponding distribution is given by

$$p(f) = \int f \, dp \quad (f \in \mathscr{C}(G))$$

Are the converses true? For the first example this is not the case even classically, although an *exponential polynomial* will define a tempered distribution if and only if the exponents are imaginary, i.e., if and only if the function has at most polynomial growth. We shall generalize this to $G = SL(2\mathbb{R})$. For the positive measures the converse is true, thus confirming that the situation is as in the classical case (where it is known as Bochner's theorem). We shall take it up first. In view of later applications we formulate it as a theorem on a *set* of positive measures.

Positive measures. We have the following.

Theorem 37. *Let P be a set of nonnegative Borel measures on G and let $B_t = \{x \mid x \in G, \sigma(x) \leqslant t\}$. Then the following are equivalent:*

(i) $\exists$ *a continuous seminorm ν on $\mathscr{C}(G)$ such that* $|\int f \, dp| \leqslant \nu(f)$ $(p \in P,$ $f \in C_c^\infty(G//K))$;

(ii) $\exists$ $C > 0$, $r \geqslant 0$ *such that* $p(B_t) \leqslant Ce^t(1+t)^r$ $(t \geqslant 0, p \in P)$;

(iii) $\exists$ $q \geqslant 0$ *such that* $\sup_{p \in P} \int_G \Xi(1+\sigma)^{-q} \, dp < \infty$.

If these are satisfied, we can choose $q \geqslant 0$ such that the integrals

$$\int_G \Xi(1+\sigma)^{-q} \, dp \quad (p \in P)$$

converge uniformly when p varies in P.

Proof. Observe that in (i) we need the estimate only for *spherical* f. The hard implication is (i)$\Rightarrow$(ii), and let us take it up first.

$(i) \Rightarrow (ii)$. Select a spherical $u \in C_c^\infty(G)$ with $u \geqslant 0$, $\sup(u) \subset B_{1/2}$, and $\int u \, dG = 1$, and put $f_t = u * 1_t * u$ where 1_t is the characteristic function of B_{t+1}. Then $f_t = 1$ on B_t and $f_t \geqslant 0$ so that

$$p(B_t) \leqslant v(f_t) \quad (t \geqslant 0, p \in P)$$

Now we assume that v is of the form

$$v(h) = \sum_{1 \leqslant j \leqslant m} \sup(\Xi^{-1}(1+\sigma)^q |a_j h b_j|)$$

where $a_i, b_i \in U(\mathfrak{g})$. For $h = f_t$ the derivatives can be pushed to act on the u's by the formula $a(u * 1_t * u)b = ub * 1_t * au$, so the $|a_j f_t b_j| \leqslant C$ everywhere and 0 outside B_{t+2}, C being a constant *independent of* $t, 1 \leqslant j \leqslant m$. So $v(f_t) \leqslant C' e^t (1+t)^q$, and we have (ii).

$(ii) \Rightarrow (iii)$ *(and the uniform convergence)*. If $x = u_{\theta_1} a_r u_{\theta_2}$ where $n < r \leqslant n+1$, $\Xi(x) \leqslant Ce^{-r}(1+r) \leqslant Ce^{-n}(n+2)$. Hence $p \in P$ and $n \geqslant 0$,

$$\int_{B_{n+1} \setminus B_n} \Xi(1+\sigma)^{-q} \, dp \leqslant Ce^{-n}(n+2)^{-(q-1)} p(B_{n+1})$$

Hence

$$\sup_p \int_{G \setminus B_n} \Xi(1+\sigma)^{-(r+3)} \, dp \leqslant \text{const.}\, n^{-1}$$

$(iii) \Rightarrow (ii)$. Take $v(f) = \sup(\Xi^{-1}(1+\sigma)^q |f|)$.

Corollary 38. *Haar measure is not tempered.*

Proof. For

$$\int_{B_t} dG = 2\pi^2 \int_0^t (\sinh 2u) \, du \geqslant \text{const.}\, e^{2t} \quad \text{for all } t \gg 0$$

Matrix elements. Let π be a representation of G in a Hilbert space $\mathscr{H}$. Assume that $\mathscr{H}$ is a Harish-Chandra G-module and u, v are K-finite vectors. Then u and v are analytic vectors and it is clear that they are $\mathfrak{Z}$-finite. Put

$$f(x) = (\pi(x)v \mid u)$$

Theorem 39. *The distribution defined by f is tempered if and only if f satisfies the weak inequality; in this case all derivatives of f satisfy the weak inequality.*

Proof. The derivatives of f are again matrix elements of the same structure. So it is enough to prove that if f is tempered it satisfies the weak inequality. I follow Harish-Chandra's argument and first establish the following lemma.

Lemma 40. *We can find $h, k \in C_c^\infty(G)$, invariant under inner automorphisms by K, such that $h*f*k=f$.*

Proof. We shall find k (even with arbitrarily small support around 1) such that $\pi(k)v = v$. The same result applied to the representation $x \mapsto \pi(x^{-1})^\dagger$ will lead to h. Given a neighbourhood V of 1 invariant under inner automorphisms by K (e.g., $\|x-1\| < \varepsilon$), let A be the algebra of $g \in C_c^\infty(V)$ with $g^{u_\theta} = g$ for all θ. There are functions $g_n \in A$ such that $\pi(g_n)v' \to v'$ as $n \to \infty$ for each $v' \in \mathscr{H}$. But v lies in a finite sum $\mathscr{H}_F = \sum_{n \in F} \mathscr{H}_n$ ($F \subset \mathbb{Z}$ a finite set) and $\dim(\mathscr{H}_F) < \infty$. As $\pi(A)$ stabilizes each $\mathscr{H}_n$, it stabilizes $\mathscr{H}_F$; and $\pi(A)\mathscr{H}_F$ is dense in $\mathscr{H}_F$. Hence $\pi(A)\mathscr{H}_F = \mathscr{H}_F$, proving $\pi(k)v = v$ for some $k \in A$.

We now come to the proof of the theorem. By assumption we can find $a_j, b_j \in U(\mathfrak{g}), q \geqslant 0$ such that, for all $g \in C_c^\infty(G)$,

$$\left| \int f(x)g(x)\,\mathrm{d}G \right| \leqslant \sum_{1 \leqslant j \leqslant m} \sup(\Xi^{-1}(1+\sigma)^q |a_j g b_j|)$$

Replace f by $h*f*k$, h, k being as in the Lemma. This is equivalent to making the change $g \mapsto h_1 * g * k_1$ so that on the right side we can replace $a_j g b_j$ by $h_1 b_j * g * a_j k_1$. By the regular nature (see Proposition 1) of the variation of Ξ and $(1+\sigma)^q$

$$\sup \Xi^{-1}(1+\sigma)^q |h_1 b_j * g * a_j k_1| \leqslant \text{const.}\, \Xi^{-1}(1+\sigma)^q |g|$$

Hence we get an estimate of the form

$$\left| \int g f\,\mathrm{d}G \right| \leqslant \text{const.} \sup(\Xi^{-1}(1+\sigma)^q |g|)$$

i.e.,

$$\left| \int f g \Xi(1+\sigma)^{-q}\,\mathrm{d}G \right| \leqslant \text{const.}\, \|g\|_\infty$$

for all $g \in C_c^\infty(G)$. This leads in an elementary way to

$$\int |f| \Xi(1+\sigma)^{-q}\,\mathrm{d}G < \infty$$

We would like to conclude from this that

$$\int_0^\infty |f(a_t)| \mathrm{e}^{-t}(1+t)^{-q}(\mathrm{e}^{2t} - \mathrm{e}^{-2t})\,\mathrm{d}t < \infty$$

If f is spherical this is immediate. If f is not spherical, we can use a trick to majorize f by a sum of spherical functions to each of which the above argument is applicable. Let L (resp. M) be the span of the K-translates of v (resp. u) and let $(\mathrm{e}_i)_{1 \leqslant i \leqslant l}$ (resp. $(f_j)_{1 \leqslant j \leqslant m}$) be an orthonormal basis of

L(resp. M) such that e_i and f_j transform according to characters of K. The matrix elements $(\pi(x)e_i|f_j) = g_{ij}(x)$ are then also tempered and hence, by the above argument,

$$\int |g_{ij}|\Xi(1+\sigma)^{-s}\,\mathrm{d}G < \infty$$

for some $s \geqslant 0$. On the other hand,

$$|f| \leqslant \text{const.} \sum |g_{ij}|$$

while each $|g_{ij}|$ is a *spherical function.* Hence

$$\int_0^\infty |f(a_t)|(\mathrm{e}^{2t} - \mathrm{e}^{-2t})\mathrm{e}^{-t}(1+t)^{-s}\,\mathrm{d}t$$

$$\leqslant \text{const.} \int_0^\infty \sum |g_{ij}(a_t)|(\mathrm{e}^{2t} - \mathrm{e}^{-2t})\mathrm{e}^{-t}(1+t)^{-s}\,\mathrm{d}t$$

$$\leqslant \text{const.} \int \sum |g_{ij}|\Xi(1+\sigma)^{-s}\,\mathrm{d}G < \infty$$

If we write $f(a_t)\mathrm{e}^t = h(t)$, we get

$$\int_0^\infty |h(t)|(1+t)^{-s}\,\mathrm{d}t < \infty$$

A similar estimate is also true for h'. Indeed, $f(x; H) = (\pi(x)\pi(H)v|u)$ is also of the same structure as f and hence by the above reasoning

$$\int_0^\infty |(\mathrm{d}/\mathrm{d}t)f(a_t)|(\mathrm{e}^{2t} - \mathrm{e}^{-2t})\mathrm{e}^{-t}(1+t)^{-s'}\,\mathrm{d}t < \infty$$

for some $s' \geqslant s$, which leads to

$$\int_0^\infty |h'(t)|(1+t)^{-s'}\,\mathrm{d}t < \infty$$

Since

$$h(t)(1+t)^{-s'} = h(1)\cdot 2^{-s'} + \int_1^t [h'(w)(1+w)^{-s'} - s'h(w)(1+w)^{-(s'+1)}]\,\mathrm{d}w$$

it is clear that

$$|h(t)(1+t)^{-s'}| \leqslant \text{const.}$$

or

$$|f(a_t)| \leqslant \text{const.}\, \Xi(a_t)(1+\sigma(a_t))^{s'} \quad (t \geqslant 0)$$

To get this estimate for all x and not merely for the a_t, we use the same trick as before and use the estimate

$$|f| \leqslant \text{const.} \sum |g_{ij}| \quad (\text{on } G)$$

For each i,j,

$$|g_{ij}(a_t)| \leqslant \text{const.}\, \Xi(a_t)(1+\sigma(a_t))^{s'}$$

for all $t \geqslant 0$. But each $|g_{ij}|$ is spherical and hence

$$g_{ij} \leqslant \text{const.}\, \Xi(1+\sigma)^{a'},$$

proving finally that f satisfies the weak inequality.

Conjugacy classes. Orbital integrals. We next consider the orbital integrals. The temperedness of these is more difficult to establish because the estimates that decide temperedness are adapted to the polar decomposition, and conjugacy classes become complicated when viewed in the polar coordinates. We begin with the hyperbolic orbital integrals.

We know (see Chapter 6, Section 2) that

$$f_{f,L}(a_t) = \tfrac{1}{4}\mathrm{e}^t \int_{\mathbb{R}} \bar{f}(a_t n_s)\,\mathrm{d}s$$

where $\bar{f}(x) = \int f(u_\theta x u_\theta^{-1})\,\mathrm{d}\theta$. It is thus a question of studying the convergence of the integrals

$$\mathrm{e}^t \int \Xi(a_t n_s)(1+\sigma(a_t n_s))^{-q}\,\mathrm{d}s$$

We now have the following.

Lemma 41. *Given $t, s \in \mathbb{R}$ with $t \geqslant 0$, let $r = r(t,s) \geqslant 0$ be defined by the relation $u_{\theta_1} a_r u_{\theta_2} = a_t n_s$. Then*

$$|r - (t + \tfrac{1}{2}\log(1+s^2))| \leqslant \tfrac{1}{2}\log 2$$

In particular, for $t \geqslant 0$, $s \in \mathbb{R}$,

$$\mathrm{e}^t \Xi(a_t n_s) \leqslant \text{const.}\,(1+s^2)^{-1/2}$$
$$(1+\sigma(a_t n_s)) \geqslant \text{const.}\,(1+t)$$
$$1+\sigma(a_t n_s) \geqslant \tfrac{1}{2}\log(1+s^2)$$

Proof. We have

$$u_{\theta_1} a_r u_{\theta_2} = \begin{pmatrix} \mathrm{e}^t & s\mathrm{e}^t \\ 0 & \mathrm{e}^{-t} \end{pmatrix}$$

Taking the Hilbert–Schmidt norm of both sides we get

$$\mathrm{e}^{2r} + \mathrm{e}^{-2r} = \mathrm{e}^{2t}(1+s^2) + \mathrm{e}^{-2t}$$

From this we have, as $t \geqslant 0$,

$$2\mathrm{e}^{2r} \geqslant \mathrm{e}^{2t}(1+s^2), \quad \mathrm{e}^{2r} \leqslant 2\mathrm{e}^{2t}(1+s^2)$$

which implies the required estimate for r. The inequalities for Ξ and σ are immediate.

Proposition 42. *If $q > 1$, the integral*

$$\int \Xi(a_t n_s)(1 + \sigma(a_t n_s))^{-q}\, \mathrm{d}s$$

converges for all real t; and for any integer $r \geqslant 0$, and all real t,

$$\int \Xi(a_t n_s)(1 + \sigma(a_t n_s))^{-(q+r)}\, \mathrm{d}s \leqslant \text{const.}\, \mathrm{e}^{-t}(1 + |t|)^{-r}$$

Proof. We have, by Lemma 41, when t is $\geqslant 0$

$$\Xi(a_t n_s)(1 + \sigma(a_t n_s))^{-(q+r)} \leqslant \text{const.}\, \mathrm{e}^{-t}(1 + t)^{-r}(1 + s^2)^{-1/2}[\log(1 + s^2)]^{-q}$$

from which the result follows. For $t < 0$ we observe (cf. the treatment of $F_{f,L}$ in Chapter 6, Section 2) that for any *spherical* continuous function $f \geqslant 0$, $t \neq 0$,

$$\frac{\pi}{2}\mathrm{e}^{t} \int f(a_t n_s)\, \mathrm{d}s = \frac{\pi}{2}\mathrm{e}^{-t} \int f(a_{-t} n_s)\, \mathrm{d}s = F_{f,L}(a_t)$$

in the sense that if one integral is finite, then all are finite and equal. This proves the proposition.

Theorem 43. *For any $q > 1$ the integral*

$$J_q(t) = \int \Xi(x a_t x^{-1})(1 + \sigma(x a_t x^{-1}))^{-q}\, \mathrm{d}(G/L) < \infty$$

t being $\neq 0$; and, for any $r \geqslant 0$, for all $t \neq 0$

$$|\mathrm{e}^{t} - \mathrm{e}^{-t}|\, J_{q+r}(t) \leqslant \text{const.}\, (1 + |t|)^{-r}$$

In particular, the invariant measures on the hyperbolic conjugacy classes are tempered. For any $f \in \mathscr{C}(G)$, $F_{f,L}(a_t)$ is well defined for all $t \neq 0$, extends to an even function in $\mathscr{S}(\mathbb{R})$ (denoted again by $F_{f,L}$), and the map $f \mapsto F_{f,L}$ is continuous from $\mathscr{C}(G)$ into $\mathscr{S}(\mathbb{R})$.

Proof. The first two assertions follow from Proposition 42. The extension of $F_{f,L}$ is defined by

$$F_{f,L}(a_t) = \tfrac{1}{4}\mathrm{e}^{t} \int \bar{f}(a_t n_s)\, \mathrm{d}s \quad (f \in \mathscr{C}(G))$$

The same proposition shows that for any integer $r \geqslant 0$

$$(\mathrm{d}^r/\mathrm{d}t^r) F_{f,L}(a_t) = \tfrac{1}{4}\mathrm{e}^{t} \int \bar{f}(b; a_t n_s)\, \mathrm{d}s$$

where $b = (H+1)^r$; this proves that

$$\sup(1+|t|)^r |F^{(r)}_{f,l}(a_t)| \leqslant \text{const.} \sup(\Xi^{-1}(1+\sigma)^{q+r} |\bar{f}b|)$$

where $q > 1$. The theorem is now clear.

Elliptic orbital integrals. These require a little more effort.

Lemma 44. *Suppose* $t \geqslant 0$, $\theta \not\equiv 0, \pi \pmod{2\pi}$, *and* $a_t u_\theta a_t^{-1} = u_{\theta_1} a_r u_{\theta_2}$ *where* $r \geqslant 0$. *Then*

$$r \leqslant \tfrac{1}{2}\log 2 + 2t, \quad -r \leqslant \tfrac{1}{2}\log 2 - 2t + \log(1/|\sin\theta|)$$

Proof. Proceeding as in Lemma 41 we have

$$2\cos^2\theta + (e^{4t} + e^{-4t})\sin^2\theta = e^{2r} + e^{-2r}$$

so that $2e^{2r} \geqslant (\sin^2\theta)e^{4t}, e^{2r} \leqslant 2e^{4t}$. The lemma is now immediate.

Proposition 45. *For any* $q > 2$, *the integral*

$$J(u_\theta) = \int \Xi(x u_\theta x^{-1})(1 + \sigma(x u_\theta x^{-1}))^{-q}\, dG$$

converges when $\theta \not\equiv 0, \pi \pmod{2\pi}$; *and there is a constant* $C = C_q > 0$ *such that, for all* θ,

$$|\sin\theta|\, J(u_\theta) \leqslant C(1 + \log(1/|\sin\theta|))^2$$

Proof. If $x = u_{\theta_1} a_t u_{\theta_2}$ and f is spherical, $f(x u_\theta x^{-1}) = f(a_t u_\theta a_t^{-1})$. Hence

$$J(u_\theta) \leqslant \text{const.} \int_0^\infty \Xi(a_t u_\theta a_t^{-1})(1 + \sigma(a_t u_\theta a_t^{-1}))^{-q} \cdot (e^{2t} - e^{-2t})\, dt$$

By the above lemma the integrand is majorized by

$$\text{const.}\, e^{-2t} \cdot (1/|\sin\theta|) \cdot (1+t) \cdot (1 + \sigma(a_t u_\theta a_t^{-1}))^{-q} \cdot (e^{2t} - e^{-2t}).$$

For σ the lemma gives the estimate

$$\sigma(a_t u_\theta a_t^{-1}) \geqslant \begin{cases} t & t \geqslant \log(1/|\sin\theta|) + \tfrac{1}{2}\log 2 \\ 0 & 0 \leqslant t \leqslant \log(1/|\sin\theta|) + \tfrac{1}{2}\log 2 \end{cases}$$

Hence, for all θ as above,

$$J(u_\theta) \leqslant \text{const.} \frac{1}{|\sin\theta|} \int_0^\infty (1+t)^{-(q-1)}\, dt + \text{const.} \frac{1}{|\sin\theta|}(1 + \log(1/|\sin\theta|))^2$$

which proves the proposition.

Write $C^\infty_{\text{bdd}}(B')$ for the space of all $h \in C^\infty(B')$ such that

$$\mu_m(h) = \sup_{B'} |(d/d\theta)^m h(u_\theta)| < \infty$$

for every integer $m \geqslant 0$. It is a Fréchet space with the seminorms μ_m.

Theorem 46. *The invariant measures on the regular elliptic conjugacy classes are tempered. For* $f \in \mathscr{C}(G)$,

$$F_{f,B}(u_\theta) = (\mathrm{e}^{\mathrm{i}\theta} - \mathrm{e}^{-\mathrm{i}\theta}) \int f(xu_\theta x^{-1})\, \mathrm{d}(G/B)$$

is well defined on B' *and lies in* $C^\infty_{\mathrm{bdd}}(B')$; *and the map* $f \mapsto F_{f,B}(u_\theta)$ *from* $\mathscr{C}(G)$ *to* $C^\infty_{\mathrm{bdd}}(B')$ *is continuous.*

Proof. The previous proposition gives the temperedness of the invariant measures and so $F_{f,B}$ is well defined on B'. We shall now study $F_{f,B} = F_f$, $f \in C_c^\infty(G)$, *but use only continuous seminorms on* $\mathscr{C}(G)$ *in estimating it.*

We begin by observing that

$$f \mapsto F'_f(1),\ f \mapsto F_f(1\pm),\ f \mapsto F_f(\gamma \pm)$$

are all tempered distributions. For the first one this is obvious, for it is a multiple of the Dirac measure at the origin by the limit formula. For $F_f(1\pm)$ we recall that

$$F_f(1\pm) = \pm \text{const.} \int_0^\infty \bar{f}(n_{\pm s})\, \mathrm{d}s$$

and the temperedness is immediate from Proposition 42. For $F_f(\gamma\pm)$ we translate by γ and come down to the previous case.

This said, we fix a $q > 2$, put $\mu(f) = \sup(\Xi^{-1}(1+\sigma)^q |f|)$, and use Proposition 45 to make the following obvious estimate:

$$|F_f(u_\theta)| \leqslant C\mu(f)(1 + \log(1/|\sin\theta|))^2$$

But we know that

$$-(\mathrm{d}^2/\mathrm{d}\theta^2)F_f = F_{\Omega f}$$

Hence

$$|F''_f(u_\theta)| \leqslant \nu_1(f)(1 + \log(1/|\sin\theta))^2$$

for all $f \in C_c^\infty(G)$; here $\nu_1(f) = C\mu(\Omega f)$ and so ν_1 is a continuous seminorm on $\mathscr{C}(G)$. The function $(1 + \log(1/|\sin\theta|))^2$ is *integrable* on B and so

$$|F'_f(u_\theta)| \leqslant F'_f(1)| + \text{const.}\ \nu_1(f)$$

for all $f \in C_c^\infty(G)$. In view of our remark above we thus obtain

$$|F'_f(u_\theta)| \leqslant \nu_2(f) \quad (f \in C_c^\infty(G))$$

for all $u \in B'$, ν_2 being a continuous seminorm on $\mathscr{C}(G)$. Integrating once again and using the temperedness of $f \mapsto F_f(1\pm)$ we get

$$|F_f(u_\theta)| \leqslant \nu_3(f) \quad (f \in C_c^\infty(G))$$

for all $u_\theta \in B'$, being a continuous seminorm on $\mathscr{C}(G)$. Using the relations $F_f^{(2k)} = (-1)^k F_{\Omega^k f}$ we thus obtain that

$$|F_f^{(k)}(u_\theta)| \leqslant \nu_k(f) \quad (f \in C_c^\infty(G))$$

for all $u_\theta \in B'$. The map $f \mapsto F_f$ is thus continuous from $C_c^\infty(G)$ to $C_{\text{bdd}}^\infty(B')$, *where* $C_c^\infty(G)$ *is given the topology inherited from* $\mathscr{C}(G)$. As $C_c^\infty(G)$ is dense in $\mathscr{C}(G)$ we thus have a continuous linear map L from $\mathscr{C}(G)$ to $C_{\text{bdd}}^\infty(B')$ extending it. It is clear that

$$(Lf)(u_\theta) = F_{f,B}(u_\theta) \quad (u_\theta \in B')$$

We now have the following immediate corollary.

Theorem 47. *The limit formula and the jump relations are valid for* $f \in \mathscr{C}(G)$;

$$f \mapsto F_f(1 \pm), \quad f \mapsto F_f(\gamma \pm)$$

are tempered distributions.

Characters. The consequences of the foregoing analysis for character theory are swift and pleasant.

Theorem 48. *The characters* $T_\chi = T_{\varepsilon, i\mu}$ $(\mu \in \mathbb{R})$ *are tempered and*

$$T_\chi(f) = \hat{F}_f(\chi) \quad (f \in \mathscr{C}(G))$$

Proof. We know this for $f \in C_c^\infty(G)$; the general case is then clear from Theorem 43.

Recall the function D which is invariant and analytic on G, and coincides with Δ_B^2 on B and Δ_L^2 on L. We know that $|D|^{-1/2}$ is locally integrable on G.

Theorem 49. *The measure* $|D|^{-1/2}\,\mathrm{d}G$ *is tempered; in particular there is an integer* $q \geqslant 0$ *such that*

$$\int |D|^{-1/2} \Xi (1+\sigma)^{-q}\,\mathrm{d}G < \infty$$

If Θ *is any invariant function such that*

$$|\Theta| \leqslant \text{const.}\, |D|^{-1/2} \quad (\text{on } G')$$

then the distribution defined by Θ *is tempered, and*

$$\Theta(f) = \int \Theta(x) f(x)\,\mathrm{d}x \quad (f \in \mathscr{C}(G))$$

the integral being absolutely convergent. This is in particular the case for the characters Θ_m $(m \neq 0)$, $\Theta_{0,\pm}$.

Proof. In view of Theorem 36 it is enough to check that the distribution defined by $|D|^{-1/2}$ is tempered. By the integration formula, we have, for any $f \in C_c^\infty(G)$,

$$\left|\int |D|^{-1/2} f \, dG\right| \leqslant \text{const.}\left(\int |F_{f,B}| d\theta + \int |F_{f,L}| dt\right)$$

and Theorem 4.6 proves what we want. The remaining assertions now follow trivially.

It is natural to call an irreducible representation in a Hilbert space *tempered* if its character is a tempered distribution on G.

Theorem 50. *The matrix elements defined by the K-finite vectors of a tempered irreducible representation satisfy the weak inequality. The only tempered irreducible representations of $G = SL(2, \mathbb{R})$ are the $\pi_{\varepsilon, i\mu}$ $(\mu \neq 0)$, $\pi_{0,0}$, D_k $(k \neq 0)$, $D_{0,\pm}$.*

Proof. Suppose π is an irreducible representation of G in a Hilbert space and (e_n) is an orthonormal basis such that $\pi(u_\theta)e_n = e^{ik_n\theta}e_n$. If

$$g_n(x) = (\pi(x)e_n \,|\, e_n)$$

then, for the character Θ of π we have

$$\Theta(f) = \sum_n \int f g_n \, dG \quad (f \in C_c^\infty(G))$$

The characters of K of course occur with multiplicity 1 and so the orthogonal projection on the span of e_n is the operator

$$E_n = \frac{1}{2\pi}\int \pi(u_\theta) e^{-ik_n\theta} \, d\theta$$

So,

$$\int f g_n \, dG = \text{tr}(\pi(f)E_n) = \text{tr}\, \pi(f_n)$$

where

$$f_n(x) = \frac{1}{2\pi}\int f(xu_\theta) e^{ik_n\theta} \, d\theta$$

In other words,

$$\int f g_n \, dG = \Theta(f_n)$$

and, as the map $f \mapsto f_n$ is a continuous endomorphism of $\mathscr{C}(G)$, it is clear that the temperedness of Θ implies that of g_n. As π is irreducible, all its matrix elements are obtained from any nonzero g_n by differentiation. Hence they are all tempered and must satisfy the weak inequality by Theorem 39. For the second assertion let π be a tempered irreducible representation and Θ its character. Then $\Omega\Theta = \mu^2\Theta$ for some $\mu \in \mathbb{C}$. We first observe that the proof of Lemma 30 in Chapter 6 goes through with only minor changes to imply that the function Φ_L associated to Θ on L' must be bounded. We now use the results of the discussion after Theorem 3 of Chapter 7. Note that $\Theta = \pm\Theta_\gamma$. If $\mu \notin \mathbb{Z}$, then Θ is a multiple of $T_{0,\mu}$ or $T_{1,\mu}$, and Φ_L cannot be bounded unless $\mu \in i\mathbb{R}$. We then use the Harish-Chandra subquotient theorem to infer that $\pi \cong \pi_{\varepsilon,\mu'}$. If $\mu = m \in \mathbb{Z}\backslash(0)$, Θ is a linear combination of $\Theta_{\pm m}, F_m, T_{\varepsilon,m}$ (ε has the parity of m); the boundedness of Φ_L excludes F_m and $T_{\varepsilon,m}$; and so, by the subquotient theorem, $\pi \cong D_{\pm m}$. Finally, if $\mu = 0$, $\pi \cong \pi_{1,0\pm}$, similarly.

Remark. It is very interesting that *a tempered irreducible representation is already unitary*. This is a theorem true in complete generality and was proved by Harish-Chandra. It is also interesting that *the trivial representation is not tempered.*

8.9 Plancherel formula in Schwartz space

We can now prove that the Plancherel formula is valid as an identity in the space of tempered invariant distributions.

Theorem 51 (Plancherel formula on $\mathscr{C}(G)$). *We have, for all $f \in \mathscr{C}(G)$,*

$$2\pi^2 f(1) = \sum_{m\neq 0} |m|\Theta_m(f) + \frac{1}{2}\int_0^\infty \mu \tanh(\mu\pi/2) T_{0,i\mu}(f)\,\mathrm{d}\mu$$
$$+ \frac{1}{2}\int_0^\infty \mu \coth(\mu\pi/2)\, T_{1,i\mu}(f)\,\mathrm{d}\mu$$

where all the characters are calculated with respect to d*G, and the series converge absolutely.*

Proof. It is a question of proving that the right side defines a tempered distribution. We shall actually prove that if

$$J(f) = \sum_{m\neq 0} |m||\Theta_m(f)| + \frac{1}{2}\int_0^\infty \mu \tanh(\mu\pi/2)|\,T_{0,i\mu}(f)|\,\mathrm{d}\mu$$
$$+ \frac{1}{2}\int_0^\infty \mu \coth(\mu\pi/2)|\,T_{1,i\mu}(f)|\,\mathrm{d}\mu$$

then J is a continuous seminorm on $\mathscr{C}(G)$. The estimates needed follow from the differential equations

$$\Omega\Theta_m = m^2\Theta_m, \quad \Omega T_{\varepsilon,i\mu} = -\mu^2 T_{\varepsilon,i\mu}$$

First we have, for all m, μ, ε, and all $f \in \mathscr{C}(G)$,

$$|T_{\varepsilon,i\mu}(f)| = |\hat{F}_{f,L}(\varepsilon, i\mu)| \leqslant \|F_{f,L}\|_1 = \nu_1(f)$$

$$|\Theta_m(f)| \leqslant \int |D|^{-1/2}|f|\,\mathrm{d}G = \nu_2(f)$$

where ν_1 and ν_2 are continuous seminorms on $\mathscr{C}(G)$. Then

$$\begin{aligned}|T_{\varepsilon,i\mu}(f)| &\leqslant (1+\mu^2)^{-2}|T_{\varepsilon,i\mu}((1-\Omega)^2 f)|\\ &\leqslant (1+\mu^2)^{-2}\nu_1((1-\Omega)^2 f)\\ |\Theta_m(f)| &\leqslant m^{-4}\nu_2(\Omega^2 f)\\ \mu\tanh(\mu\pi/2) + \mu\coth(\mu\pi/2) &\leqslant c\mu \quad (\mu > 0)\end{aligned}$$

where $c > 0$ is a constant. Hence

$$J(f) \leqslant \nu_2(\Omega^2 f)\cdot \sum_{m\neq 0} |m|^{-3} + c\nu_1((1-\Omega)^2 f)\cdot \int_0^\infty \mu(1+\mu^2)^{-2}\,\mathrm{d}\mu$$

proving that J is a continuous seminorm on $\mathscr{C}(G)$. Since the formula has been proved for $C_c^\infty(G)$, it is now true for $\mathscr{C}(G)$ by continuity.

8.10 Completeness theorems

Once the Plancherel formula has been obtained on $\mathscr{C}(G)$, the completeness theorem in the context of the Harish-Chandra transform is immediate. We use the following simple lemma.

Lemma 52. *Let f be an element of $\mathscr{C}_{mn}$, $f^0(x) = f(x^{-1})$, and let π be any irreducible unitary representation whose character Θ is tempered. Then $\Theta(\bar{f} * f^0)$ is zero unless χ_m, χ_n occur in $\pi|_K$; if e_m, e_n are unit vectors transforming according to χ_m and χ_n respectively and $g_{mn}(x) = (\pi(x)e_n|e_m)$,*

$$\Theta(\bar{f} * f^0) = \left|\int f\bar{g}_{mn}\,\mathrm{d}G\right|^2$$

Proof. We know from Theorem 50 that g_{mn} must satisfy the weak inequality. Hence both sides of the above relations are continuous on $\mathscr{C}(G)$ and so we need only establish it when $f \in C_c^\infty(G)$. Now $\pi(f^0) = \pi(\bar{f})^\dagger$ and so, if e_k, e_l are unit vectors transforming according to χ_k and χ_l under K and $(\pi(\bar{f})e_l|e_k) = a_{kl}$, then $(\pi(f^0)e_l|e_k) = \bar{a}_{lk}$. It is obvious that $a_{kl} = 0$ unless $k = m$, $l = n$. Thus, if χ_m and χ_n do not occur in $\pi|_K$, $\pi(\bar{f}) = 0$ and so $\Theta(\bar{f} * f^0) = \mathrm{tr}(\pi(\bar{f})\pi(f^0)) = 0$. If they do occur, then the matrix of $\pi(\bar{f})$ has the sole nonzero entry a_{mn} and hence $\Theta(\bar{f} * f^0) = |a_{mn}|^2$. It is obvious that $a_{mn} = \int \bar{f} g_{mn}\,\mathrm{d}G$.

Theorem 53. *For any fixed* (m, n), ${}^{00}\mathscr{C}_{mn} = {}^{0}\mathscr{C}_{mn}$. *More precisely,* $\mathscr{C}^{\#}_{mn}$ *is of finite codimension in* $\mathscr{C}_{mn}$, *and its orthogonal complement in* $\mathscr{C}_{mn}$ *is the span of the discrete-series matrix elements of type* (m, n). *Moreover,* $\mathscr{C}_{mn} = {}^{0}\mathscr{C}_{mn} + \mathscr{C}^{\#}_{mn}$, *the sum being direct and smooth in the topology of* $\mathscr{C}$. *In particular, for* $m, n = 0, \pm 1$, $\mathscr{C}_{mn} = \mathscr{C}^{\#}_{mn}$.

Proof. We compute $\Theta(\bar{f} * f^0)$ for $f \in \mathscr{C}_{mn}$, $\Theta = \Theta_m$ or $T_{\varepsilon,i\mu}$, using Lemma 50. We find $T_{\varepsilon,i\mu}(\bar{f} * f^0) = |(\mathscr{H} f)(\mu)|^2$ if ε has the parity of m, n, and zero otherwise. Similarly, $\Theta_k(\bar{f} * f^0) = |(f \mid \phi_{kmn})|^2$, where ϕ_{kmn} is the matrix element of D_k defined by e_m and e_n, and the scalar product is in $L^2(\mathrm{d}G)$. Hence the Plancherel formula applied to $\bar{f} * f^0$ (Theorem 51) gives

$$2\pi^2 \int |f|^2 \, \mathrm{d}G = \sum_{k \neq 0} |k| |(f \mid \phi_{kmn})|^2 + \frac{1}{2} \int_0^\infty |\mathscr{H} f|^2 b_\varepsilon \, \mathrm{d}\mu$$

where $b_\varepsilon = \mu \tanh(\mu\pi/2)$ or $\mu \coth(\mu\pi/2)$ according as $\varepsilon = 0$ or 1. Now, the orthogonality relations for the discrete series representations and the formula for the formal degree of D_k (Theorem 27, Chapter 6) show that $\{(2\pi^2)^{-1/2} |k|^{1/2} \phi_{kmn}\}$ is an orthonormal family. Hence if ${}^{00}f$ is the orthogonal projection of f on ${}^{00}\mathscr{C}_{mn}$

$$\sum |k| |(f \mid \phi_{kmn})|^2 = (2\pi^2) \| {}^{00}f \|^2$$

On the other hand, by Theorem 35,

$$\frac{1}{2} \int_0^\infty |\mathscr{H} f|^2 b_\varepsilon \, \mathrm{d}\mu = (2\pi^2) \| f^{\#} \|^2$$

where $f^{\#}$ is the projection of f on $\mathscr{C}^{\#}_{mn}$. Hence

$$\| f \|^2 = \| {}^{00}f \|^2 + \| f^{\#} \|^2$$

showing that $f = {}^{00}f + f^{\#}$. The statements of Theorem 53 are now clear except for the last. For that it is enough to notice that the characters $\chi_0, \chi_{\pm 1}$ never occur in $D_k, k \neq 0$.

8.11 The Harish-Chandra transform on $\mathscr{C}(G)$

The restrictions to functions of a fixed type (m, n) in the definition of the Harish-Chandra transform is somewhat unsatisfactory. In order to understand the structure of the *whole* of $\mathscr{C}(G)$ it is necessary to consider all the types at once, and this can be done only with the help of the complete Fourier transform. All the machinery for doing this is already in place, and so we can obtain the main results with almost no extra effort. The only additional estimates that we need are the decay properties of the discrete series matrix elements *where the types as well as the parameter k are allowed to vary.*

Estimates for the matrix elements of discrete series. For any integer $k \neq 0$ and $m, n \geqslant k+1$ (for $k>0$) or $\leqslant k-1$ (for $k<0$), we write ϕ_{kmn} for the matrix element

$$\phi_{kmn}(x) = (\pi_k(x)e_n | e_m)$$

where π_k is the discrete series representation corresponding to k and (e_r) is an orthonormal basis in the space of π_k; $\pi_k(u_\theta)e_r = e^{ir\theta}e_r$. Given k, m, n, this determines ϕ_{kmn} uniquely if $m=n$, and in all cases up to a constant factor of absolute value 1.

Theorem 54. *Fix $a, b \in U(\mathfrak{g}), l \geqslant 1$. Then there are an integer $q \geqslant 0$ and a constant $c>0$ such that, for all $x \in G, m, n, k$ with $|k| \geqslant l$,*

$$|\phi_{kmn}(b; x; a)| \leqslant c[k, m, n]^q e^{-(l+1)\sigma(x)}$$

In particular, for all k, m, n,

$$|\phi_{kmn}(b; x; a)| \leqslant c[k, m, n]^q \Xi(x)^2$$

We will prove this by the method of differential equations. Note that all these derivatives are matrix elements of π_k and hence themselves derivatives of the matrix elements corresponding to the highest- (or lowest-) weight vector of the module D_k. Thus it is not difficult to show using the formula for this special matrix element (cf. Chapter 6, Section 5) that the above estimate is valid for *fixed* k, m, n; the point of the theorem is that we can control the growth in k, m, n to be at most polynomial. We work with $k>0$; the case $k<0$ is similar.

We need a lemma.

Lemma 55. *For any fixed $a \in U(\mathfrak{g})$ of degree $\leqslant r$, the entries of the matrix of $\pi_k(a)$ are majorized by*

$$c \cdot [k, m, n]^r$$

and they are zero unless $|m-n| \leqslant 2r$.

Proof. Since this property is preserved under multiplication it is enough to check it for $a = H', X', Y', b = 1$. For H' it is obvious; for X' and Y', if we write

$$\pi_k(X')e_m = c(k, m)e_{m+2}, \quad \pi_k(Y')e_n = d(k, n)e_{n-2}$$

the relation $\pi_k(X')^\dagger = -\pi_k(Y')$ shows that $c(k, m) = -\bar{d}(k, m+2)$; on the other hand, $\pi_k(Y')\pi_k(X')e_m = c(k, m)d(k, m+2)e_m$, and by Proposition 31 of Chapter 5 we have $c(k, m)d(k, m+2) = (k^2 - (m+1)^2)/4$. (Warning: the e_n of that proposition are not the *normalized* e_n we are working with here.) The lemma is now clear.

Since $a\ \phi_{kmn}b$ are the entries of the matrix $\pi(b)(\phi_{kmn})\pi_k(a)$, this lemma shows that it is enough to prove the estimate of the theorem for just the ϕ_{kmn}. For this purpose we go over to

$$\Phi_{kmn}(t) = \begin{pmatrix} \tilde{\phi}_{kmn}(t) \\ \tilde{\phi}'_{kmn}(t) \end{pmatrix}$$

where

$$\tilde{\phi}_{kmn}(t) = (2\sinh 2t)^{1/2}\,\phi_{kmn}(a_t)$$

We have

$$(\mathrm{d}/\mathrm{d}t)\Phi_{kmn} = M_k\Phi_{kmn} + P_{mn}\Phi_{kmn}$$

where

$$M_k = \begin{pmatrix} 1 & 1 \\ k^2 & 0 \end{pmatrix} \quad P_{mn} = \begin{pmatrix} 0 & 0 \\ q_{mn} & 0 \end{pmatrix}$$

and

$$|q_{mn}(t)| \leqslant c[m, n]^2 \mathrm{e}^{-2t} \quad (t \geqslant 1)$$

Our method of analyzing this equation is almost the same as before. The eigenvalues of M_k are $\pm k$ and the associated spectral projections are

$$E_k^{\pm} = \begin{pmatrix} \frac{1}{2} & \pm 1/2k \\ \pm k/2 & \frac{1}{2} \end{pmatrix} \quad \| E_k^{\pm} \| \leqslant c[k]$$

Since the eigenvalues are *real*, the solutions of the equation $\mathrm{d}W/\mathrm{d}t = M_k W$ are *not* bounded and so greater care is now required. The initial estimates become very crucial and we formulate them as follows.

Lemma 56(a). *For any $a, b \in U(\mathfrak{g})$,*

$$\sup_x |\phi_{kmn}(a; x; b)| \leqslant c[k, m, n]^{\deg(a) + \deg(b)}$$

where $c = c(a, b) > 0$ is a constant.

(b) *For some constant $c > 0$, we have for all k, m, n*

$$\| \Phi_{kmn}(t) \| \leqslant c[k, m, n]\mathrm{e}^t \quad (t \geqslant 1)$$

(c) *For fixed k, m, n,*

$$\| \Phi_{kmn}(t) \| \leqslant \mathrm{const}.\mathrm{e}^{-(|k|+1)t} \quad (t \geqslant 1)$$

Proof. (a) As before we may assume $a = b = 1$. But then $|\phi_{kmn}(x)| \leqslant 1$.

(b) Since $(\mathrm{d}/\mathrm{d}t)\phi_{kmn}(a_t) = (\pi_k(a_t)\pi_k(H)e_n | e_m)$ it is clear that

$$|(\mathrm{d}/\mathrm{d}t)\phi_{kmn}(a_t)| \leqslant c[k, m, n] \quad (t \geqslant 1)$$

The required estimate for Φ_{kmn} is immediate.

(c) For fixed k, m, n and $a, b \in U(\mathfrak{g})$, $a\phi^{kmn}b = \psi$ is itself a matrix element of π_k, and hence of the form $a_1 g b_1$ where g is the matrix element defined by the lowest weight vector of D_k, and $a_1, b_1 \in U(\mathfrak{g})$. By the results of

Chapter 6, Section 5,

$$|g(a_t)| \leqslant \text{const.}\, e^{-(k+1)t} \quad (t \geqslant 1)$$

and so

$$|g| \leqslant \text{const.}\, \Xi^{k+1}$$

On the other hand, $g = h_1 * g * h_2$ for suitable $h_1, h_2 \in C_c^\infty(G)$ (Lemma 38) and so

$$|a_1 g b_1| = |h_1 b_1 * g * a_1 h_2| \leqslant \text{const.}\, \Xi^{k+1}$$

This gives (c).

Proof of Theorem 54. We write

$$E_k^\pm \Phi_{kmn} = \Phi_{kmn}^\pm, \quad \Psi_{kmn}^\pm = E_k^\pm P_{mn} \Phi_{kmn}$$

Then

$$(d/dt)\Phi_{kmn}^\pm = \pm k \Phi_{kmn}^\pm + \Psi_{kmn}^\pm$$

The initial estimates now are ($t \geqslant 1$)

$$\| \Phi_{kmn}^\pm(t) \| \leqslant c[k, m, n]^2 e^t \tag{A}$$

$$\| \Psi_{kmn}^\pm(t) \| \leqslant c[k, m, n]^2 \| \Phi_{kmn}(t) \| e^{-2t} \tag{B}$$

$$\| \Phi_{kmn}(t) \| \leqslant c_{kmn} e^{-(k+1)t} \tag{C}$$

We write the differential equation as

$$(d/dt)(e^{\mp kt} \Phi_{kmn}^\pm(t)) = e^{\mp kt} \Psi_{k:mn}^\pm(t)$$

So

$$\Phi_{kmn}^+(t) = -e^{kt} \int_t^\infty e^{-ku} \Psi_{kmn}^+(u)\, du$$

the integral being convergent in view of (C) and (B). Hence, using (A) and (B),

$$\begin{aligned} \| \Phi_{kmn}^+(t) \| &\leqslant c[k, m, n]^4 e^{kt} \int_t^\infty e^{-(k+1)u}\, du \\ &\leqslant c[k, m, n]^4 e^{-t} \end{aligned}$$

On the other hand, for any $t_0 \geqslant 1$,

$$\Phi_{kmn}^-(t) = e^{-kt + kt_0} \Phi_{kmn}^-(t_0) + e^{-kt} \int_{t_0}^t e^{ku} \Psi_{kmn}^-(u)\, du$$

Using (A) and (B) again,

$$\begin{aligned} \| \Phi_{kmn}^-(t) \| &\leqslant c[k, m, n]^2 e^{-kt} + c[k, m, n]^4 e^{-kt} \int_{t_0}^t e^{(k-1)u} du \\ &\leqslant c[k, m, n]^4 e^{-t} \end{aligned}$$

So we get

$$\|\Phi_{kmn}^{\pm}(t)\| \leqslant c[k,m,n]^4 \mathrm{e}^{-t} \tag{A1}$$

which is an improvement of (A) since the exponential factor has gone down from e^t to e^{-t}. We now use (A_1), (B) and (C) and proceed as before. We shall suppose $k \geqslant l$ where l is a fixed integer $\geqslant 1$. Then we can use this iterative *bootstrap* method of estimation $[l/2]+1$ times to find

$$\|\Phi_{kmn}^{\pm}(t)\| \leqslant c[k,m,n]^q \mathrm{e}^{-lt}$$

Then

$$|\phi_{kmn}(x)| \leqslant c[k,m,n]^q \mathrm{e}^{-(l+1)\sigma(x)}$$

for $\sigma(x) \geqslant 1$, and we are through if we use (a) of Lemma 54 for the compact set $\sigma(x) \leqslant 1$.

Remark. It is remarkable that as $|k|$ keeps increasing the ϕ_{kmn} decay faster and faster. This is true in complete generality.

The Harish-Chandra transform on $\mathscr{C}(G)$. We are now ready to introduce the Harish-Chandra transform without any type restrictions. Let $\mathscr{A}$ be the set of all complex infinite matrices

$$a = (a_{mn})_{m,n \in \mathbb{Z}}$$

such that

(i) $a_{mn} = 0$ if m, n are of opposite parity,
(ii) a_{mn} is rapidly decreasing in m, n, i.e., for each $q \geqslant 0$, $|a_{mn}| \leqslant$ const. $[m,n]^{-q}$.

It is easy to check that $\mathscr{A}$ is an algebra; the series encountered in multiplying two elements of $\mathscr{A}$ are absolutely convergent. We also introduce the 'dual' algebra of complex matrices

$$a^* = (a^*_{mn})_{m,n \in \mathbb{Z}}$$

such that

(i) $a^*_{mn} = 0$ if m and n have opposite parity,
(ii) $|a^*_{mn}| \leqslant c[m,n]^q$ (for some q),
(iii) $a^*_{mn} = 0$ unless $|m-n| \leqslant p$ (for some p),

where p and q are integers independent of m and n but depend on a^* in general. It is again easy to verify that $\mathscr{A}^*$ operates on $\mathscr{A}$ by multiplication:

$$\mathscr{A}^*\mathscr{A} \subset \mathscr{A}, \quad \mathscr{A}\mathscr{A}^* \subset \mathscr{A}$$

We view $\mathscr{A}$ as a Fréchet algebra with respect to the seminorms

$$|a|_p = \sup_{m,n} |a_{mn}|[m,n]^p$$

Then $\mathscr{A}^*$ operates continuously on $\mathscr{A}$.

The Schwartz space

$$\mathscr{S}(\mathbb{R}:\mathscr{A})$$

is the space of $\mathscr{A}$-valued Schwartz functions on $\mathbb{R}$, i.e., the space of matrices

$$a = (a_{mn})_{m,n\in\mathbb{Z}}$$

where the $a_{mn} \in \mathscr{S}(\mathbb{R})$ are such that for any continuous seminorm μ on $\mathscr{S}(\mathbb{R})$ the matrix

$$\mu(a) = (\mu(a_{mn})) \in \mathscr{A}$$

i.e., for any $p \geqslant 0$

$$\mu_p(a) = \sup_{m,n} \mu(a_{mn})[m, n]^p < \infty$$

It is clear that $\mathscr{S}(\mathbb{R}:\mathscr{A})$ is a Fréchet algebra, the multiplication being *pointwise* on $\mathbb{R}$. Similarly we speak of the Schwartz space $\mathscr{S}(\mathbb{Z}':\mathscr{A})$, where $\mathbb{Z}' = \mathbb{Z}\backslash(0)$; it is the space of all functions

$$a : \mathbb{Z}' \to \mathscr{A}$$

such that

$$\sup_{m,n,k} |a_{mn}(k)| [m, n, k]^q < \infty$$

for every $q > 0$. It is also a Fréchet algebra. We may introduce also the algebra $\mathscr{T}(\mathbb{R}:\mathscr{A}^*)$ of moderately growing smooth functions on $\mathbb{R}$ with values in $\mathscr{A}^*$ i.e., matrices

$$a^* = (a^*_{mn})_{m,n\in\mathbb{Z}}, \quad a^*_{mn} \in C^\infty(\mathbb{R})$$

such that

(i) $a^*_{mn} = 0$ if m and n have opposite parity,
(ii) $a^*_{mn} = 0$ unless $|m - n| \leqslant p = p(a^*)$,
(iii) for any integer $r \geqslant 0$, $\exists q \geqslant 0 \ni$

$$|(\mathrm{d}^r/\mathrm{d}\mu^r) a^*_{mn}(\mu)| \leqslant c[m, n, \mu]^q$$

for all $\mu \in \mathbb{R}$, $m, n \in \mathbb{Z}$.

It is clear that $\mathscr{T}(\mathbb{R}:\mathscr{A}^*)$ operates continuously on $\mathscr{S}(\mathbb{R}:\mathscr{A})$.

Suppose $f \in \mathscr{C}(G)$. We define its *Harish-Chandra transform* $\mathscr{H} f$ to be the pair

$$\mathscr{H} f = (\mathscr{H}_B f, \mathscr{H}_L f)$$

where $\mathscr{H}_B f$ and $\mathscr{H}_L f$ are defined as follows. $\mathscr{H}_B f$ is an (infinite) matrix-valued function on $\mathbb{Z}'$; its value at $k \in \mathbb{Z}'$ is the matrix whose mnth entry is

$$\int f\bar{\phi}_{kmn}\,\mathrm{d}G$$

for all m, n with $\phi_{kmn} \neq 0$, and 0 otherwise. $\mathscr{H}_L f$ is the (infinite) matrix-valued function whose value at $\mu \in \mathbb{R}$ is

$$\int f\,\overline{f_{mn}(\mu:x)}\,\mathrm{d}G$$

***Theorem* 57.** *For any* $f \in \mathscr{C}(G)$, $\mathscr{H}_B f$ *lies in* $\mathscr{S}(\mathbb{Z}':\mathscr{A})$ *and* $\mathscr{H}_L f$ *lies in* $\mathscr{S}(\mathbb{R}:\mathscr{A})$. $\mathscr{H}_B$ *and* $\mathscr{H}_L$ *are continuous homomorphisms,* $\mathscr{C}(G)$ *being viewed as a convolution algebra. If* t_n *are defined as in Lemma 21 and* $U = \operatorname{diag}(\ldots, t_n, \ldots)$, *then* U *and* U^{-1} *are in* $\mathscr{T}(\mathbb{R}:\mathscr{A}^*)$, $U^{-1} = U^0(U^0(\mu) = U(-\mu))$, *and*

$$U\mathscr{H}_L f U^{-1} = (\mathscr{H}_L f)^0.$$

Proof. It is useful to introduce the projections

$$E_{mn}:\mathscr{C} \to \mathscr{C}_{mn}, \quad E_{mn}f = f_{mn}$$

where

$$f_{mn}(x) = \int\int f(u_{\theta_1}xu_{\theta_2})\mathrm{e}^{-\mathrm{i}(m\theta_1 + n\theta_2)}\frac{\mathrm{d}\theta_1}{2\pi}\frac{\mathrm{d}\theta_2}{2\pi}$$

It is easy to check that, if v is a continuous seminorm on $\mathscr{C}$, there is a continuous seminorm $\bar{v}$ on $\mathscr{C}$ such that

$$v(f_{mn}) \leqslant \bar{v}(f) \quad (f \in \mathscr{C}, m, n \in \mathbb{Z})$$

This said, we take up proof. From Theorem 54 we know that

$$|\phi_{kmn}(x)| \leqslant c[k, m, n]^q \Xi(x)^2$$

and so

$$|(\mathscr{H}_B f)(k:m:n)| \leqslant c[k, m, n]^q v(f)$$

where $v(f) = \int |f| \Xi^2\,\mathrm{d}G$. On the other hand, if

$$\Box h = \Omega\Omega_K h \Omega_K \quad (h \in \mathscr{C}, \Omega_K = 1 - (X - Y)^2)$$

we have

$$\Box^r \phi_{kmn} = [k^2(1 + m^2)(1 + n^2)]^r \phi_{kmn}$$

Hence

$$|(\mathscr{H}_B f)(k:m:n)| \leqslant c[k, m, n]^{q-2r} v(\Box^r f)$$

This implies $\mathscr{H}_B f \in \mathscr{S}(\mathbb{Z}':\mathscr{A})$. For $\mathscr{H}_L$ we use the obvious relation

$$(\mathscr{H}_L f)_{mn} = \mathscr{H} f_{mn}$$

By Theorem 20, if ν is a continuous seminorm on $\mathscr{S}(\mathbb{R})$, we can find a continuous seminorm ν' on $\mathscr{C}$ such that

$$\nu((\mathscr{H}_L f)_{mn}) \leqslant \nu'(f_{mn})$$

for all f, m, n, and hence

$$\nu((\mathscr{H}_L f)_{mn}) \leqslant \bar{\nu}'(f)$$

for all f, m, n. If

$$\square_1 h = \Omega_K h \Omega_K$$

we have, as usual,

$$\mathscr{H}_L \square_1^r f = (1+m^2)^r (1+n^2)^r \mathscr{H}_L f$$

and so

$$\nu((\mathscr{H}_L f)_{mn}) \leqslant c[m,n]^{-2r} \bar{\nu}'(\square_1^r f)$$

It is thus seen that $\mathscr{H}_L f \in \mathscr{S}(\mathbb{R}:\mathscr{A})$. Further, Lemma 21 and Proposition 22 show that $U \in \mathscr{T}(\mathbb{R}:\mathscr{A}^*)$ and Theorem 23 shows that $\mathscr{H}_L f$ satisfies the functional equation

$$(\mathscr{H}_L f)^0 = U(\mathscr{H}_L f) U^{-1}$$

It only remains to check that

$$\mathscr{H}(f * g) = \mathscr{H} f \cdot \mathscr{H} g.$$

It is enough to do this for $f, g \in C_c(G)$. But then $(\mathscr{H}_L f)(\mu)$(resp. $(\mathscr{H}_B f)(k)$) is the complex conjugate matrix of the operator $\pi_{i\mu}(\bar{f})$(resp. $\pi_k(\bar{f})$) and the required relation follows from the relation $\pi(a*b) = \pi(a)\pi(b)$ valid for any representation (warning: for $f \in \mathscr{C}$ and π tempered, $\pi(f)$ may not be a bounded operator in the space of π).

The obvious thing to do now is to invert $\mathscr{H}$. This requires the introduction of the wave packet map, as well as the projection on the discrete part. As a consequence of all the work on estimates that has been done so far we can work over $\mathscr{C}(G)$ and not just at the L^2-level, and obtain really beautiful results.

First let us make some remarks on convergence of series in a Fréchet space V. A family $(f_j)(j \in J$, a denumerable set) in V defines a *series* $\sum_j f_j$; it is said to be *absolutely convergent* if, for *any* continuous seminorm μ on V, $\sum \mu(f_j) < \infty$. It is then convergent in the usual sense; the sum is independent of the order of summation. We shall use this for families $f_{mn} \in \mathscr{C}_{mn}$ such that, for any continuous seminorm ν on $\mathscr{C}$, $\nu(f_{mn}) \leqslant$ const. $[m,n]^{-q}$ for every $q \geqslant 0$. Then $\sum_{m,n} f_{mn}$ is absolutely convergent, and, if f is its sum, $E_{mn} f = f_{mn}$ for all f. For any $h \in \mathscr{C}(G)$,

$$E_{mn} \Omega_K^r h \Omega_K^r = (1+m^2)^r (1+n^2)^r E_{mn} h$$

from which it is easy to see that

$$\sum_{m,n} E_{mn} h$$

is absolutely convergent to h.

Theorem 58. *For any $b \in \mathscr{S}(\mathbb{Z}':\mathscr{A})$, the series*

$$(2\pi^2)^{-1} \sum |k| b(k:m:n) \phi_{kmn} = {}^0Wb$$

converges absolutely in $\mathscr{C}$. If ${}^0\mathscr{F}$ is the set of all $b \in \mathscr{S}(\mathbb{Z}':\mathscr{A})$ such that $b(k:m:n) = 0$ unless m, n k are such that $\phi_{kmn} \neq 0$, then ${}^0\mathscr{F}$ is a closed subalgebra and ${}^0W$ is a linear topological isomorphism of ${}^0\mathscr{F}$ with a closed subspace ${}^0\mathscr{C}$ of $\mathscr{C}$; and its inverse is the restriction of $\mathscr{H}_B$ to ${}^0\mathscr{C}$. Moreover ${}^0W\mathscr{H}_B = {}^0E$ is the restriction to $\mathscr{C}$ of the orthogonal projection from $L^2(\mathrm{d}G)$ to its discrete part ${}^0L^2(\mathrm{d}G)$,

$$\mathscr{H}_B {}^0Wb = b$$

$$\int |{}^0Wb|^2 \,\mathrm{d}G = \frac{1}{2\pi^2} \sum_{k,m,n} |k| \, |b(k:m:n)|^2$$

In particular, ${}^0\mathscr{C} = {}^0L^2(\mathrm{d}G) \cap \mathscr{C}$.

Proof. It is immediate from Theorem 54 that ${}^0Wb$ is well defined, the series defining it is absolutely convergent, and $b \mapsto {}^0Wb$ is continuous. It is also clear that ${}^0\mathscr{F}$ is a closed subalgebra of $\mathscr{S}(\mathbb{Z}':\mathscr{A})$ and that $\mathscr{H}_B\mathscr{C} \subset {}^0\mathscr{F}$. Further we see from Theorem 27 of Chapter 6 that $\mathscr{H}_B {}^0Wb = b$ and $\int |{}^0Wb|^2 \,\mathrm{d}G$ equals

$$\frac{1}{2\pi^2} \sum_{k,m,n} |k| \, |b(k:m:n)|^2$$

At this stage we know that ${}^0\mathscr{C} = {}^0W {}^0\mathscr{F}$ is closed in $\mathscr{C}$ and that ${}^0W({}^0\mathscr{F} \to {}^0\mathscr{C})$ is a linear topological isomorphism with inverse $\mathscr{H}_B$. If now $f \in \mathscr{C}(G)$, its orthogonal projection on the discrete part of $L^2(\mathrm{d}G)$ is precisely

$$\frac{1}{2\pi^2} \sum |k| (f \,|\, \phi_{kmn}) \phi_{kmn}$$

the series being convergent in $L^2(\mathrm{d}G)$. This is just

$$\frac{1}{2\pi^2} \sum |k| (\mathscr{H}_B f)(k:m:n) \phi_{kmn}$$

which is convergent in $\mathscr{C}$ to ${}^0W\mathscr{H}_B f$. This proves everything.

For any $b \in \mathscr{S}(\mathbb{R}:\mathscr{A}), b = (b_{mn})$, let

$$W^\# b_{mn} = \frac{1}{4\pi^2} \int b_{mn} |c_{mn}^+|^{-2} f_{mn}(\mu:\cdot) \,\mathrm{d}\mu$$

We have the following.

Theorem 59. *Let $\mathscr{F}^{\#}$ be the subspace of all $b \in \mathscr{S}(\mathbb{R}:\mathscr{A})$ such that $b^0 = UbU^{-1}$. Then $\mathscr{F}^{\#}$ is a closed subalgebra of $\mathscr{S}(\mathbb{R}:\mathscr{A})$, and $b \mapsto \frac{1}{2}(b + (UbU^{-1})^0)$ is a projection of $\mathscr{S}(\mathscr{R}:\mathscr{A})$ onto $\mathscr{F}^{\#}$. For any $b \in \mathscr{F}^{\#}$ the series $\sum_{m,n} W^{\#} b_{mn}$ is absolutely convergent in $\mathscr{C}$ to an element $W^{\#}b$, and the map $W^{\#}(b \mapsto W^{\#}b)$ is a linear topological isomorphism of $\mathscr{F}^{\#}$ with a closed subspace $\mathscr{C}^{\#}$ of $\mathscr{C}$, with $\mathscr{H}_L$ (on $\mathscr{C}^{\#}$) as its inverse. Moreover, $W^{\#}\mathscr{H}_L = E^{\#}$ is the orthogonal projection $\mathscr{C} \to \mathscr{C}^{\#}$, and*

$$\mathscr{H}_L W^{\#} b = b$$

$$\int |W^{\#}b|^2 \, dG = \frac{1}{2\pi^2} \sum_{m,n\,\text{even}} \frac{1}{2} \int_0^{\infty} |b_{mn}|^2 \mu \tanh(\mu\pi/2) \, d\mu$$

$$+ \frac{1}{2\pi^2} \sum_{m,n\,\text{odd}} \frac{1}{2} \int_0^{\infty} |b_{mn}|^2 \mu \coth(\mu\pi/2) \, d\mu$$

Proof. By Theorem 33, if v is any continuous seminorm on $\mathscr{C}$, we can find a continuous seminorm v' on $\mathscr{S}(\mathbb{R})$ and an integer $q \geqslant 0$ such that $\forall m, n$

$$v(W^{\#} b_{mn}) \leqslant [m, n]^q v'(b_{mn})$$

for all. By the definition of $\mathscr{S}(\mathbb{R}:\mathscr{A})$ we now have $\sum_{m,n} v(W^{\#} b_{mn}) < \infty$. Moreover, if $W^{\#}b = \sum W^{\#} b_{mn}$, we have

$$v(W^{\#}b) \leqslant \sum_{m,n} [m, n]^q v'(b_{mn})$$

and the right side is a continuous seminorm on $\mathscr{S}(\mathbb{R}:\mathscr{A})$. Since the $\mathscr{C}_{mn}$ are mutually orthogonal, everything else follows from the theory of the Harish-Chandra transform on the $\mathscr{C}_{mn}$. The relation $\mathscr{H}_L W^{\#} b = b$ shows that $\mathscr{C}^{\#}$ is closed and $W^{\#}(\mathscr{F}^{\#} \to \mathscr{C}^{\#})$ is a linear topological isomorphism with $\mathscr{H}_L$ as its inverse.

Theorem 60. *The subspaces ${}^0\mathscr{C}$, $\mathscr{C}^{\#}$ are closed in $\mathscr{C}$; we have*

$$\mathscr{C} = {}^0\mathscr{C} + \mathscr{C}^{\#}$$

where the sum is orthogonal and the two projections are continuous on $\mathscr{C}$. The Harish-Chandra transform on $\mathscr{C}$ is an isomorphism of topological algebras of $\mathscr{C}$ with ${}^0\mathscr{F} \oplus \mathscr{F}^{\#}$, and its inverse is ${}^0W \oplus W^{\#}$. In particular, ${}^0\mathscr{C}$ and $\mathscr{C}^{\#}$ are closed two-sided ideals in $\mathscr{C}$ and are respectively the kernels of $\mathscr{H}_L$ and $\mathscr{H}_B$.

Proof. The orthogonality of ${}^0\mathscr{C}$ and $\mathscr{C}^{\#}$, as well as the fact that for any $f \in \mathscr{C}$ one has $f = {}^0\!f + f^{\#}$, follows by Theorem 53 for $f \in \mathscr{C}_{mn}$, and in the general case by continuity. The rest is trivial.

If one operates at a Hilbert-space level, there is just one thing to do: prove that the matrix elements of the discrete series and the wave packets of the unitary principal series span a *dense subspace of* $L^2(G)$. The theory we have developed has led to the *unexpected* conclusion that there is, *even at the level of Schwartz space, a splitting into the 'discrete' part and the 'continuous' part.* As mentioned at the outset, for $SL(2, \mathbb{R})$ such results already go back to Ehrenpreis and Mautner. I have followed the method that I had worked out with *R.* Ranga Rao in 1968 in Princeton. It is one of the wondrous achievements of Harish-Chandra that he was able to establish the splitting of the various series at the Schwartz-space level for all connected semisimple groups with finite centre.

8.12 Coda

Let me begin with Harish-Chandra's beautiful characterization of ${}^0\mathscr{C}$. For a finite group F, if $F_1 \subset F$ is a subgroup, a function f on F is orthogonal to all the matrix coefficients of the representations of F induced by the trivial representation of F_1 if and only if

$$\sum_{n \in F_1} f(xny) = 0$$

for all $x, y \in F$. If one extends this formally to G and takes F_1 to be N, then the condition that $\pi(f) = 0$ for all representations of the unitary principal series becomes

$$\int_{\mathbb{R}} f(xn_s y)\, \mathrm{d}s = 0 \quad (x, y \in G)$$

In the theory of automorphic forms such a condition defines a *cusp form.* So Harish-Chandra called functions in $\mathscr{C}(G)$ *cusp forms* on G. Observe that in view of Proposition 42 the measure $\mathrm{d}N$ is tempered, i.e., the map

$$f \mapsto \int f(n_s)\, \mathrm{d}s$$

is a well-defined continuous linear functional on $\mathscr{C}(G)$. The space of cusp forms is thus a well defined closed subspace of $\mathscr{C}$.

Theorem 61. *${}^0\mathscr{C}$ and $\mathscr{C}^{\#}$ are translation invariant; ${}^0\mathscr{C}$ is precisely the space of cusp forms.*

Note that f is a cusp form if and only if all the $E_{mn} f = f_{mn}$ are cusp forms. We need a lemma.

Lemma 62. *If $f \in \mathscr{C}_{mn}$, then*

$$\int f \overline{f_{mn}(\mu)}\, \mathrm{d}G = \pi \hat{g}_f(-\mathrm{i}\mu)$$

where ^ is Fourier transform and

$$g_f(a_t) = \mathrm{e}^t \int f(a_t n_s)\,\mathrm{d}s$$

Proof. It is enough to prove this for $f \in C_c^\infty(G)$; for, in view of Proposition 42, the map $f \mapsto g_f$ is well defined and continuous from $\mathscr{C}$ to $\mathscr{S}(\mathbb{R})$. If f has compact support,

$$\int f\overline{f_{mn}(\mu:\cdot)}\,\mathrm{d}G = \iint f(x)\mathrm{e}^{(\mathrm{i}\mu-1)t(x^{-1}u_\theta)}\chi_n(x^{-1}[u_\theta])\chi_{-m}(u_\theta)\,\mathrm{d}G\,\mathrm{d}\theta$$

Substituting $x^{-1}u = y$ this becomes $(k(y) = u_\varphi)$

$$\begin{aligned}&\iint f(u_\theta y^{-1})\mathrm{e}^{(\mathrm{i}\mu-1)t(y)}\mathrm{e}^{\mathrm{i}n\varphi}\mathrm{e}^{-\mathrm{i}m\theta}\,\mathrm{d}G\,\mathrm{d}\theta\\ &= \frac{1}{2}\iint f(u_\theta n_{-s}a_{-t}u_\varphi^{-1})\mathrm{e}^{(\mathrm{i}\mu+1)t}\mathrm{e}^{\mathrm{i}n\varphi-\mathrm{i}m\theta}\,\mathrm{d}t\,\mathrm{d}s\,\mathrm{d}\theta\,\mathrm{d}\varphi\\ &= \pi\iint f(a_t n_s)\mathrm{e}^{-\mathrm{i}\mu t}\mathrm{e}^t\,\mathrm{d}t\,\mathrm{d}s\\ &= \pi\hat{g}_f(-\mathrm{i}\mu).\end{aligned}$$

Proof of Theorem 61. Since ${}^0\mathscr{C} = \mathscr{C} \cap {}^0L^2(G)$, translation invariance of ${}^0\mathscr{C}$ is immediate. Since $\mathscr{C}^\#$ is the orthogonal complement of ${}^0\mathscr{C}$ in $\mathscr{C}$, its translation invariance is also clear. If $f \in {}^0\mathscr{C}$, then $f_{mn} \in {}^0\mathscr{C}$ for all m, n, so that $g_{f_{mn}} = 0$ for all m, n, by the above lemma. Hence $g_f = 0$, and in particular

$$\int f(n_s)\,\mathrm{d}s = 0$$

Changing f to the function $z \mapsto f(xzy)$ it follows that f is a cusp form. Conversely if f is a cusp form, so is f_{mn} and hence $g_{f_{mn}} = 0$. Lemma 60 shows that $\int f\overline{f_{mn}(\mu:\cdot)}\,\mathrm{d}G = 0$. Thus $\mathscr{H}_L f = 0$ or $f \in {}^0\mathscr{C}$.

By evaluating the Harish-Chandra transform at arbitrary points we obtain continuous representations of $\mathscr{C}$ into $\mathscr{A}$. This procedure allows us to recover the representations $\pi_{\varepsilon,\mathrm{i}\mu}$, π_k; it moreover reveals the splitting of $T_{1,0}$.

Theorem 63. *If* $\mu \neq 0$ *(resp.* $k \neq 0$*), the map* $f \mapsto (\mathscr{H}_L f)(\mu)$ *(resp.* $f \mapsto (\mathscr{H}_B f)(k)$*) is a continuous homomorphism of* $\mathscr{C}$ *onto* $\mathscr{A}$. If $\mu = 0$ *it maps* $\mathscr{C}$ *onto the subalgebra centralized by* $U(0)$ *where* $U(0) = (\delta_{mn}u_n)$*, and the* u_n *are* *given by*

$$u_n = \begin{cases} 1 & n \geqslant 0 \\ (-1)^n & n < 0 \end{cases}$$

Remark. The matrices in $\mathscr{F}^{\#}$ are already of the form

$$\begin{array}{c} \\ \text{odd} \\ \text{even} \end{array} \begin{array}{c} \begin{array}{cc} \text{odd} & \text{even} \end{array} \\ \left(\begin{array}{c|c} * & 0 \\ \hline 0 & * \end{array}\right) \end{array}$$

The functional equation $b^0 = UbU^{-1}$ then becomes (at $\mu = 0$)

$$b(0) = U(0)b(0)U(0)^{-1}$$

so that $b(0)$ has the structure

$$\begin{array}{c} \\ \text{odd} \begin{array}{c} >0 \\ <0 \end{array} \\ \text{even} \end{array} \begin{array}{c} \begin{array}{cc} \text{odd} & \text{even} \end{array} \\ \left(\begin{array}{c|c} \begin{array}{c|c} * & 0 \\ \hline 0 & * \end{array} & 0 \\ \hline 0 & * \end{array}\right) \end{array}$$

The irreducibility of $T_{0,0}$ and splitting of $T_{1,0}$ will be clear if we prove that the evaluation map is *surjective* onto the subalgebra in question.

Proof. Let $\mu \neq 0$ and $\beta \in \mathscr{A}$. We may assume $\mu > 0$. Choose $\alpha \in C_c^\infty((0, \infty))$ such that $\alpha(\mu) = 1$, and let $b'_{mn} = \beta_{mn}\alpha$. Then $b' \in \mathscr{S}(\mathbb{R}:\mathscr{A})$ and $b = b' + (Ub'U^{-1})^0$ is an element of $\mathscr{F}^{\#}$ with $b(\mu) = \beta$. The argument is the same for $k \neq 0$. If $\mu = 0$ and $\beta \in \mathscr{A}$ is such that $U(0)\beta U(0)^{-1} = \beta$, then the same construction yields $b \in \mathscr{F}^{\#}$ with $b(0) = \beta$; we take $\alpha \in C_c^\infty(\mathbb{R})$ with $\alpha(0) = \frac{1}{2}$.

Let π be any tempered unitary irreducible representation, (e_n) an orthonormal basis of vectors transforming according to characters χ_{k_n} of K and let $\mathscr{M}$ be the algebra of matrices (a_{mn}) such that

$$\sum (1 + k_n^2)^r (1 + k_m^2)^r |a_{mn}| < \infty$$

for all $r \geqslant 0$. If $f \in \mathscr{C}(G)$, as f is not in general in $L(G)$, the operator $\pi(f)$ is not well defined as a bounded operator on the Hilbert space of π. If $f_{mn}(x) = (\pi(x)e_n | e_m)$ the f_{mn} satisfy the weak inequality and so $\int f f_{mn} \, dG$ exists for all m, n. We may thus associate to f the matrix $\pi(f) = (\int f f_{mn} \, dG)_{m,n}$ which lies in $\mathscr{M}$. Since π is one of π_k or $\pi_{\varepsilon, i\mu}$, Theorem 63 leads to the following.

Theorem 64. *π maps $\mathscr{C}(G)$ onto $\mathscr{M}$.*

The reader should view this as a strong generalization of the classical fact that, if π is an irreducible representation of a finite group, it maps the group algebra *onto* the full endomorphism algebra of the space of π.

The isomorphism

$$\mathscr{H}:\mathscr{C} \cong {}^0\mathscr{F} \oplus \mathscr{F}^\#$$

leads to an identification of their topological duals. Since the duals of ${}^0\mathscr{F}$ and $\mathscr{F}^\#$ are explicitly computable, one obtains a very explicit description of the *Harish-Chandra transform* $\mathscr{H}T = \hat{T}$ of any tempered distribution on G: if T is a tempered distribution on G, $\mathscr{H}T$ is the element of $({}^0\mathscr{F})^* \oplus (\mathscr{F}^\#)^*$ such that

$$T(f) = (\mathscr{H}T)(\mathscr{H}f)$$

The Plancherel formula may then be interpreted as the determination of $\mathscr{H}\delta$, δ being the Dirac delta measure at the identity element of the group. We shall not go into this in any detail here.

The result that the transform of a Schwartz function on G is a *matrix valued* Schwartz function shows that transform theory in general is difficult. However, there is a part of it which reduces to *scalar functions*, namely the so-called *invariant transform theory*. It is especially suited for studying *tempered invariant distributions*. It was introduced by Harish-Chandra [H12]. Given any $f \in \mathscr{C}(G)$ we define its *invariant Harish-Chandra transform* $\hat{f}$ as the pair of functions

$$\hat{f} = (\hat{f}_B, \hat{f}_L)$$

where (with $\hat{B} \cong \mathbb{Z}$ and $\hat{L} \cong \{0, 1\} \times i\mathbb{R}$)

(i) $\hat{f}_B$ and $\hat{f}_L$ are respectively defined on $\hat{B}$ and $\hat{L}$,
(ii) $\hat{f}_B(m) = \Theta_m(f)(m \neq 0)$, $\hat{f}_B(0) = \Theta_{0,+}(f) - \Theta_{0,-}(f)$, $\hat{f}_L(\varepsilon, i\mu) = T_{\varepsilon, i\mu}(f)$.

Clearly $\hat{f}$ is obtained by taking the trace of $\mathscr{H}f$; indeed,

$$\hat{f}_L(\varepsilon, i\mu) = \mathrm{tr}\,(\mathscr{H}_L f)(\varepsilon, i\mu) \quad \hat{f}_B(m) = \mathrm{tr}\,(\mathscr{H}_B f)(m) \quad (m \neq 0)$$

However, $\mathscr{H}_B f$ is defined only on $\mathbb{Z}'$ while the definition of $\hat{f}_B$ specifies its value at the trivial character 0 of B. The point is that when the *full* transform $\mathscr{H}f$ is given one is also given at $\mu = 0$ the two submatrices corresponding to the splitting of $T_{1,0}$; but, on taking traces, this information is lost, and we obtain only $\Theta_{0,+}(f) + \Theta_{0,-}(f)$. The definition of $\hat{f}_B$ at 0 is designed to recover $\Theta_{0,\pm}(f)$ from a knowledge of $\hat{f}_B$ and $\hat{f}_L$.

Let $\mathscr{S}(\hat{B})$ and $\mathscr{S}(\hat{L})$ be the Schwartz spaces of $\hat{B}$ and $\hat{L}$; $\mathscr{S}(\hat{B})$ is the space of rapidly decreasing sequences on $\hat{B} \cong \mathbb{Z}$ while $\mathscr{S}(\hat{L}) \cong \mathscr{S}(\mathbb{R}) \oplus \mathscr{S}(\mathbb{R})$. Given $(a, b) \in {}^0\mathscr{F} \oplus \mathscr{F}^\#$ we define $\mathrm{tr}\,(a, b) = (\alpha, \beta)$ as follows. Let $P^\varepsilon(\varepsilon = 0, 1)$ be the projection matrices corresponding to the partition of $\mathbb{Z}$ into integers of parity ε, $\varepsilon = 0, 1$; let $P^1_\pm$ be the projections corresponding to odd integers >0 and odd integers <0. Then

$$\alpha(m) = \mathrm{tr}\,(a(m)) \quad (m \neq 0), \quad \alpha(0) = \mathrm{tr}\,(b(0)P^1_+) - \mathrm{tr}\,(b(0)P^1_-)$$
$$\beta(\varepsilon{:}\mu) = \mathrm{tr}\,(b(\mu)P^\varepsilon)$$

Since $b^0 == UbU^{-1}$, it is clear that $\beta^0 = \beta$, i.e., β is symmetric, i.e., $\beta(\varepsilon, \mu) = \beta(\varepsilon, -\mu)$.

Theorem 65. *The map* tr *is a continuous open surjection of* ${}^0\mathscr{F} \oplus \mathscr{F}^\#$ *onto* $\mathscr{S}(\hat{B}) \oplus \mathscr{S}(\hat{L})^{\mathrm{symm}}$ *where* symm *denotes the subalgebra of symmetric functions. Its kernel is the closure of the span of the commutators in* ${}^0\mathscr{F} \oplus \mathscr{F}^\#$. *Moreover*

$$\mathrm{tr}\,(\mathscr{H} f) = \hat{f} \quad (f \in \mathscr{C}(G))$$

Proof. It is enough to prove the surjectivity and the statement on the kernel; the continuity of tr is obvious, its openness is immediate from the open mapping theorem for Fréchet spaces, and the last relation is immediate from the definition of tr and $\hat{f}$.

Surjectivity. Suppose $\alpha \in \mathscr{S}(\hat{B})$, $\beta \in \mathscr{S}(\hat{L})^{\mathrm{symm}}$. We define $a \in \mathscr{S}(\mathbb{Z}'{:}\mathscr{A})$, $b \in \mathscr{S}(\mathbb{R}{:}\mathscr{A})$ as follows. Let e_{mn} be the matrix units; then

$$a(\pm m) = \alpha(\pm m)e_{\pm(m+1), \pm(m+1)} \quad (m \in \mathbb{Z}, m > 0)$$
$$b(\mu) = \beta(0{:}\mu)e_{0,0} + \tfrac{1}{2}(u(\mu) + \beta(1{:}\mu))e_{1,1} + \tfrac{1}{2}(-u(\mu) + \beta(1{:}\mu))e_{-1,-1}$$

where $u \in \mathscr{S}(\mathbb{R})^{\mathrm{symm}}$ is such that $u(0) = \alpha(0)$. Since b is diagonal and symmetric, we have $b = UbU^{-1} = b^0$, i.e., $b \in \mathscr{F}^\#$. That $a \in {}^0\mathscr{F}$ is obvious.

Determination of the kernel. Let $\mathscr{D}$ be the closure of the linear span of the commutators in the algebra ${}^0\mathscr{F} \oplus \mathscr{F}^\#$. It is clear that tr is 0 on $\mathscr{D}$. We shall now prove the converse. Suppose $(a, b) \in {}^0\mathscr{F} \oplus \mathscr{F}^\#$ and we are given that $\mathrm{tr}\,(a, b)$ is zero. We wish to show that $(a, b) \in \mathscr{D}$. We may argue separately for a and b.

We start with a. Define $\gamma_n \in \mathscr{S}(\mathbb{Z}')$ by

$$\gamma_n = \sum_{j \prec n} a_{jj}$$

where $j \prec n$ means $j \leqslant n$ and has the same parity as n. Then, as $\sum_{j \equiv n(2)} a_{jj} = 0$, we have

$$\gamma_n = -\sum_{j \succ n+2} a_{jj}$$

We may therefore conclude that, for any continuous seminorm ν on $\mathscr{S}(\mathbb{Z})$, $\nu(\gamma_n) \to 0$ rapidly in n as $n \to \pm\infty$. Hence

$$\sum_n \gamma_n(e_{nn} - e_{n+2,n+2})$$

is an absolutely convergent series in $\mathscr{S}(\mathbb{Z}':\mathscr{A})$, with sum equal to

$$\sum_n (\gamma_n - \gamma_{n-2})e_{n,n} = \sum_n a_{nn}e_{n,n}$$

In other words,

$$a = \sum_n \gamma_n(e_{n,n} - e_{n+2,n+2}) + \sum_{m \neq n} a_{mn}e_{mn}$$

both series being absolutely convergent. The argument for a will be complete if we can show that, for *fixed* m, n, both $\gamma_n(e_{nn} - e_{n+2,n+2})$ and $a_{mn}e_{mn}$ are in $\mathscr{D}$. To see this for the off-diagonal term, let $\alpha_{mn}(k) = \beta_{mn}(k) = (a_{mn}(k))^{1/2}$ (some choice of square root for each k). Then α_{mn}, β_{mn} are in $\mathscr{S}(\mathbb{Z}')$ and

$$[\alpha_{mn}e_{mm}, \quad \beta_{mm}e_{mn}] = a_{mn}e_{mn}$$

Moreover $\alpha_{mn}(k)e_{mm}$ and $\beta_{mn}(k)e_{mn}$ are zero unless $a_{mn}(k) \neq 0$, so that $\alpha_{mn}e_{mm}$ and $\beta_{mn}e_{mn}$ are both in ${}^0\mathscr{F}$. For the diagonal term, write $\gamma_n'(k) = \gamma_n(k)^{1/2}$; then $\gamma_n' \in \mathscr{S}(\mathbb{Z}')$ and

$$\gamma_n(e_{n,n} - e_{n+2,n+2}) = [\gamma_n' e_{n,n+2}, \gamma_n' e_{n+2,n}]$$

If $k > 0$ and $n < k+1$, $a_{jj}(k) = 0$ for $j \prec n$, showing $\gamma_n(k) = 0$, thus $\gamma_n'(k) = 0$; if $k < 0$ and $n + 2 > k - 1$, $a_{jj}(k) = 0$ for $j \succ n + 2$, so that $\gamma_n(k) = 0$ once again (using the alternative representation of γ_n), showing $\gamma_n'(k) = 0$. This proves that $\gamma_n' e_{n,n+2}$ and $\gamma_n' e_{n+2,n}$ are both in ${}^0\mathscr{F}$. Thus, $\gamma_n(e_{n,n} - e_{n+2,n+2}) \in \mathscr{D}$.

We now take up $b \in \mathscr{F}^\#$ with

$$\sum_{n \equiv \varepsilon} b_{nn} = 0 \quad (\varepsilon = 0, 1)$$

$$\sum_{n \text{ odd} \geqslant 1} b_{nn}(0) = \sum_{\substack{n \text{ odd}, \\ \leqslant -1}} b_{nn}(0) = 0$$

The symmetry condition on b is

$$b^0_{mn} = (t_m/t_n)b_{mn}$$

As in the previous case write

$$\gamma_n = \sum_{j \prec n} b_{jj} = -\sum_{j \succ n+2} b_{jj}$$

Then $\gamma_n \in \mathscr{S}(\mathbb{R})^{\text{symm}}$, $\nu(\gamma_n) \to 0$ rapidly as $n \to \pm\infty$ for any continuous seminorm ν on $\mathscr{S}(\mathbb{R})$, so that

$$b = \sum_n \gamma_n(e_{n,n} - e_{n+2,n+2}) + \sum_{m \neq n} b_{mn}e_{mn},$$

the series converging absolutely in $\mathscr{S}(\mathbb{R}:\mathscr{A})$. Once again we are reduced to proving that the individual terms of the series are in $\mathscr{D}$. In doing this we shall use the following easily proved result: we can find a sequence

$(h_r)_{r\geqslant 1}$ of elements of $\mathscr{S}(\mathbb{R})^{\mathrm{symm}}$ such that, for any $f\in\mathscr{S}(\mathbb{R}), h_r f\to f$ in $\mathscr{S}(\mathbb{R})$. We discuss the various cases. Fix $(h_r)_{r\geqslant 1}$ as above.
(a) ge_{mn}, *where* $m\neq n, g^0=(t_m/t_n)g$. Here $h_r^0=h_r$ and $(h_r g)^0=(t_m/t_n)h_r g$, so that $h_r g e_{mn}\in\mathscr{F}^{\#}$ and $h_r e_{mm}\in\mathscr{F}^{\#}$. Moreover,

$$ge_{mn}=\lim_r (h_r g)e_{mn}=\lim_r [h_r e_{mm}, ge_{mn}]\in\mathscr{D}$$

(b) $\gamma\cdot(e_{n,n}-e_{n+2,n+2})$, *where* $\gamma\in\mathscr{S}(\mathbb{R})^{\mathrm{symm}}, n\neq -1$. We define α_r, β_r by

$$\alpha_r(\mu)=h_r(\mu)(n+1+\mathrm{i}\mu),\quad \beta_r(\mu)=\gamma(\mu)(n+1+\mathrm{i}\mu)^{-1}$$

The condition on n guarantees that $\alpha_r, \beta_r\in\mathscr{S}(\mathbb{R})$, and $\gamma=\lim_r(\alpha_r\beta_r)$. Moreover, $\alpha_r^0=\alpha_r(t_n/t_{n+2})$ and $\beta_r^0=\beta_r(t_{n+2}/t_n)$, so that $\alpha_r e_{n,n+2}$ and $\beta_r e_{n+2,n}$ are both in $\mathscr{F}^{\#}$. Then

$$\gamma(e_{n,n}-e_{n+2,n+2})=\lim_r [\alpha_r e_{n,n+2}, \beta_r e_{n+2,n}]\in\mathscr{D}$$

(c) $\gamma\cdot(e_{-1,-1}-e_{1,1})$, $\gamma\in\mathscr{S}(\mathbb{R})^{\mathrm{symm}}$, $\gamma(0)=0$. This is exactly the case for $\gamma=\gamma_{-1}$ because

$$\gamma_{-1}(0)=\sum_{j\,\mathrm{odd}\leqslant -1} b_{jj}(0)=0$$

The symmetry of γ now shows that γ has *a double zero* at the origin so that $\gamma(\mu)=\mu^2\delta(\mu)$ where $\delta\in\mathscr{S}(\mathbb{R})^{\mathrm{symm}}$. We then define

$$\alpha_r(\mu)=\mu h_r(\mu),\quad \beta_r(\mu)=\mu\delta(\mu)$$

Then $\alpha_r\beta_r\to\gamma$, moreover, $\alpha_r^0=-\alpha_r$, $\beta_r^0=-\beta_r$, so that (as $t_{-1}/t_1=-1$) $\alpha_r e_{-1,1}$ and $\beta_r e_{1,-1}$ are both in $\mathscr{F}^{\#}$. Thus

$$\gamma\cdot(e_{1,-1}-e_{1,1})=\lim_r [\alpha_r e_{-1,1}, \beta_r e_{1,-1}]\in\mathscr{D}.$$

This completes the proof of the theorem.

Let

$$[\mathscr{C},\mathscr{C}]=\begin{cases}\text{closure of linear span of commutators}\\ f*g-g*f,\quad f,g\in\mathscr{C}\end{cases}$$

Then

$$\mathscr{C}^{\mathrm{inv}}:=\mathscr{C}/[\mathscr{C},\mathscr{C}]$$

is a Fréchet algebra; the product is the one induced from $\mathscr{C}$ by * and converts $\mathscr{C}^{\mathrm{inv}}$ into a *commutative algebra.* Theorem 63 leads at once to the following theorem.

Theorem 66. *The invariant Harish-Chandra transform* $f\mapsto\hat{f}$ *induces a topological algebra isomorphism*

$$\mathscr{C}^{\mathrm{inv}}\cong\mathscr{S}(\hat{B})\oplus\mathscr{S}(\hat{L})^{\mathrm{symm}}$$

where the right side is viewed as an algebra under pointwise multiplication.

Theorem 67. *If $f \in \mathscr{C}$, then $f \in [\mathscr{C}, \mathscr{C}]$ if and only if $T_{\varepsilon, i\mu}(f) = 0$ $(\mu \in \mathbb{R})$, $\Theta_m(f) = 0$ $(m \in \mathbb{Z} \backslash (0))$, $\Theta_{0,\pm}(f) = 0$. If T is a tempered distribution on G, T is invariant if and only if T vanishes on all commutators, i.e.,*

$$T(f * g) = T(g * f) \quad (f, g \in \mathscr{C})$$

In this case there are unique tempered distribution $\hat{T}_B$ on $\hat{B}$ and $\hat{T}_L$ on $\hat{L}$, with $\hat{T}_L$ symmetric, such that

$$T(f) = \hat{T}_B(\hat{f}_B) + \hat{T}_L(\hat{f}_L)$$

Proof. A tempered distribution on $\hat{B}$ is just a function on $\hat{B} \approx \mathbb{Z}$ which grows at most polynomially in the variable $n \in \mathbb{Z}$. If $f \in \mathscr{C}$ and is killed by $T_{\varepsilon, i\mu}, \Theta_m, \Theta_{0,\pm}, \hat{f} = 0$ and so Theorem 66 shows that $f \in [\mathscr{C}, \mathscr{C}]$. If a tempered distribution T vanishes on $[\mathscr{C}, \mathscr{C}]$, we can find unique tempered distributions $\hat{T}_B$, $\hat{T}_L$ as described such that

$$T(f) = \hat{T}_B(\hat{f}_B) + \hat{T}_L(\hat{f}_L).$$

Since f does not change when f is changed to $f^x(f^x(y) = f(x^{-1}yx))$, it is clear that T is invariant.

We shall call

$$\hat{T} = (\hat{T}_B, \hat{T}_L)$$

the *Fourier transform of* T. It is obvious that the map

$$T \mapsto \hat{T}$$

is a *bijection* from the space of tempered invariant distributions on G to the space of distributions $\hat{T} = (\hat{T}_B, \hat{T}_L)$ where $\hat{T}_B$ is tempered and $\hat{T}_L$ is tempered and symmetric. Many of the central questions in harmonic analysis depend on the computation of $\hat{T}$ for various T.

In a discussion in Princeton in 1968 Harish-Chandra asked me whether the characterization of an invariant distribution given in Theorem 67 was true. The proofs of Theorems 65, 66, 67 given here are essentially the ones worked out at that time by R. Ranga Rao and myself.

Finally, the reader should compare Theorem 64 with the compact case where $\mathscr{C} = C^\infty(G)$ (G compact semisimple) and,

$$\mathscr{C} = [\mathscr{C}, \mathscr{C}] \oplus \mathscr{C}^{\text{inv}}$$

$\mathscr{C}^{\text{inv}}$ being the subalgebra of invariants of $C^\infty(G)$. For noncompact G, $\mathscr{C}^{\text{inv}}$ can only be defined as a quotient.

For a deep study of function spaces and their Fourier transforms on $SL(2, \mathbb{R})$ the reader should look at the papers of Ehrenpreis and Mautner

[EM]. For a treatment of the operatorial Fourier transform the reader should consult Arthur's work [A]. I have used in my approach mostly those ideas which can be formulated in the general context. I have not treated the L^1-analogue of $\mathscr{C}$, namely the space $\mathscr{C}^1(G)$. It is another story, long (as this one is), and I may take it up later.

Problems

1. Let $F_{mn}(\mu)$ be as in the text (see just before Theorem 3). Prove that $F_{mn}(\mu) = \mathbb{C}\cdot f_{mn}(\mu:\cdot)$.
2. Let $G = SL(2, \mathbb{R})$ and $\mathscr{C}^1(G) = \mathscr{C}^1$ be the space of all C^∞ functions f on G such that, for $a, b \in U(\mathfrak{g})$, $r \geqslant 0$,

$$\nu_{a,b:r}(f) = \sup_G \Xi^2 |(afb)|(1+\sigma)^r < \infty$$

Prove that $\mathscr{C}^1 \subset L^1(G)$, and extend to $\mathscr{C}^1$ the elementary properties of $\mathscr{C}$.
3. Prove that characters of irreducible unitary representations of $G = SL(2, \mathbb{R})$ define continuous linear functionals on $\mathscr{C}^1$.
4. Prove the following formulae ([H13]) which compute the invariant Fourier transforms of the regular elliptic conjugacy classes:

$$\begin{aligned} i\pi[F_{f,B}(u_\theta) - F_{f,B}(u_{-\theta})] &= \sum \Theta_m(f) \sin|m|\theta - \frac{1}{2}\int_0^\infty T_{0,i\mu}(f)\frac{\cosh(\pi/2-\theta)\mu}{\cosh(\pi\mu/2)}\,d\mu \\ &\quad - \frac{1}{2}\int_0^\infty T_{1,i\mu}(f)\frac{\sinh(\pi/2-\theta)\mu}{\sinh(\pi\mu/2)}\,d\mu, \end{aligned}$$

$$-\pi[F_{f,B}(u_\theta) + F_{f,B}(u_{-\theta})] = -\Theta_0(f) - \sum_{m\neq 0} (\operatorname{sgn}(m)\cos m\theta \cdot \Theta_m(f))$$

Here $0 < \theta < \pi$, $\Theta_0 = (\Theta_{0,+} - \Theta_{0,-})/2$, and $f \in \mathscr{C}(G)$.
5. Calculate the invariant Fourier transforms of $F dG$ where $F = |D|^{-1/2}$, $|D_L|^{-1/2} = |D|^{-1/2} \cdot 1_{\text{hyp}}$, and 1_{ell}; here 1_{ell} (resp. 1_{hyp}) is the characteristic function of the elliptic set (resp. hyperbolic).
6. Prove that for $G = SL(2, \mathbb{C})$ the unitary principal series are precisely all the tempered irreducible representations.
7. Prove that for $G = SL(2, \mathbb{R})$ the matrix elements of the supplementary series representations L_s $(0 < s < 1)$ decay asymptotically like e^{-st} (hint: use differential equations).

Appendix 1

Functional analysis

A1.1 Generalities The reader should be familiar with the elements of the theory of Hilbert and Banach spaces. Good standard references are [RS], [Y]. The reader may also refer to a beautiful survey by Gårding and Lions [GL] written for physicists. For many problems one has to go beyond Banach spaces and work with topological vector spaces, of which the most important are the *locally convex* spaces. Let V be a vector.space over $\mathbb{C}$; a function $|\cdot|(V \to \mathbb{R})$ is called a *seminorm on* V if (a)$|v| \geqslant 0$ $(v \in V)$, (b) $|av| = |a||v|$ $(a \in \mathbb{C},\ v \in V)$, (c) $|u+v| \leqslant |u| + |v|$, $(u, v \in V)$. A collection $\mathcal{N} = \{|\cdot|_n | n \in N\}$ of seminorms on V is called *separating* if $|v|_n = 0$ for all $n \in \mathbb{N} \Rightarrow v = 0$. If $\mathcal{N}$ is a separating collection of seminorms on V, one can convert V into a topological vector space by requiring that, for any $v_0 \in V$, a basis for the topology at v_0 consist of sets of the form $v_0 + U(a, F)$ where $a > 0$, $F \subset N$ is a finite set, and

$$U(a, F) = \{v | v \in V, |v|_n < a \text{ for } n \in F\}$$

The separating condition guarantees that V is Hausdorff. The sets $U(a, F)$ are themselves open, and, as they are all covnvex, the topology of V has a basis of convex neighborhoods at every point. Conversely, if V is a Hausdorff topological vector space whose topology has a basis of convex neighbourhoods at the origin, it can be obtained in the above manner; in fact, we can take $\mathcal{N}$ to be the collection of *all* seminorms on V that are continuous. These are the *locally convex* spaces. A *Fréchet space* is a locally convex space defined by a *countable* separating set of seminorms which is *complete*, i.e., in which Cauchy sequences always converge. In particular Banach and Hilbert spaces are Fréchet.

The basic theorems of linear analysis – the open mapping theorem, the closed graph theorem, and the uniform boundedness principle – are all valid in Fréchet spaces.

A1.2 Spectral theory The core of linear functional analysis is the spectral theory of linear operators in Banach, especially Hilbert spaces. If V is a Banach space and $A(D(A) \to V)$ is a linear operator defined on a dense linear subspace $D(A)$ of V, A is called *closed* if its graph is closed in $V \oplus V$. The *resolvent set* of $A, \rho(A)$, is the set of all $\lambda \in \mathbb{C}$ such that $(\lambda I - A)^{-1}$ exists as a bounded operator, i.e., $\lambda I - A$ is a bijection of $D(A)$ with dense range and $(\lambda I - A)^{-1}$ extends to a bounded operator on V; the *spectrum* of $A, \sigma(A)$, is the set $\mathbb{C}\backslash\rho(A)$. The spectrum is always closed; if A is a bounded operator, it is nonempty. If $\dim(V) < \infty$, $\sigma(A)$ is the set of eigenvalues of A.

Let $\mathscr{H}$ be a Hilbert space and $A(\mathscr{H} \to \mathscr{H})$ a bounded linear operator. A is *self-adjoint* if $(Au, v) = (u, Av)$ for all $u, v \in \mathscr{H}$. The diagonalizability theorem for self-adjoint, i.e., Hermitian matrices has a far-reaching extension for bounded self-adjoint operators in a Hilbert space, namely, Hilbert's spectral theorem. To formulate it we need new concepts. Let X be a space and $\mathscr{B}$ a σ-algebra of subsets of X. A *projection valued measure* (pvm) on $(X, \mathscr{B})$ (or X) is a map $P(E \to P(E))$ from $\mathscr{B}$ to the set of orthogonal projections in $\mathscr{H}$ such that (a) $P(\varnothing) = 0$, $P(X) = 1$, (b) $P(E)u = \sum_n P(E_n)u$ for all $u \in \mathscr{H}$ if E is the disjoint union of the E_n. If P is a pvm, the spectral integrals

$$A(f) = \int_X f(x)\,\mathrm{d}P(x)$$

can be defined in a straightforward way for all $\mathscr{B}$-measurable functions that are essentially bounded, i.e., for some set E, $P(E) = 1$ and f is bounded on E. The $A(f)$ are bounded linear operators defined by

$$(A(f)u, v) = \int_X f(x)\,\mathrm{d}P_{u,v}(x) \quad (u, v \in \mathscr{H}, P_{u,v}(E) = (P(E)u, v))$$

where the right side is the usual integral with respect to the complex measure $P_{u,v}$. If X is a topological space satisfying the second axiom of countability and $\mathscr{B}$ is the σ-algebra of its Borel sets, there is a smallest closed set C such that $P(C) = 1$; C is called the *support* of P. Hilbert's spectral theorem for a bounded self-adjoint A in $\mathscr{H}$ asserts the existence of a unique pvm P on $\mathbb{R}$ with compact support such that

$$A = \int_{\mathbb{R}} \lambda\,\mathrm{d}P(\lambda)$$

The support of P is then precisely the spectrum of A. P is called the *spectral measure* of A, the $P(E)$ the *spectral projections* of A; and the subspaces on which the $P(E)$ project are called the *spectral subspaces* of A. If $\lambda_0 \in \mathbb{R}$, it is an eigenvalue of A if and only if $P(\{\lambda_0\}) \neq 0$, and then the range of $P(\{\lambda_0\})$ is the corresponding eigensubspace. The point of introducing the pvm's is of course to take care of the cases where A has has no eigenvalues, for instance, when A is multiplication by the coordinate variable x in $L^2(0, 1)$. If $\dim(\mathscr{H}) < \infty$ and $\mathscr{H}(\lambda)$ $(\lambda \in \sigma(A))$ are the eigenspaces, and we write $\mathscr{H}(E) = \bigoplus_{\lambda \in E \cap \sigma(A)} \mathscr{H}(\lambda)$, then $P(E)$ is the orthogonal projection $\mathscr{H} \to \mathscr{H}(E)$. If $\dim(\mathscr{H}) = \infty$ but A is compact (i.e., maps norm bounded sets into sets with compact closure), the spectral theorem is still very close to its finite dimensional origins and one does not need to go beyond the concepts of eigenvalues and eigenspaces. In this case $\sigma(A)$ is the closure of the set of eigenvalues of A; we still define $\mathscr{H}(E)$, $P(E)$ as before. P is then the spectral measure of A and the $\mathscr{H}(\lambda)$ are finite-dimensional for $\lambda \neq 0$. The classical examples of compact A are the *integral operators* A_K defined by compactly supported continuous kernels K on $X \times X$ where X is a locally compact Hausdorff space with Borel measure μ:

$$(A_K f)(x) = \int_X K(x, y) f(y) \, \mathrm{d}\mu(y) \quad (f \in L^2(\mu))$$

For A_K to be self-adjoint the condition is of course $K(x, y) = K(y, x)^{\mathrm{conj}}$. For a general bounded self-adjoint A on $\mathscr{H}$, the spectrum $\sigma(A) \subset \mathbb{R}$ so that $R(\lambda) = (A - \lambda 1)^{-1}$ is a bounded operator defined for all $\lambda \in \mathbb{C} \backslash \mathbb{R}$ and depending holomorphically on λ; and the spectral projections may be calculated explicitly from Stone's formula: for any $a, b \in \mathbb{R}$ with $a < b$,

$$\operatorname*{s-lim}_{\varepsilon \to 0} (1/2\pi \mathrm{i}) \int [R(\lambda + \mathrm{i}\varepsilon) - R(\lambda - \mathrm{i}\varepsilon)] \, \mathrm{d}\lambda = P((a, b)) + \tfrac{1}{2}(P(\{a\}) + P(\{b\})$$

Here s-lim means limit in the strong operator topology (see Section 5 below for the definition of strong topology). Finally, $A \geqslant 0$ if and only if the support of A is contained in $[0, \infty)$, i.e., $P([0, \infty)) = 1$.

The spectral theorem allows us to develop a functional calculus. If f is a polynomial, we have

$$f(A) = \int_{\mathbb{R}} f(\lambda) \, \mathrm{d}P(\lambda)$$

We now define $f(A)$ for any essentially bounded Borel measurable f by the right side.

The needs of quantum mechanics highlighted the necessity of extending the spectral theorem to operators which are not everywhere defined, such as multiplication by x in $L^2(-\infty, \infty)$, or $(1/\mathrm{i})\mathrm{d}/\mathrm{d}x$ in $L^2(-\infty, \infty)$, which are the operators of position and momentum in the x-direction. The von Neumann spectral theorem does exactly this. A closed operator A with a dense domain of definition $D(A)$ is called *symmetric* if $(Au, v) = (u, Av)$ for all $u, v \in D(A)$. If A and B are linear operators with domains $D(A)$ and $D(B)$ respectively, we say that B is an extension of A if $D(A) \subset D(B)$ and B coincides with A on $D(A)$. Any symmetric A has a unique smallest closed extension (= closure), and this extension is symmetric. One can then define the adjoint A^* of A as follows: $D(A^*)$ is the linear space of all u for which $v \to (Av, u)$ is a continuous linear functional on $D(A)$; and $(v, A^*u) = (Av, u)$. The symmetry of A means that A^* is an extension of A. Von Neumann defined a symmetric A to be *self-adjoint* if $A = A^*$, i.e. if $D(A^*) = D(A)$. The von Neumann spectral theorem asserts that there is a canonical bijection between the set of self-adjoint operators A and the set of pvm's P on $\mathbb{R}$, related by the equation

$$A = \int_{\mathbb{R}} \lambda \, \mathrm{d}P(\lambda)$$

The spectral integral on the right side is to be interpreted carefully, but essentially in the same way as before; $D(A)$ is the set of u for which

$$\int_{\mathbb{R}} \lambda^2 \, \mathrm{d}P_{u,u}(\lambda) < \infty$$

and

$$(Au, v) = \int_{\mathbb{R}} \lambda \, \mathrm{d}P_{u,v}(\lambda) \quad (u \in D(A), v \in \mathscr{H})$$

In particular

$$\| Au \|^2 = \int_{\mathbb{R}} \lambda^2 \, \mathrm{d}P_{u,u}(\lambda)$$

Of course A is bounded if and only if P has compact support. If A is the operator of multiplication by x in $L^2(-\infty, \infty)$, $P(E)$ is multiplication by the characteristic function of E.

In order to get a little more understanding of von Neumann's delicate definition of a self-adjoint operator it is useful to look at extensions of symmetric operators. For a closed, symmetric, densely defined A, the subspaces $C_{\pm} = \{Au \pm \mathrm{i}u \mid u \in D(A)\}$ are closed and

$$U_A : Au - \mathrm{i}u \to Au + \mathrm{i}u$$

is an isometry of C_- with C_+. U_A is the *Cayley transform* of A, and the correspondence $B \leftrightarrow U_B$ is a bijection between the set of closed symmetric extensions of A and the set of isometric extensions of U_A. In particular, A is self-adjoint if and only if U_A is unitary, i.e., $C_- = C_+ = \mathscr{H}$. The dimensions of $C_{\pm}^{\perp}$ are called the *deficiency indices* of A, and their equality is necessary and sufficient for A to have a self-adjoint extension. Let $\mathscr{H} = L^2(0, \infty)$ and $A = (1/\mathrm{i})\mathrm{d}/\mathrm{d}x$, $D(A)$ being the set of all $u \in L^2(0, \infty)$ such that u is absolutely continuous and $\mathrm{d}u/\mathrm{d}x \in L^2(0, \infty)$; then A is symmetric, closed, and its deficiency indices are 0, 1, so that A is already maximally symmetric and does not admit any self-adjoint or even symmetric extension. The corresponding operator on $L^2(-\infty, \infty)$ is of course self-adjoint, and its spectral measure can be computed using Fourier transforms.

The spectral theory of self-adjoint operators can be extended without difficulty to one-parameter unitary groups. If $t \to U_t$ is a unitary representation of $\mathbb{R}$ in $\mathscr{H}$, there is a unique pvm P on $\mathbb{R}$ such that

$$U_t = \int_{\mathbb{R}} e^{it\lambda} \, dP(\lambda) \quad (t \in \mathbb{R})$$

This is Stone's theorem, and (symbolically)

$$U_t = e^{itH}, \quad H = \int_{\mathbb{R}} \lambda \, dP(\lambda)$$

This theorem goes over to unitary representations $U(g \to U_g)$ of locally compact (second countable, as always for us) abelian groups; the corresponding pvm is now defined on the dual group $\hat{G}$, and

$$U_g = \int_{\hat{G}} \chi(g) \, dP(\chi) \quad (g \in G)$$

The correspondence $U \leftrightarrow P$ is bijective.

The spectral theory of differential operators is a vast subject by itself. The case of singular ordinary differential operators of order 2 on a half-line already goes back to Hermann Weyl [W4]; its modern treatment began with Stone [St], and was completed by Kodaira [Kd] and Titchmarsh [Ti]; see the monumental [DS], Part II.

A1.3 Operators of Hilbert–Schmidt and trace class In analogy with L^p spaces we can introduce certain subspaces of $B(\mathscr{H})$, the algebra of all bounded linear operators of $\mathscr{H}$. If $A \in B(\mathscr{H})$, A is said to be of *Hilbert–Schmidt* (H.S.) *class* if

$$|||A|||_2^2 = \sum_n \|Ae_n\|^2 < \infty$$

for some orthonormal (ON) basis (e_n) of $\mathscr{H}$; this is then true of all ON bases and the value of $|||A|||_2^2$ is the same for all such bases. The operators of H.S. class form a Hilbert space under the scalar product

$$(A, B)_2 = \sum_n (Ae_n, Be_n)$$

where (e_n) is any ON basis, the value of the series being independent of the choice of the basis; $|||\cdot|||_2$ is the corresponding norm. The operators of H.S. class form a two-sided ideal in $B(\mathscr{H})$.

An operator $A \in B(\mathscr{H})$ is said to be of *trace class* if the series $\sum_n (Ae_n, e_n)$ is convergent for all ON bases (e_n) of $\mathscr{H}$. The value of the sum is then independent of the ON basis and is called the *trace* of A, written $\mathrm{tr}(A)$. The operators of trace class form a two-sided ideal in $B(\mathscr{H})$; $\mathrm{tr}(AB) = \mathrm{tr}(BA)$ if A is of trace class and $B \in B(\mathscr{H})$; $\mathrm{tr}(SAS^{-1}) = \mathrm{tr}(A)$ for A of trace class and S any invertible element of $B(\mathscr{H})$. An $A \in B(\mathscr{H})$ is of trace class if and only if the nonnegative operator $|A| = (A^*A)^{1/2}$ is of trace class. If A is nonnegative, it is of trace class if and only if $A^{1/2}$ is of H.S. class, i.e., if and only if $\sum_n (Ae_n, e_n) < \infty$ for some ON basis (e_n). Finally, the operators of trace class form a Banach space under the norm $|||A|||_1 = \mathrm{tr}(|A|)$. If $A \in B(\mathscr{H})$ and (e_n) is an ON basis, A is called *summable* (with respect to this basis) if $\sum_{m,n} |(Ae_m, e_n)| < \infty$; it is then of trace class. If A, B are of H.S. class, AB is of trace class.

If $(X, \mathscr{B}, \mu)$ is a measure space and $K \in L^2(X \times X, \mu \times \mu)$, the integral operator A_K defined by the kernel K is of H.S. class. In general it may not be of trace class. Let X be a C^∞ manifold, K a C^∞ function with compact support, and let μ be C^∞, i.e., on any open set with coordinates $x_1, \ldots, x_n$, $\mathrm{d}\mu = g(x_1, \ldots, x_n)\,\mathrm{d}x_1 \cdots \mathrm{d}x_n$ where g is C^∞. Then A_K is summable and

$$\mathrm{tr}(A_K) = \int_X K(x, x)\,\mathrm{d}\mu(x)$$

The proof of this result may be obtained along the following lines. By using a partition of unity we may assume that X is covered by a single coordinate chart, and, as the support of K is compact, we may go over to the case when X is a torus and μ is the Lebesgue measure. The summability of A_K with respect to the usual trigonometric basis is immediate because the Fourier series of K is rapidly convergent. The formula for the trace is immediate from the convergence of the Fourier series of K. This proof shows that it is enough to assume that K is of classs C^r, for some large r ($r > \frac{1}{2}\dim(X)$ will do). The reader should complete the details left out in these remarks.

A1.4 Distributions Distributions or generalized functions have had a long and fascinating history. The fundamental ideas were slow in evolving and important contributions came from Hadamard, M. Riesz, Weyl, Dirac, Sobolev and many others. The formalization and synthesis that have made distributions one of the most powerful tools in modern analysis are due to L. Schwartz. His books [Sch] as well as the book of Gel'fand and Shilov [GS] are standard references.

Let Ω be an open subset of $\mathbb{R}^n$. A *distribution* on Ω is a linear functional T on the space $C_c^\infty(\Omega)$ of smooth compactly supported functions on Ω with the following continuity property: if $K \subset \Omega$ is compact, and (f_n) is a sequence of elements of $C_c^\infty(\Omega)$ such that $\mathrm{supp}(f_n) \subset K$ for all n (supp = support) and f_n, as well as all its derivatives, converges to 0 uniformly as $n \to \infty$, then $T(f_n) \to 0$. It is possible to give $C_c^\infty(\Omega)$ a locally convex topology so that this condition becomes equivalent to the continuity of

T. It is clear from the definition that the linear functional

$$f \to -T((\partial/\partial x_j)f)$$

is also a distribution, denoted by $(\partial/\partial x_j)T$. In this way distributions can be differentiated any number of times. If F is a measurable locally integrable function on Ω, we identify F with the distribution

$$T_F : f \to \int F f \, \mathrm{d}x$$

F is determined almost everywhere by T_F, and hence everywhere if it is a continuous function. If F is smooth, this identification takes $(\partial/\partial x_j)F$ to the distribution $(\partial/\partial x_j)T_F$, as may be seen by the formula for integration by parts. The concept of distributions and their derivatives is thus a generalization of that of a smooth function and its derivatives. It is a far-reaching notion that has revolutionized modern analysis so much that it is impossible to think back to the times when it was not available.

The possibility of unrestrictedly differentiating any distribution allows us to speak of distribution solutions of linear partial differential equations. For any distribution T on Ω and $g \in C^\infty(\Omega)$, let us define gT to be the distribution $f \to T(gf)$. Then, for any linear partial differential operator L with smooth coefficients, we have a map $T \to L(T)$ of the space of distributions on Ω. Its kernel is then the space of distribution solutions of the equation $Lu = 0$ on Ω. Let us write L as

$$L = \sum a_\alpha \partial_\alpha \quad (\alpha = (\alpha_1, \ldots, \alpha_n), |\alpha| = \alpha_1 + \cdots + \alpha_n, \partial_\alpha = (\partial/\partial x_1)^{\alpha_1} \cdots (\partial/\partial x_n)\alpha_n)$$

and recall that L is *elliptic* if the polynomial

$$\sum a_\alpha(x)\xi^\alpha \quad (\xi^\alpha = \xi_1^{\alpha_1} \cdots \xi_n^{\alpha_n})$$

does not vanish when $x \in \Omega$ and ξ varies on the unit sphere of $\mathbb{R}^n$. The *regularity theorem* for elliptic operators says that, if L has smooth coefficients and is elliptic, then any distribution solution $L(T) = 0$ is necessarily a classical solution, i.e., T is a smooth function and satisfies $L(T) = 0$ in the usual sense. Moreover, if L has analytic coefficients, T is analytic also. Historically, when $n = 3$, and $L =$ the Laplacian, this result (weak harmonicity $\Rightarrow$ harmonicity) goes back to Hermann Weyl (Weyl's lemma), and was used by him as a cornerstone for the construction of meromorphic functions on Riemann surfaces. These ideas are now an indispensable part of the modern Hodge theory of compact manifolds.

A distribution of Ω is a localizable object in the sense that, if $\Omega' \subset \Omega$ is an open subset of Ω, it makes sense to restrict any distribution T on Ω to a distribution T' on Ω'; T' is the restriction of T to $C_c^\infty(\Omega')$ which is a subspace of $C_c^\infty(\Omega)$. If $\Omega = \bigcup_i \Omega_i$ where $\Omega_i \subset \Omega$ are open, and T_i are distributions on Ω_i such that T_i and T_j coincide on $\Omega_i \cap \Omega_j$ for all i, j, there is a unique distribution T on Ω which restricts to T_i on Ω_i. The support of a distribution T is the complement in Ω of the largest open subset of Ω of which T is 0.

The most famous example of a distribution is the Dirac delta function. If $p \in \mathbb{R}^n$, the Dirac delta function δ_p at p is the distribution given by $\delta_p = f(p)$, $f \in C_c^\infty(\mathbb{R}^n)$. It is just the probability measure whose entire mass is concentrated at the point p. Its derivatives are the distributions $f \to (\partial_\alpha f)(p)$, $f \in C_c^\infty(\mathbb{R}^n)$; however, these are not measures. As examples of differentiation we mention the following: in $\mathbb{R}^1$,

$$(\mathrm{d}/\mathrm{d}x)H = \delta_0$$

where H is the function on $\mathbb{R}^1$ which is 0 for $x < 0$ and 1 for $x > 0$, and we identify it with the distribution $f \to \int_{\mathbb{R}^1} H(x) f(x)\,\mathrm{d}x$; in $\mathbb{R}^2$,

$$(\partial^2/\partial x^2 + \partial^2/\partial y^2)\log r = 2\pi\delta_0$$

where $r = (x^2 + y^2)^{1/2}$, $\log r$ is the distribution $f \to \iint \log r \cdot f\,\mathrm{d}x\,\mathrm{d}y$, and δ_0 is the Dirac delta function at the origin of $\mathbb{R}^2$. The most general distribution in $\mathbb{R}^n$ with support contained in $\{p\}$ is a linear combination of derivatives of δ_p. Similarly but more generally, the distributions supported by $x_n = 0$ are those that are (uniquely) represented, locally on $x_n = 0$, in the form

$$T_0 + (\partial/\partial x_n)T_1 + \cdots + (\partial/\partial x_n)^m T_m \qquad (*)$$

where $T_0, \ldots, T_m$ are distributions on $\mathbb{R}^{n-1}$ identified with the subspace $x_n = 0$ of $\mathbb{R}^n$. Note that, for computing the value of $(*)$ at an $f \in C_c^\infty(\mathbb{R}^n)$, it is not enough to know the restriction of f to the hyperplane $x_n = 0$; one should know the restrictions of $(\partial/\partial x_n)^r f$ also, $0 \leqslant r \leqslant m$. The integer m (where $T_m \neq 0$) is the *local transversal order of* T; we use this idea decisively in Chapter 7 in proving the Harish-Chandra regularity theorem for $SL(2, \mathbb{R})$.

The localizability of a distribution allows one to define the concept of distributions on any smooth manifold. If Ω is a smooth manifold and T is a linear functional on $C_c^\infty(\Omega)$, T is a distribution if and only if, on any open subset $\Omega' \subset \Omega$ on which one can define coordinates $x_1, \ldots, x_n$, T restricts to a distribution in $x_1, \ldots, x_n$.

For global problems, for instance those involving Fourier analysis, it is necessary to work with more restrictive notions. For Fourier analysis one introduces the Schwartz space $\mathscr{S}(\mathbb{R}^n)$ of smooth functions f such that

$$\mu_{\alpha,m}(f) = \sup(1 + r^2)^m |(\partial_\alpha f)(x)| < \infty$$

for all $m \geqslant 0$, $\alpha = (\alpha_1, \ldots, \alpha_n)$, r^2 being $x_1^2 + \cdots + x_n^2$. The $\mu_{\alpha,m}$ are seminorms converting $\mathscr{S}(\mathbb{R}^n)$ into a Fréchet space, and *tempered distributions* are by definition continuous linear functionals on $\mathscr{S}(\mathbb{R}^n)$. We have $C_c^\infty(\mathbb{R}^n) \subset \mathscr{S}(\mathbb{R}^n)$, and it is easy to verify that a tempered distribution defines by restriction a distribution on $C_c^\infty(\mathbb{R}^n)$. Moreover, $C_c^\infty(\mathbb{R}^n)$ is dense in $\mathscr{S}(\mathbb{R}^n)$ so that given any distribution on $C_c^\infty(\mathbb{R}^n)$ there is at most one tempered distribution of which it is the restriction; the original distribution is called *tempered* if there is such an extension. We make no distinction between the linear functional on $C_c^\infty(\mathbb{R}^n)$ and its extension to $\mathscr{S}(\mathbb{R}^n)$. For any $f \in \mathscr{S}(\mathbb{R}^n)$ its Fourier transform $\mathscr{F} f$ is the function

$$(\mathscr{F} f)(u) = (2\pi)^{-n/2} \int f(x) \exp(\mathrm{i}(x_1 u_1 + \cdots + x_n u_n)) \mathrm{d}x_1 \cdots \mathrm{d}x_n$$

The fundamental theorem of Fourier analysis is the assertion that $\mathscr{F} f \in \mathscr{S}(\mathbb{R}^n)$ and $f \to \mathscr{F} f$ is a topological linear isomorphism with

$$f(x) = (2\pi)^{-n/2} \int (\mathscr{F} f)(u) \exp(-\mathrm{i}(x_1 u_1 + \cdots + x_n u_n)) \, \mathrm{d}x_1 \cdots \mathrm{d}x_n$$

The concept of Fourier transform can now be extended by duality: for any tempered distribution T, its Fourier transform $\mathscr{F} T$ is the tempered distribution

$$(\mathscr{F} T)(f) = T(\mathscr{F} f)$$

The Plancherel formula is the statement that

$$\mathscr{F} \delta_0 = (2\pi)^{-n/2} \cdot 1$$

A1.5 Von Neumann algebras Except for a brief encounter in Chapter 3 we do not use these. For additional insight into the matters discussed in Chapter 3 the reader may consult the treatise of Takesaki [*T*].

For any Hilbert space $\mathscr{H}$, $B(\mathscr{H})$ can be given several topologies. The *norm topology* is the one induced by the operator norm $\|\cdot\|$ The *strong topology* is the one induced by the collection of seminorms

$$A \to \|Ax\| \quad (x \in \mathscr{H})$$

The *weak topology* is the one induced by the collection of seminorms

$$A \to (Ax, y) \quad (x, y \in \mathscr{H})$$

These are progressively weaker (= coarser). A *von Neumann algebra* is a subalgebra $\mathscr{A} \subset B(\mathscr{H})$ which is closed under taking adjoints and is a closed subset of $B(\mathscr{H})$ in the weak operator topology. The *commutant of* $\mathscr{A}$, $\mathscr{A}'$ in symbols, is the von Neumann algebra of all operators that commute with everything in $\mathscr{A}$. Central to the basic theory is von Neumann's *double commutant theorem*: $\mathscr{A} = \mathscr{A}''$. If $\mathscr{A} \cap \mathscr{A}' = \mathbb{C} \cdot 1$, $\mathscr{A}$ is called a *factor*. The reason for this terminology is that, if $\mathscr{H}$ is finite-dimensional, factors are the algebras that arise by factorizing $\mathscr{H}$ as $\mathscr{H}_1 \otimes \mathscr{H}_2$, taking $\mathscr{A} = B(\mathscr{H}_1) \otimes 1$ and $\mathscr{A}' = 1 \otimes B(\mathscr{H}_2)$. If $\mathscr{H}$ is infinite-dimensional, this construction will still give factors; but, as Murray and von Neumann discovered, there are other examples. Factors of the former type are said to be of type I. The theory of von Neumann algebras has made great strides in recent years and a knowledge, at least of its basic themes, is very useful in building up a general picture of representation theory.

Appendix 2

Topological groups

A2.1 Locally compact groups and Haar measure The concept of a topological group is obtained by combining those of a group and a topological space. Thus, a topological group G is at the same time a group and a Hausdorff topological space; the group structure and the topological structure are related by the requirement that the maps

$$x, y \to xy$$

$$x \to x^{-1}$$

are continuous. Our interest is entirely in the case when the topology is locally compact and second-countable. From now on G will denote such a topological group.

The fundamental fact is the existence, on G, of a Borel measure $d_l x$ that is invariant under all left translations. Thus

$$\int_G f(x)\, d_l x = \int_G f(yx)\, d_l x$$

for all $y \in G$ and $f \in C_c(G)$(= the space of continuous functions on G with compact support). This measure, called a *left Haar measure*, is unique up to a positive multiplicative constant. If $G = \mathbb{R}^n$, it is a multiple of Lebesgue measure; in general, there is no natural way to normalize it. If G is compact, it is usual to normalize it by the condition

$$\int_G d_l x = 1$$

Similarly we have *right Haar measures* $d_r x$. In general the two are not the same. Groups for which left Haar measures are also right-invariant are called *unimodular*. Compact groups and abelian groups are unimodular. The book of Halmos [Ha] is a good reference for a thorough treatment of Haar measure.

Appendix 3

Lie groups and Lie algebras

A3.1 Basics For the basic definitions and concepts involving C^∞ and analytic manifolds as well as a general reference for results concerning Lie groups and Lie algebras the reader should consult [V2].

The concept of a Lie group is obtained by combining the notions of smooth manifolds and groups. Thus a C^∞ (resp. (real) analytic) Lie group is a group G, which is at the same time a C^∞ (resp. analytic) manifold, such that the maps

$$x, y \to xy, \quad x \to x^{-1}$$

are C^∞ (resp. analytic). We shall always work with analytic Lie groups in view of the classical result that on any C^∞ Lie group G one can introduce in a unique way the structure of a real analytic manifold on G that converts G into an analytic Lie group.

To any Lie group G is associated its Lie algebra Lie(G). Recall that a Lie algebra over a field k of characteristic 0 is a vector space $\mathfrak{g}$ over k equipped with a bilinear map $[\cdot,\cdot]$ of $\mathfrak{g} \times \mathfrak{g}$ into $\mathfrak{g}$ such that

(a) $[X, Y] + [Y, X] = 0 (X, Y \in \mathfrak{g})$,
(b) $[[X, Y], Z] + [[Y, Z], X] + [[Z, X], Y] = 0 (X, Y, Z \in \mathfrak{g})$.

A typical example of a Lie algebra over $\mathbb{R}$ or $\mathbb{C}$ is the space of smooth vector fields on a smooth manifold; the bracket $[X, Y]$ of vector fields X and Y is the usual one, namely $[X, Y] = XY - YX$ where X and Y are viewed as derivations of the algebra of smooth functions on the manifold. In local coordinates $x_1, \ldots, x_n$, if

$$X = \sum_j a_j \partial/\partial x_j, \quad Y = \sum b_j \partial/\partial x_j$$

Then

$$[X, Y] = \sum_j \left(\sum_k (a_k(\partial b_j/\partial x_k) - b_k(\partial a_j/\partial x_k)) \right) \partial/\partial x_j$$

This shows that, if the manifold is analytic and X, Y are analytic, so is $[X, Y]$. Returning to the Lie group G, Lie(G) is the space of all analytic vector fields that are invariant under all left translations. It is clear that the bracket of two left-invariant vector fields is also left-invariant so that Lie(G) is a Lie algebra, in fact a subalgebra of the Lie algebra of analytic vector fields on G. For any $X \in \mathrm{Lie}(G)$ and $g \in G$, let X_g be the tangent vector defined by X at g; then $X \to X_1$ is a linear bijection of $\mathrm{Lie}(G)$ with the tangent space $T_1(G)$ to G at 1. In other words, $\mathrm{Lie}(G)$ has the same dimension as G. We shall usually write $\mathfrak{g}$ for $\mathrm{Lie}(G)$ and identify $\mathfrak{g}$ with $T_1(G)$ via the map $X \to X_1$.

One knows from the theory of differentiable manifolds that to any vector field on a manifold one can associate its integral curves: if M is a C^∞ manifold and X a smooth vector field on M, then for any $m \in M$ we can find $a > 0$ and a smooth map $u((-a, a) \to M)$ such that $u(0) = m$ and $\mathrm{d}u/\mathrm{d}t = X_{u(t)}$. The number a depends on m, and it may not be possible to take $a = \infty$. If G is a Lie group and $X \in \mathfrak{g}$, the integral curves are defined for all t. The integral curve through 1 is defined as $t \to \exp tX$. Thus $\exp tX$ is defined for all t, is 1 when $t = 0$, and

$$(\mathrm{d}/\mathrm{d}t)(\exp tX) = X_{\exp tX}$$

We write $\exp X$ for $(\exp tX)_{t=1}$ and get the *exponential map* exp of $\mathfrak{g}$ into G. It is easy to see that

$$((\mathrm{d}/\mathrm{d}t)(\exp t)(X + Y)))_{t=0} = X + Y$$

The reason for calling this the exponential map is that, when $G = GL(n, \mathbb{R})$, one has $\mathfrak{g} = \mathfrak{gl}(n, \mathbb{R})$, and $\exp X$ is the usual exponential of the matrix X.

If $a \in G$, a acts on G by $g \to aga^{-1}$. So we have an action of a on the vector fields, and it is clear that left invariance is preserved under this action. If $X \in \mathfrak{g}$, we write $X^a = \mathrm{Ad}(a)X$ for the transformed vector field. Since the integral curves of X^a are a(integral curves of X)a^{-1}, we have

$$\exp tX^a = a \exp tX \cdot a^{-1}$$

In particular $X^{ab} = (X^b)^a$ or

$$\mathrm{Ad}(ab) = \mathrm{Ad}(a)\mathrm{Ad}(b)$$

Since $X \to X^a$ is a linear in X, we see that $a \to \mathrm{Ad}(a)$ is a representation of G in $\mathfrak{g}$, the so-called *adjoint representation.* Ad is an analytic homomorphism of G into $GL(\mathfrak{g})$, in fact into $\mathrm{Aut}(\mathfrak{g})$, the group of automorphisms of $\mathfrak{g}$.

Using left translations we shall identify the tangent spaces at all the points of G with $\mathfrak{g}$. This is similar to the way by which the tangent spaces to a vector space V at each of its points get identified with V itself. Thus if $X \in \mathfrak{g}$, and $g \in G$, X is identified with X_g. In particular, if f is a smooth function,

$$(Xf)(g) = (\mathrm{d}/\mathrm{d}t)_{t=0} f(g \exp tX)$$

There is a natural correspondence between Lie subgroups of G and subalgebras of $\mathfrak{g}$; if $\mathfrak{h} \subset \mathfrak{g}$ is the subalgebra, its corresponding subgroup is the subgroup H of G generated by $\exp\mathfrak{h}$. However, care must be exercised in defining the analytic structure on H (see [V2], Ch. 2, §5). The point is that H need not be closed in G; the classic example of this phenomenon arises when G is the torus $\mathbb{T}^2$ with Lie algebra $\mathbb{R}^2$ and $\mathfrak{h}$ is the one-dimensional subspace defined by a line through the origin with an irrational slope. On the other hand, any closed subgroup H of G is necessarily a Lie group; this means that the connected component of H containing 1 is open in H and is a closed analytic submanifold of G, so that it acquires the structure of an analytic group automatically.

Similarly, if G_1, G_2 are Lie groups with respective Lie algebras $\mathfrak{g}_1, \mathfrak{g}_2$, there is a correspondence between analytic homomorphisms of G_1 and G_2 and Lie algebra homomorphisms of $\mathfrak{g}_1$ into $\mathfrak{g}_2$: to $\pi(G_1 \to G_2)$ corresponds $d\pi(\mathfrak{g}_1 \to \mathfrak{g}_2)$ such that

$$\exp((d\pi)(X)) = \pi(\exp X) \quad (X \in \mathfrak{g})$$

Given π, $d\pi$ is uniquely determined by this; moreover, given $d\pi$, π is also determined uniquely provided G is connected. However, not every Lie algebra homomorphism of $\mathfrak{g}_1$ to $\mathfrak{g}_2$ is of the form $d\pi$ for some π $(G_1 \to G_2)$; this is true if G is *simply connected*. Finally any continuous homomorphism of G_1 to G_2 is analytic; here it is essential to remember that it is a question of real-analytic structure. For the complex-analytic additive group $\mathbb{C}$, $z \to z^{\text{conj}}$ is a continuous automorphism that is not complex-analytic. For the adjoint representation $\mathrm{Ad}(G \to GL(\mathfrak{g}))$ the corresponding representation $\mathfrak{g} \to \mathfrak{gl}(\mathfrak{g})$ is the adjoint representation ad of Lie algebras defined by

$$\mathrm{ad}(X)(Y) = [X, Y]$$

This follows from the formula

$$\mathrm{Ad}(\exp X) = \exp(\mathrm{ad}\, X)$$

(see [V2], Ch.2, for more details concerning these remarks).

A3.2 Universal enveloping algebra Each element of $\mathfrak{g}$ may be viewed as a derivation of $C^\infty(G)$. So it makes sense to speak of the algebra of endomorphisms of $C^\infty(G)$ generated by the elements of $\mathfrak{g}$. Clearly the elements of this algebra may be viewed as analytic differential operators on G which are left-invariant. One can prove that all left-invariant differential operators are obtained this way. This algebra is denoted by $U(\mathfrak{g})$ and is called the *universal enveloping algebra of* $\mathfrak{g}$ (or G).

We have

$$\mathfrak{g} \subset U(\mathfrak{g})$$
$$[X, Y] = XY - YX \quad (X, Y \in \mathfrak{g})$$

where the multiplication on the right side is the one in $U(\mathfrak{g})$. The word universal refers to the following universal property possessed by the pair $(\mathfrak{g}, U(\mathfrak{g}))$: if A is any associative algebra and $f(\mathfrak{g} \to A)$ is a linear map such that

$$f([X, Y]) = f(X)f(Y) - f(Y)f(X) \quad (X, Y \in \mathfrak{g})$$

then there is a unique extension of f to a homomorphism $f(U(\mathfrak{g}) \to A)$.

Given any Lie algebra, say $\mathfrak{n}$, over a field k (of characteristic 0) we may define the *universal enveloping algebra* $U(\mathfrak{n})$ *of* $\mathfrak{n}$ by the requirements that it contain $\mathfrak{n}$, be generated by $\mathfrak{n}$, and possess the above universal property. It is basic theorem that the algebra $U(\mathfrak{n})$ exists and is essentially unique ([V2], Ch. 3, §2). The universal property allows us to identify Lie algebra representations with representations of associative algebras. In fact, if π is a representation $\mathfrak{n}$ in V, we view π as a linear map of $\mathfrak{n}$ into the associative algebras End (V) such that $\pi([X, Y]) = \pi(X)\pi(Y) - \pi(Y)\pi(X)$, and so obtain a unique extension of π to a representation of the associative algebra $U(\mathfrak{n})$ in V. The theory of representations of Lie algebras acquires its special character because it is in substance the theory of representations of certain special associative algebras which are infinite-dimensional.

If $X_1, \ldots, X_n$ is a basis for $\mathfrak{n}$, monomials in them will span $U(\mathfrak{n})$ since $\mathfrak{n}$ generates $U(\mathfrak{n})$. However, there are relations between them because we can always replace $X_i X_j$ by $X_j X_i + \sum_k c_{ijk} X_k$ where $\sum_k c_{ijk} X_k = [X_i, X_j]$. But if we *fix* the enumeration $X_1, \ldots, X_n$ and consider only the *standard* monomials

$$X_1^{r_1} \cdots X_n^{r_n} \quad (r_1, \ldots, r_n \geqslant 0)$$

we obtain a *basis of* $U(\mathfrak{n})$. This is the Poincaré–Birkhoff Witt theorem ([V2], Theorem 3.2.2).

Finally we note that the interpretation of elements of $U(\mathfrak{g})$ as left invariant differential operators on G is made very direct by the formula

$$(\partial(L_1 \cdots L_m)f)(g) = (\partial^m/\partial t_1 \cdots \partial t_r)_{t_1 = \cdots = t_r} f(g \exp t_1 L_1 \cdots \exp t_m L_m)$$

$(L_1, \ldots, L_m \in \mathfrak{g})$. For the action $\partial_r(L_1 \cdots L_m)$ of $L_1, \ldots, L_m \in U(\mathfrak{g})$ as a *right invariant* differential operator we use

$$(\partial_r(L_1 \cdots L_m)f)(g) = (\partial^m/\partial t_1 \cdots \partial t_m)_{t_1 = \cdots = t_r = 0} f(\exp t_1 L_1 \cdots \exp t_m L_m g)$$

A3.3 The fundamental theorem of Lie This asserts that, for any Lie group G, the *local* structure of G is completely determined by Lie(G). To explain the meaning of this statement we introduce the concept of local isomorphism between Lie groups. If G_1 and G_2 are Lie groups, a *local isomorphism* of G_1 with G_2 is an analytic diffeomorphism φ of an open neighbourhood U_1 of 1 in G_1 with an open neighbourhood U_2 of 1 in

G_2 such that (a) $\varphi(1) = 1$ and (b) if $x, y \in U_1$, then $xy \in U_1$ if and only if $\varphi(x)\varphi(y) \in U_2$, and then $\varphi(xy) = \varphi(x)\varphi(y)$. For instance, if $\varphi(G_1 \to G_2)$ is a *surjective* analytic homomorphism with a *discrete* kernel, then φ is a local isomorphism. A more geometric way to define a local isomorphism is to require the existence of a Lie group G and two surjective analytic homomorphisms $\pi_j(G \to G_j)$ with discrete kernels; this is equivalent to the definition given earlier. Then Lie's fundamental theorem asserts that G_1 and G_2 are locally isomorphic if and only if Lie(G_1) and Lie(G_2) are isomorphic as Lie algebras. In particular, if G_1 and G_2 are simply connected, we can replace local by global isomorphism. Actually the result is much stronger. The *functor* $G \to$ Lie(G) is an *equivalence of categories* from the category of connected, simply connected, Lie groups to the category of Lie algebras (of finite dimension).

In down-to-earth terms this means that the multiplicative structure of G near the identity can be recovered from the structure of $\mathfrak{g}$ as a Lie algebra. Classically this was done by exhibiting a formula

$$\exp X \cdot \exp Y = \exp Z$$

for X, Y near $0, Z$ being given explicitly as a convergent series

$$Z = X + Y + \tfrac{1}{2}[X, Y] + \sum_{n \geqslant 3} c_n(X:Y)$$

where c_n is a homogeneous polynomial map of $\mathfrak{g} \times \mathfrak{g}$ into $\mathfrak{g}$ of degree n, expressed entirely in terms of the brackets in $\mathfrak{g}$ ([V2], Ch. 2, §15). This is the *Campbell–Baker–Hausdorff formula.* The c_k may be determined formally by a recursion formula which reveals how its structure is entirely determined by the Lie algebra structure of $\mathfrak{g}$(cf.[V2], equation (2.15.15)).

A3.4 Integration. Distributions Integration on analytic (or smooth) manifolds requires the use of exterior differential forms. We shall be only concerned with volume integrals, hence only with forms of the highest degree ([V2], Ch. 1).

Let M be an analytic manifold of dimension n and assume that M is oriented. This means that at each point $m \in M$ the class of positive exterior forms on $T_m(M)$ has been singled out, and that the choices are coherent. Coherence means that if $x_1, \ldots, x_n$ are coordinates at m and $(dx_1)_m \wedge \cdots \wedge (dx_n)_m$ is positive, then $(dx_1)_p \wedge \cdots \wedge (dx)_p$ is positive for p near m also. Such a coordinate system is called *positive.* If ω is now any n-form with compact support, $\int_M \omega$ may be defined in a natural manner so that, if $x_1, \ldots, x_n$ form a positive system of coordinates on an open set U and $g \in C_c^\infty(U)$,

$$\int_M g\omega = \int_{\mathbb{R}} g'(x_1, \ldots, x_n)\omega'(x_1, \ldots, x_n)\, dx_1 \cdots dx_n$$

where $g = g'((x_1,\ldots,x_n), \omega = \omega'(x_1,\ldots,x_n)\,dx_1 \wedge \cdots \wedge dx_n$, and $g' \in C_c^\infty(\mathbb{R}^n)$. In particular, if ω is a smooth positive n-form on M the integrals $\int_M f\omega$ are well-defined for all $f \in C_c^\infty(M)$.

The basic formula is that of integration by parts. Let M be an oriented smooth manifold and ω a positive n-form on M, n being $\dim(M)$. Then, given any smooth differential operator D on M, one can define its *formal transpose* D^t in such a way that

$$\int_M Df \cdot g\omega = \int_M f \cdot D^t g\omega$$

for all f, $g \in C^\infty(M)$ with at least one of f and g having compact support. The classical expression for D^t in local coordinates $x_1,\ldots,x_n$ is computed by writing D as $\sum a_\alpha \partial_\alpha$, $\omega = \omega'\,dx_1 \wedge \cdots \wedge dx_n$, and integrating by parts the integral

$$\int Df \cdot g\omega'\,dx_1 \cdots dx_n$$

Then

$$D^t = \sum(-1)^{|\alpha|}\partial_\alpha \circ (a_\alpha \omega')$$

If $M = G$ is a Lie group we usually take ω to be left invariant and orient G to make ω positive. Then the above integration is just integration with respect to a left Haar measure. Suppose now G is unimodular, and dg a Haar measure. Then, for $X \in \mathfrak{g}$.

$$\int_G f(g \exp tX)\,dg = \int_G f(g)\,dg \quad (f \in C_c^\infty(G))$$

and so, differentiating with respect to t, we get

$$\int_G (Xf)(g)\,dg = 0$$

Replacing f by fh we see that

$$\int_G Xf \cdot h\,dg = -\int_G f \cdot Xh\,dg$$

In other words, the formal transpose of $\partial(X)$ is $-\partial(X)$:

$$\partial(X)^t = -\partial(X) \quad (X \in \mathfrak{g})$$

Hence

$$\partial(L_1 \cdots L_m)^t = (-1)^m \partial(L_m \cdots L_1)$$

In other words there is a unique *antiautomorphism* $a \to a^t$ of $U(\mathfrak{g})$ such that $X^t = -X$ for $X \in \mathfrak{g}$, and

$$\partial(a)^t = \partial(a^t)$$

These remarks lead naturally to the action of elements of $U(\mathfrak{g})$ on distributions on G. If $T(C_c^\infty(G) \to \mathbb{C})$ is a distribution on G and $a \in U(\mathfrak{g})$, the distribution $aT := \partial(a)T$ is defined by

$$(aT)(f) = T(a^t f) \quad (f \in C_c^\infty(G))$$

If F is a smooth function, we identify it with the distribution

$$T_F : f \to \int_G F f \, dg$$

Then

$$T_{\partial(a)F} = \partial(a) T_F$$

A3.5 Semisimple Lie groups and Lie algebras The original programme of Sophus Lie was to obtain a classification of all smooth actions of all (local) Lie groups. It was Killing who seems to have realized the fruitfulness of classifying the Lie algebras themselves, and had the great insight that the *simple* Lie algebras over $\mathbb{C}$ may be classified. Killing's work was quite incomplete at many points and the full classification was carried out by Elie Cartan in his 1894 thesis. Cartan's work was simplified further by Dynkin in the 1940s ([V2], Ch. 4). For an absorbing historical account of the development of the ideas of Killing and Cartan, see [Haw].

A Lie algebra $\mathfrak{g}$ over $\mathbb{C}$ is *simple* if it is *not* one-dimensional and if 0 and $\mathfrak{g}$ are its only ideals. The classical Lie algebras

$$A_l(l \geqslant 1), \quad B_l(l \geqslant 2), \quad C_l(l \geqslant 3), D(l \geqslant 4)$$

are all simple. Here, in time honoured notation,

$A_l =$ Lie algebra of $(l+1) \times (l+1)$ complex matrices of trace 0,
$B_l =$ Lie algebra of $(2l+1) \times (2l+1)$ complex skew symmetric matrices,
$C_l =$ Lie algebra of $(2l) \times (2l)$ complex matrices L such that

$$L^t F + FL = 0$$

where

$$F = \begin{pmatrix} 0 & I \\ -I & 0 \end{pmatrix} \quad (l \times l \text{ blocks})$$

$D_l =$ Lie algebra of $(2l) \times (2l)$ complex skew-symmetric matrices.

The corresponding complex groups are respectively $SL(l+1, \mathbb{C})$, $SO(2l, +1, \mathbb{C})$, $Sp(l, \mathbb{C})$ (group of all $g \in SL(2l, \mathbb{C})$ such that $g^t F g = F$) and $SO(2l+1, \mathbb{C})$. The Cartan–Killing classification simply says that the *classical* Lie algebras $A_l(l \geqslant 1)$, $B_l(l \geqslant 2)$, $C_l(l \geqslant 3)$ and $D_l(l \geqslant 4)$ are all simple and are mutually nonisomorphic, and that apart from these there are only five other simple algebras over $\mathbb{C}$, namely, the so-called *exceptional* Lie algebras

$$G_2, F_4, E_6, E_7, E_8$$

([V2], Ch. 4). Cartan also classified the irreducible representations of the simple Lie algebras by their highest weights.

A *semisimple* Lie algebra over $\mathbb{C}$ may be defined as a direct sum of simple Lie algebras over $\mathbb{C}$. This is not, however, the preferred definition. For any Lie algebra $\mathfrak{g}$ over a field k of characteristic 0, one defines its *Cartan–Killing form* as the symmetric bilinear form $\langle \cdot, \cdot \rangle$ on $\mathfrak{g} \times \mathfrak{g}$ given by

$$\langle X, Y \rangle = \mathrm{tr}(\mathrm{ad}\, X\, \mathrm{ad}\, Y) \quad (X, Y \in \mathfrak{g})$$

A semisimple Lie algebra $\mathfrak{g}$ is by definition a Lie algebra whose only solvable (or abelian) ideal is 0. Cartan's famous criterion for semisimplicity is that $\mathfrak{g}$ is semisimple if and only if its Cartan–Killing from is non-degenerate. It can then be proved that, over $\mathbb{C}$, $\mathfrak{g}$ is semisimple if and only if it is a direct sum of ideals which are simple Lie algebras. Both the notion of semisimplicity and Cartan's criterion make sense over any field of characteristic 0, and are preserved when passing to any extension of the base field.

The Cartan–Killing classification, in the simplified form given to it by Dynkin, proceeds by associating to each simple Lie algebra over $\mathbb{C}$ a combinatorial graph called its *Dynkin diagram*, and then classifying the possible diagrams. By explicit calculations going back to Cartan and Killing one then shows that all the possible diagrams arise from suitable simple Lie algebras. These calculations are tedious, especially for the exceptional Lie algebras, and a general existence proof was obtained only in 1948–50 when Chevalley and Harish-Chandra established it independently. The modern theory of semisimple Lie algebras may be said to have begun with these articles of Chevalley and Harish-Chandra. Their results were subsequently sharpened by Serre. In the formulation of Chevalley, Harish-Chandra, and Serre, the Dynkin diagram is used to obtain a *presentation* of the semisimple Lie algebra as the *unique* Lie algebra with generators and relations that are completely specified by the diagram ([V2], Ch. 4, §§5–8).

In representation theory one deals with *real* groups. A Lie group G will be called semisimple if $\mathfrak{g} = \mathrm{Lie}(G)$ is semisimple. The complexification $\mathfrak{g}_{\mathbb{C}}$ of $\mathfrak{g}$ will still be semisimple, and its structure theory will lead to the structure theory for $\mathfrak{g}$. These case of compact G goes back to Hermann Weyl. For arbitrary real G this method leads to the Iwasawa decomposition and other related results on structure of real semisimple Lie algebras, such as Cartan subalgebras, parabolic subalgebras, and so on.

References

[A] J. Arthur, Harmonic analysis of the Schwartz space of a reductive Lie group. I, II. (Preprint, 1973).

[Ba] V. Bargmann, Irreducible unitary representations of the Lorentz Group, *Ann. Math.*, **48** (1947), 568–640.

[Bo] A. Borel, Linear algebraic groups. Algebraic groups and discontinuous subgroups. *Proc. Symp. Pure Math., vol. IX* (1966), ed. A Borel and G.D. Mostow, 3–19.

[Br] F. Bruhat, Sur les représentations induites des groupes de Lie, *Bull. Soc. Math. France*, **84** (1956), 97–205.

[C] C. Chevalley, *Theory of Lie groups I*, Princeton University Press, 1946.

[Da] H. Davenport, *Multiplicative number theory* (ed. & rev. H.L. Montgomery), Springer Verlag, 1980.

[Di] J. Dixmier, *Algèbres enveloppantes*, Gauthier-Villars, 1974.

[DS] N. Dunford and J. Schwartz, *Linear operators, Part II*, Wiley-Interscience, 1963.

[EM] L. Ehrenpreis and F. Mautner, Some properties of the Fourier transform on semisimple Lie groups, I, II, III. I. *Ann. Math.*, **61** (1965), 406–39; II. *Trans. Amer. Math. Soc.*, **84** (1955), 1–55; III. *Trans. Amer. Math. Soc.*, **90** (1959), 431–84.

[GGV] I.M. Gel'fand and M.I. Graev, *Generalized functions* (translated from Russian), vol. 5, Academic Press, 1966.

[GL] L. Gårding and J.L. Lions, Functional analysis, Supplement to vol. XIV, Series X, *Nuovo Cimento*, **N. 1** (1959), 9–66.

[GN] I.M. Gel'fand and M.A. Naimark, *Unitary representations of the classical groups* (in Russian), Publications of the Stekhlov Mathematics Institute, 1950.

[GP] I.M. Gel'fand and V.A. Ponomarev, The category of Harish-Chandra modules over the Lie algebra of the Lorentz group, *Dok. Akad. Nauk. SSSR*, **176** (1967), 1114–17; Indecomposable representations of the Lorentz group, *Uspekhi Mat. Nauk*, **23** (1968), 3–60 (*Russ. Math. Surveys*, **23** (1968), 1–58).

[GS] I.M. Gel'fand and G.E. Shilov, *Generalized functions* (Translated from Russian), Vol. 1, Academic Press, 1964.

[Gå] L. Gårding, Note on continuous representations of Lie groups, *Proc. Nat. Acad. Sci. U.S.A.*, **33** (1947) 331–2.

[H1] Harish-Chandra, On some applications of the universal enveloping algebra of a semisimple Lie algebra, *Trans. Amer. Math. Soc.*, **70** (1951), 28–96.

[H2] Harish-Chandra, Representations of a semisimple Lie group on a Banach space. I, *Trans. Amer. Math. Soc.*, **75** (1953), 185–243.

[H3] Harish-Chandra, Representations of semisimple Lie groups. II, *Trans. Amer. Math. Soc.*, **76** (1954), 26–65.

[H4] Harish-Chandra, Representations of semisimple Lie groups. V, *Proc. Nat. Acad. Sci. U.S.A.*, **40** (1954), 1076–7.

[H5] Harish-Chandra, Discrete series for semisimple Lie groups. I, *Acta Math.*, **113** (1965), 241–318.

[H6] Harish-Chandra, Discrete series for semisimple Lie groups. II, *Acta Math.*, **116** (1966), 1–111.

[H7] Harish-Chandra, Harmonic analysis on real reductive groups, I, II, III. I, *J. Funct. Anal.*, **19** (1975), 104–204; II, *Invent. Math.*, **36** (1976), 1–55; III, *Ann. Math.*, **104** (1976), 117–201.

[H8] Harish-Chandra, Plancherel formula for the 2 × 2 real unimodular group, *Proc. Nat. Acad. Sci. U.S.A.*, **38** (1952), 337–42.

[H9] Harish-Chandra, Plancherel formula for complex semisimple Lie groups, *Proc. Nat. Acad. Sci. U.S.A.*, **37** (1951), 813–18.

[H10] Harish-Chandra, The Plancherel formula for complex semisimple Lie groups, *Trans. Amer. Math. Soc.*, **76** (1954), 485–528.

[H11] Harish-Chandra, Representations of semisimple Lie groups. V, *Amer. J. Math.*, **78** (1956), 1–41.

[H12] Harish-Chandra, Harmonic analysis on semisimple Lie groups, *Bull. Amer. Math. Soc.*, **76** (1970), 529–51.

[H13] Harish-Chandra, Harmonic analysis on semisimple Lie groups. Some recent advances in the basic sciences, *Belfer Graduate School of Science Annual Science Conference Proceedings*, vol. 1 (1962, 1963, 1964) (ed. A. Gelbart), 35–40.

[H14] Harish-Chandra, *Collected papers* (ed. V.S. Varadarajan), Springer Verlag, 1984.

[Ha] P.R. Halmos, *Measure theory*, Van Nostrand, 1950.

[Haw] T. Hawkins, Wilhelm Killing and the structure of Lie algebras, *Archive for History of Exact Sciences*, **26** (1982), 127–92.

[Hb] W.J. Haboush, Reductive groups are geometrically reductive, *Ann. Math.*, **102** (1975), 67–85.

[K] V.G. Kac, *Infinite dimensional Lie algebras*, Birkhäuser, Boston, 1983. Second edition, Cambridge University Press, 1985.

[Kd] K. Kodaira, The eigenvalue problem for ordinary differential equations of the second order and Heisenberg's theory of *S*-matrices, *Amer. J. Math.* **71** (1949), 921–45.

[Kn] A.W. Knapp, *Representations of semisimple Lie groups*, Princeton University Press, 1986.

[Ko] B. Kostant, On the existence and irreducibility of certain series of representations, *Lie groups and their representations*, ed. I.M. Gel'fand, John Wiley 1975, 231–329.

[L1] R.P. Langlands, *Euler products*, Yale University Press, 1967.

[L2] R.P. Langlands, Problems in the theory of automorphic forms, *Lectures in modern analysis and applications*, Springer Lecture Notes, no. **170** (1970), 18–86.

[Ma1] G.W. Mackey, *Unitary group representation in physics, probability and number theory*, Benjamin, 1978.

[Ma2] G.W. Mackey, Harmonic analysis as the exploration of symmetry, a historical survey, *Bull. Amer. Math. Soc.*, **3** (1980), 543–698.

[Ma3] G.W. Mackey, Induced representations of locally compact groups, I, II. I, *Ann. Math.*, **55** (1952), 101–39; II, *Ann. Math.*, **58** (1953), 193–221.

[MF] D. Mumford and J. Fogarty, *Geometric invariant theory*, Springer Verlag, 1982.

[MZ] D. Montgomery and L. Zippin, *Topological transformation groups*, Wiley-Interscience, 1955.

[P] L.S. Pontryagin, *Topological groups*, Princeton University Press, 1939.

[PRV] K.R. Parthasarathy, R. Ranga Rao and V.S. Varadarajan, Representations of complex semisimple Lie groups and Lie algebras, *Ann. Math.*, **85** (1967), 383–429.

[PS] A.N. Pressley and G.B. Segal, *Loop groups*, Oxford University Press, 1986.

[RS] M. Reed and B. Simon, *Methods of modern mathematical physics*, vol. I, Academic Press, 1980.

[S1] J.P. Serre, *Représentations linéaires des groupes finis*, Hermann, 1971.

[S2] J.P. Serre, *A course in arithmetic*, Springer Verlag, 1973.

[S3] J.P. Serre, *Algèbres de Lie semi-simple complexes*, Benjamin, 1966.

[Sc] I. Schur, *Vorlesungen über Invarienteniheorie* (ed. Helmut Grunsky), Springer Verlag, 1968.

[Sch] L. Schwartz, *Théorie des distributions*, Hermann, 1973.

[St] M.H. Stone, Linear transformations in Hilbert spaces and their applications to analysis, *Amer. Math. Soc. Colloq. Pub.*, vol. **15**, 1932.

[T] M. Takesaki, *Theory of operator algebras*, Springer Verlag, 1979.

[Ta] J. Tate, Fourier analysis in number fields and Hecke's zeta functions (Thesis), *Algebraic number theory*, ed. J.W.S. Cassels and A. Fröhlich, Thompson Book Company (1967), 305–47.

[Te] A. Terras, *Harmonic analysis on symmetric spaces and applications I, II*. I Springer Verlag, 1985. II, Springer Verlag, 1987.

[Ti] E.C. Titchmarsh, *Eigenfunction expansions associated with second order differential equations*, Oxford University Press, 1946.

[V1] V.S. Varadarajan, *Geometry of quantum theory*, Springer Verlag, 1985.

[V2] V.S. Varadarajan, *Lie groups, Lie algebras and their representations*, Springer Verlag, 1984.

[V3] V.S. Varadarajan, *Harmonic analysis on real reductive groups*, Springer Lecture Notes, no. **576** (1977).

[W1] H. Weyl, *The classical groups*, Princeton University Press, 1946.

[W2] H. Weyl, Theorie der Darstellung kontinuerlicher halbeinfacher Gruppen durch lineare Transformationen, I, II, III and Supplement. I, *Math. Zeitsch.*, **23** (1925), 271–309; II, *Math. Zeitsch.*, **24** (1926), 328–76; III, *Math. Zeitsch.*, **24** (1926), 377–95; Supplement, *Math. Zeitsch.*, **24** (1926), 789–91.

[W3] H. Weyl, *The theory of groups and quantum mechanics*, Dover Publications.

[W4] H. Weyl, *Gesammelte Abhandlungen*, Band I, nos. **6**, **7**, **8**.

[Wa1] N.R. Wallach, *Real reductive groups*, Academic Press, 1988.

[Wa2] N.R. Wallach, Cyclic vectors and irreducibility for principal series representations, *Trans. Amer. Math. Soc.*, **158** (1971), 107–12.

[We1] A. Weil, *L'Intégration dans les groupes topologiques et ses applications*, Hermann, 1940.

[We2] A. Weil, *Basic number theory*, Springer Verlag, 1967.

[We3] A. Weil, (a) Sur certains groupes d'opérateurs unitaires, *Acta Math.*, **111** (1964), 143–211, (b) Sur la formule de Siegel dans la théorie des groupes classiques, *Acta Math.*, **113** (1965), 1–87.

[Y] K. Yoshida, *Functional analysis*, Springer Verlag, 1968.

[Z] D.P. Zhelebenko, *Compact Lie groups and their representations*, Translations of Mathematical Monographs, vol. **40**, *Amer. Math. Soc.*, 1973.

Subject index